ENVIRONMENTAL CHEMISTRY RESEARCH PROGRESS

Environmental Chemistry Research Progress

Paul Robinson
and
Richard Gallo
Editors

Nova Science Publishers, Inc.
New York

For permission to use material from this book please contact us:
Telephone 631-231-7269; Fax 631-231-8175
Web Site: http://www.novapublishers.com

LIBRARY OF CONGRESS CATALOGING-IN-PUBLICATION DATA

Environmental chemistry research progress / editors, Paul Robinson and Richard Gallo.
p. cm.
Includes index.
ISBN 978-1-60741-055-3 (hardcover)
1. Environmental chemistry. I. Robinson, Paul, 1958- II. Gallo, Richard, 1959-
TD193.E5528 2009
628.5--dc22 2009001827

Published by Nova Science Publishers, Inc. ✢ New York

CONTENTS

PREFACE

Photodissociation reactions play a vital role in initiating many chemical reactions in the atmosphere. This book presents a review on the photochemistry of alkyl nitrites, nitro compounds and carboxylic acid, with an emphasis on the recent progress in photodissociation dynamics by using laser induced fluorescent (LIF) technique. The use of artificial neural networks in predicting pollutants concentrations is also evaluated. Biodegradable materials are widely used in the biomedical field because there is no post-operative surgery after implantation. The use of nanofibers for tissue engineering is described, as well as the electrospinning technique as a way to fabricate controllable continuous nanofiber scaffold mimicking the ECM structure. Organochlorine pesticides, which have been used for decades as insecticides and for the control of vector-borne diseases, is looked at as well. Specifically, the use of organophosphorus pesticides as a replacement, their long-term persistance and their metabolites presented in water, sediments, soil and vegetables, as well as the modern strategies for the extraction of these compounds from environmental samples. Poly (ethylene terephthalate)(PET) is one of the most eco-friendly polymeric materials. It is largely collected and consumed in the form of PET bottles in Japan. This trend will most likely become a worldwide phenomenon, however, there are also problems attributed to this material recycling system. These problems, as well as the new technology used to improve processability for PET, are examined in this book. The innovative 3R "Recycle-Reuse-Reduce" AGROCARBON technology, which provides recycling of agricultural organic and mineral by-products, is also discussed, as well as the process that upgrades by-products to high added-value plant growth promotion and natural fertilization products for environmentally friendly vegetable cultivation.

Chapter 1 - The innovative 3R “Recycle-Reuse-Reduce” AGROCARBON technology provides recycling of agricultural organic and mineral by-products provides carbon products for soil amendment and restoration of soil natural balance. The input feed stream is refuse grain, food processing and/or other by-products. The innovative technology providing surface modified charcoals and minerals for plant availability. The process is upgrading by-products to high added-value plant growth promotion and natural fertilization products for environmentally friendly vegetable cultivation, with carbon sequestration potential. The 3R is a horizontally arranged and indirectly heated low temperature zero emission carbonization system (operating under vacuum, up to 850 °C±50°C material core temperature) and directly integrated novel agro biotechnological processing units of agrocarbon specific solid state fermentation and formulations.

Chapter 2 - In recent years, the photodissociation dynamics of alkyl nitrites, nitro compounds and carboxylic acid has been a subject of intensive studies, which is closely related to the atmospheric chemistry. Here we present a review on the photochemistry of alkyl nitrites, nitro compounds and carboxylic acid, with emphasis on the recent progress in photodissociation dynamics at 266 and 355 nm by using laser induced fluorescence (LIF) technique. The OH photofragments were observed in these photolysis processes. Internal state distributions of the nascent OH from the photodissociation were measured. The OH spin-orbit state and Λ-doublet state were observed. The *ab initio* calculations have also been employed to investigate those photodissociation processes. Based on the experimental and theoretical results, the photodissociation dynamics of these molecules are revealed.

Chapter 3 - The use of Artificial Neural Networks (ANNs) for emission estimation and forecasting has been recently proven to be promising. For instance, a neural network approach can account for the synergistic effects that arise from the complex interactions of several variables that affect atmospheric dispersion of gases. Unlike other modeling approaches, artificial neural networks are able to provide reliable estimates of emission rates based solely on limited information that is not known with as much certainty. The deterministic models, for instance, would require information regarding atmospheric stability, dispersion coefficients in the lateral and vertical directions, reaction mechanisms, and kinetic data. These information are usually either not well understood or only known with a certain degree of uncertainty. The objective of the present chapter is to offer a background on the use of artificial neural networks in predicting pollutants concentrations. The chapter will start with an introduction to ANNs and will discuss different algorithms used for training them. The chapter will then present several examples that will illustrate the predictive performance of the prepared network models when compared against linear and non-linear regression models, commercial simulators, and measured data.

Chapter 4 - Organochlorine pesticides have been used during decades as insecticides and for the control of vector-borne diseases. In the 1980s, these pesticides were banned in developed countries and were replaced by other kind of compounds like organophosphorus pesticides; however, they are usually more toxic than organochlorine pesticides. Nowadays these compounds are present in the environment; thus the long-term persistence and their metabolites are presented in water, sediments, soils and vegetables.

The methods usually used for the extraction of these compounds from environmental samples are often too time consuming, involve multi-step procedures that are prone to loss of analytes and use high amounts of toxic organic solvents.

Modern strategies in analytical chemistry have been developed towards the simplification of sample's preparation, as well as the minimization of organic solvent used on the extraction process. An alternative to organic solvents is the use of micellar systems as extractants of organic pollutants in liquid and solid environmental samples. Micellar media facilitate the solubilization of different compounds present in different kind of samples. Benefits obtained when micellar media are used, are lesser toxicity and cost than organic solvents.

Chapter 5 - Poly(ethylene terephthalate) (PET) is one of the most eco-friendly polymeric materials, since a large amount of PET bottles, more than 65% in Japan, is collected and consumed as recycled PET. This trend will spread worldwide in the near future because of the rapid growth of the interest in global environment as well as the excellent cost-performance. Although the material recycling system seems to be successful, it has some problems attributed to the depression of molecular weight leading to low level of melt elasticity. The lack of the melt elasticity leads to sagging, i.e., downward- deformation of the extrudate by gravitational force, which prohibits extrusion processing, and inhomogeneity of the thickness of the products at film processing and blow-molding. As a result, the application of recycled PET is quite limited, such as spinning to produce fibers and injection-molding.

In this chapter, a new method employing reactive polymer blend technique is proposed to improve the processability for polyester resins by rheological modification. The technique would be applicable to recycled PET. Further, appropriate processing condition is also demonstrated for industrial application. The technique enhances the melt elasticity, i.e., the most serious problem for recycled PET and thus improves the extrusion processability.

Short Communication - Manganese doped nano-TiO_2 was synthesized with easy and available inorganic salts by hydrothermal methods. TEM results show that the sample was sphere shape, XRD patterns show that it was anatase structure. Photo-degradation of acid red 3R was performed under high pressure Xe lamp with the photo-catalyst. The effects of pH value, dopant concentration, catalyst's dosage, crystal structure, calcinations on the photo-catalytic reaction were studied. The results indicated that nano-particle doped with 0.5 at. % Mn has higher activity than pure TiO_2, and mineralization percent of organic molecules approaches 100% within 120min.

In: Environmental Chemistry Research Progress
Editors: P. Robinson and R. Gallo
ISBN 978-1-60741-055-3

Chapter 1

FOOD CROP MINERAL DEFICIENCY AND DISTURBANCE STRESS MITIGATION IN TEMPERATE CLIMATIC REGIONS BY ECONOMICAL AND ENVIRONMENTAL VALORIZATION OF AGRICULTURAL BY-PRODUCTS "PROTECTOR"

Edward Someus*
3R Environmental Technologies Ltd, Budapest, Hungary

ABSTRACT

The innovative 3R "Recycle-Reuse-Reduce" AGROCARBON technology provides recycling of agricultural organic and mineral by-products that provide carbon products for soil amendment and restoration of soil natural balance. The input feed stream is comprised of refuse grain, food processing and/or other by-products. The innovative technology provides surface-modified charcoals and minerals for plant availability. The process upgrades by-products to high added-value plant growth promotion and natural fertilization products for environmentally-friendly vegetable cultivation with carbon sequestration potential. The 3R technology is a horizontally-arranged and indirectly-heated low-temperature zero-emission carbonization system (operating under vacuum, up to 850°C±50°C material core temperature) and directly-integrated novel agro biotechnological processing units of agrocarbon specific solid state fermentation and formulations. Performance features the following:

1. *Food crop mineral deficiency and disturbance stress mitigation in temperate climatic regions by restoration of soil natural balance.*
2. *Input feed streams:* low value organic and/or inorganic by-products such as refuse grain, sawdust and/or high phosphorous content animal bone meal, and/or

* Environmental engineer.Tel: +(36-20) 201 7557;Fax: +(36-1) 424 0224. Web: http://www.terrenum.net Email1: edward@terrenum.net Email2: edward.someus@gmail.com H-1222 Budapest, Szechenyi str. 59, Hungary

other by-products, which can be valorization transformed by added-value integrated thermal and biotechnological means.
3. *The 3R* biotechnology integrated industrialized biochar production technology is a modern *zero emission solution*, whereby all and any output products are recycled and reused, aiming at prevention-protection-preservation approaches.
4. *The output products* are different types of soil biotechnology-specific solid carrier composites and adapted microbiological fungus and/or bacteria strain consortiums. Depending on the soil and climate application scenario conditions, different types of relevant 3R products can be made.
5. *The application objective* of the products are the natural balance and functionality restorations of degraded temperate agriculture soils with controlled microbiological activity and precision farming nutrient supply, promotes humus building and mineral mobilization towards plant availability, for sustainable, improved, economical and ecological food crop production. Carbon sequestration is also targeted.
6. *The application targets* combined effects, such as plant growth promotion, biological control against soil-borne plant pathogens and natural fertilization, especially sequenced mobilized phosphorus supply and improved nutrient use efficiency.
7. *The application sectors* are the organic farming and/or low input farming for environmentally-friendly vegetable cultivation and other food crop productions.
8. *Status*: A "product-like" field demonstration plant has been developed and successfully tested for scale-up optimization. The comprehensive industrialized engineering design is made for 30,000 m^3/year input feed stream as of modern US/EU industrial standards. Patented original solution. Available for licensing.

1. ABBREVATIONS

ABC	Animal Bone Charcoal Mineral, which has been produced by a specific 3R carbonization technology surface modification method from animal bone meal, carbon/ mineral-based agricultural and food industrial waste and by-products, aiming for agro biotechnological added-value re-use in food production cultivation applications. Different types of ABC are produced and optimized to meet the regional application need for different types of soils under different worldwide climatic conditions, including scenarios for drought under climate change conditions. Scheduled dose from 400 kg/ha to 1250 kg/ha.
ADE	Amazonian Dark Earth, also known as terra preta
AMF	Arbuscular Mycorrhiza Fungi
Anthrosols	The anthrosols are soils that have been formed or profoundly modified through long-term human activities, such as addition of organic materials or household wastes, irrigation or cultivation, where human and time factors impacted the genesis. It has been identified in tropical/

	subtropical weathered soils (Brazil, Japan, China), but also associated with infertile soils such as arenosols, podzols and albeluvisols, as well as with wetland soils (fluvisols, gleysols). Hortic anthrosols occur everywhere where long-term intensive cultivation has taken place [90].
3R AGROCARBON	The agricultural product family of the 3R technology outputs, mineral/charcoal product containing abundant microorganisms. The specific biochar-charcoal may be of plant and/or animal bone origin.
3R technology	Recycle-Reduce-Reuse, integrated carbonization and agro biotechnology for added "carbon economy" value processing of agricultural and food industrial waste and by-products, done by specific pyrolysis and SSFF technologies. Inventor: Edward Someus, a Swedish environmental engineer. Patented.
3R group	3R Environmental Technologies Ltd, a holding company for rights, knowledge, know-how and engineering design for the 3R technology and PROTECTOR products, with subsidiary of the Terra Humana Ltd. (since 1989) and other organizations. Owner and general manager: Edward Someus
3R-T7	*Trichoderma harzianum* sp. soil fungus strain selected by the 3R group
Cd	Cadmium
cfu	Micro-organism colony-forming units
EU	European Union
ft^3	Cubic feet
EPA	United States Environmental Protection Agency
GHG	Green House Gas
GWP	Global Warming Potential
kBq	Kilo Becquerel
MJ	Mega Joule
OECD	Organisation for Economic Co-operation and Development
PR	Phosphate rock
P	Phosphorus
PROTECTOR product	Brand name of the 3R technology output product
PROTECTOR project	Applied scientific development project acronym, contracted 2005–2008 with the European Commission, DG RTD E.4, SDME 08/16 Directorate E–Biotechnology, Agriculture, Food Unit 4, Contarct No. Food-2005-514082. Full title: Recycling and upgrading of bone meal for environmentally friendly crop protection and nutrition. www.terrenum.net/protector
RTD	Applied Scientific Research and and Technical Development

SSFF	Solid State Fermentation and Formulation, specifically developed for ABC biotechnological processing, including formulations of individual stains or formulations of strain consortium or co-composting with ABC mineral/charcoal containing abundant microorganisms
SOM	Soil Organic Material
SRM	Specified Risk Material, food safety standard and classification for handling of waste materials from food processing
Terra Preta	Very dark and fertile anthropogenic soils found in the Amazon Basin with high charcoal content. It is also known as “Amazonian Dark Earth” ADE or “Indian black earth” or “Terra preta de índio”
Tg	10^{12} g or million metric tons
Temperate climatic	The temperate latitudes of the globe lie between the tropics and the polar circles and including the European Union, USA, Canada, Japan, China, Russia. The north temperate zone extends from the Tropic of Cancer (at about 23.5 degrees north latitude) to the Arctic Circle (at approximately 66.6 degrees north latitude). The south temperate zone extends from the Tropic of Capricorn (at approximately 23.5 degrees south latitude,) to the Antarctic Circle (at approximately 66.5 degrees south latitude)
Temperate agriculture soil	Different types of human influenced and intensive cultivated soils for food crop and energetic biomass production in temperated climatic regions. Usual characteristics is that many of these soils the erosion, degradation after long time of intensive agricultural cultivations and depleted carbon stocks and nutrient deficiency, compared to the native ecosystems
TOC	Total Organic Carbon
VOC	Volatile Organic Compounds
WSP	Water-soluble Phosphorus

2. Introduction

The innovative 3R AGROCARBON technology and PROTECTOR product are patented and IPR-protected original solutions for food crop mineral deficiency and disturbance stress mitigation in temperate climatic regions by economical and environmental valorization of agricultural by-products. The 3R technology is performed by *integrated low temperature carbonization and biotechnological recycling of mineral and/or carbon* biotechnological carriers from organic and inorganic agricultural by-products. The input feed stream is comprised of high phosphorus content animal bone meal, refuse grain, manganese carbonate, food processing and/or other by-products. The innovative technology provides surface modified charcoals and minerals for plant available microbiological carriers and food industrial by-product streams as nutrients during the solid substrate fermentation/formulation, the innovative design of which is a zero-emission industrial process. Natural soil-borne microbiological fungus and/or bacteria aerobic strains are selected for solid carrier entrapment and formulated for ambient temperature long-term storage. The process upgrades by-products to high added-value and safe biological crop protection, plant growth promotion and natural fertilization products for environmentally-friendly vegetable cultivation with carbon sequestration potential. The biological-control effect targets primarily combat against crown rot and damping off plant pathogens with preventive measures, enhancement of plant natural resistance and soil natural balance restoration.

The advanced science/engineered design includes the following: added-value by-product transformation by a specific thermal desorption/low temperature carbonization system; selection of micro-organisms for solid state fermentation and formulation with combined biological-control/mineral-mobilization effects; solid-state-fermentation-formulation; process optimization and scale-up; mineral-carbon soil kinetics evaluation; comprehensive risk assessment; validation/field demonstration under several Europe-wide climatic/soil conditions; and, Authority permitting, cost-benefit-analysis consumer acceptance and EU/USA market evaluation. The final achievements are prevention-protection-preservation approaches, aiming for environmentally-friendly and healthy food production for the low-input and organic farming horticultural production sectors.

3. General Considerations: The Multidimensional Concerns of the Intensive Crop Production

Agribusiness is an industry dependent upon manufactured chemical inputs for high yields. For every additive beneficial to productivity, there are by-products requiring extensive environmental services to render them inert. Radioactive gypsum stacks, residual pesticides and increasing nutrient loading to lakes and rivers are examples [10].

In the last 50 years, mankind has modified the environment more than during the whole history of humanity, in which the artificial modification process by the intensive agriculture production played decisive roll. The increased population pressure, reduced length of fallow, deforestation, biodiversity decreases and improper agricultural practices have led to widespread soil degradation in many parts of the world, whereas the biodiversity loss disrupts the functioning of ecosystems. An important manifestation of this environmental damage is the inadequate replenishment of soil nutrients and organic matter. In particular, P deficiency is becoming critical in many soils. Moreover, because of complementarities in the uptake of plant nutrients, this deficiency threatens to disturb the viability of applying other nutrients. In order to preserve the sustainability of agriculture and safeguard the livelihood of large segments of the rural population, there is an urgent need to rebuild soil fertility and so maintain and improve current levels of productivity and farm income [38].

A doubling in global food demand projected for the next 50 years poses huge challenges for the sustainability both of food production and of terrestrial and aquatic ecosystems and the services they provide to society. New incentives and policies for ensuring the sustainability of agriculture and ecosystem services will be crucial if we are to meet the demands of improving yields without compromising environmental integrity or public health. Modern agriculture now feeds 6,000 million people. By 2050, global population is projected to be 50% larger than at present and global grain demand is projected to double [97].

Global food stocks are running low and rich nations should not take security of supplies for granted. There are *food supply uncertainties* we face in the 21st century for which there are several reasons:

1. *increasing use of crops for bio-energy*
2. *increasing demand for meat and milk products in the developing world*
3. *poor harvests around the world following droughts and floods*

The global area for food production will decline as farmland is lost to housing, bio-energy cropping and, ultimately, sea level rise as well. However, it will not be acceptable to increase production without regard for the environment, and we will increasingly demand food that is safe and contributes to healthy diets, while it is predicted that an excessive supply of food just exports environmental problems. About half of global usable land is already in pastoral or intensive agriculture [96].

Furthermore, only recently have we started to think about how agriculture should contribute to *managing climate change by controlling the release of greenhouse gases and by storing carbon in the soil,* while we also need diversity in agriculture and restoration of soil natural balance after decades of high input farming. There will be no single path to sustainability; organic, low input farming, hi-tech plant and animal breeding will be part of the mix, possibly concentrated in different regions and serving different markets. Recycling, reduction and reuse of waste in the food chain is another issue that needs to importantly be addressed. There is no point producing more food from the land without trying to use what we already have more efficiently [26]. Although we can produce enough food to feed today's population, that achievement has come at the cost of an ever-increasing impact on Earth's sustainability. The striking increase in the use of nitrogen (N) and phosphorus (P) fertilizers between 1960 and 2000 by intensive agricultural practices has led to degradation of air and water quality [12] [98]. In addition to causing the loss of natural ecosystems, agriculture adds globally significant and environmentally detrimental amounts of nitrogen and phosphorus to terrestrial ecosystems. [110] As currently agriculture will require an additional 40 and 20 Tg (10^{12} g or million metric tons) of N and P fertilizer, respectively, applied to agricultural soils to meet food production needs in 2040 [12].

These considerations mean that we will need to produce more healthy food with higher nutritional content per hectare from the farmland that will remain, but based on long termed sustainable environmental and ecological systems.

In this context we define sustainable agriculture as practices that meet current and future societal needs for food and fibre, for ecosystem services, for healthy lives, and that do so by maximizing the net benefit to society when all costs and benefits of the practices are considered [97]. The main *environmental impacts of agriculture come from the conversion of natural ecosystems to agriculture*, from agricultural nutrients that pollute aquatic-terrestrial habitats-groundwater, and from pesticides, especially bio-accumulating or persistent organic agricultural pollutants. Agricultural nutrients enter other ecosystems through leaching, volatilization and the waste streams of livestock and humans. Pesticides can also harm human health, as can pathogens, including antibiotic-resistant pathogens associated with certain animal production practices [97].

Continuous cropping and inadequate replacement of nutrients removed in harvested materials or lost through erosion, leaching or gaseous emissions deplete fertility and cause soil organic matter levels to decline, often to half or less of original levels [63].

The effects of land degradation on productivity can be compensated for by increased fertilization, irrigation, and disease control, which increase production costs [70].

The increased urbanisation of populations creates greater dependence upon mineral fertilizers as a greater proportion of wastes from food distribution and consumption is generated remote from farms [93]. Soil nutrient management needs to satisfy the multiple agricultural goals of increasing productivity, being economically viable, and sustaining environmentally healthy soil and crop systems. The vegetables belong to the group of intensive humus consumers, taking up to approx. 760 kg Humus-C per ha and year. Regulation on soil states that humus has to be compensated at least every third year to zero losses [13]. Sustainable agriculture is that which is managed toward greater resource efficiency and conservation while maintaining an environment favourable for the evolution of all species [36].

Table 1. Agriculture production and resource use, the recent past to the near future [107]

Item	**1960**	**2000**	**2030-2040**
Food production (Mt)	1.8 x 10^9	3.5 x 10^9	5.5 x 10^9
Population (billions)	3	6	8 (maybe 10)
Irrigated land (% of arable)	10	18	20
Cultivated land (hectares)	1.3 x 10^9	1.5 x 10^9	1.8 x 10^9
Water-stressed countries	20	28	52
N fertilizer use (Tg)	10	88	120
P fertilizer use (Tg)	9	40	55-60

4. Trends in Fertilizer Use

Until about 1950, farmers throughout the world relied on animal wastes and/or crop rotation with nitrogen-fixing crops to keep the soil fertile. Though these practices are still common in some locations, much of the world's agriculture now uses inorganic (chemical) fertilizers to supplement and restore soil nutrients. Between 1920 and 1960, phosphate fertilizer consumption was greater than that of either nitrogen or potash fertilizer. After 1960, nitrogen fertilizer consumption quickly surpassed that of phosphate and potash fertilizer. Today, nitrogen fertilizer consumption is about 2.5 times higher than that of phosphate fertilizer, and almost four times higher than that of potash fertilizer. Total fertilizer consumption is about 80 times its 1920 rate [29].

Fertilizer consumption reached its maximum in the early 1980s for developed areas, such as Western Europe and North America, and has since levelled off somewhat. Asia has seen a significant rise in fertilizer consumption, increasing from about 1 million to 38 million metric tons of nutrients between 1960 and 1999. The consumption rise in South Asia is also notable—increasing from less than 1 to over 22 million metric tons of nutrients. Other developing regions, including Latin America, East Asia, Near East, Africa, and Oceania, have increased fertilizer consumption steadily since the middle of the twentieth century [29]. Developing Asia uses an enormous quantity of nitrogen fertilizer; in fact, it consumes more than three times as much nitrogen fertilizer as North America. The nitrogen–phosphate–potash ratio of fertilizer consumption for Developing Asia is roughly 6–2–1; for North America, the ratio is roughly 2½–1–1. Latin America consumes the three types of fertilizers in more equal quantities, with nitrogen use only slightly higher than phosphate use, and phosphate use only slightly higher than potash use. Worldwide, the current nitrogen–phosphate–potash ratio of fertilizer consumption is about 8–3–2 [29].

5. The Energy Demand of the Intensive Farming

Direct energy consumption in the agricultural sector includes use of gas, diesel, liquid petroleum (LP), natural gas, and electricity. Indirect energy use involves agricultural inputs, such as nitrogen fertilizer, which have a significant energy component associated with their production. Since 1992, direct fuel and electricity expenses for U.S. farms have averaged around 7 percent of total operating costs. Diesel fuel and gasoline are widely used for tillage, planting, transportation, and harvesting. Electricity, gas, and natural gas are primarily used in drying; irrigation; operation of livestock, poultry, and dairy facilities; and on farm processing and storage of perishable commodities. Expenses from indirect energy use increase total energy expenditures to 15 percent of operating costs and the accelerating energy prices in 2008 and onward will further increase the energy consumption expenditures in the agriculture sector. Fertilizers embody the most energy among production inputs because natural gas is the primary input (70-90 percent of the cost of producing nitrogen fertilizer) [85].

Energy consumed in primary agricultural production is classified as either direct or indirect use. Direct energy use is the consumption of fossil fuels for various farm operations such as fuel for tractors or fuel for crop drying. Indirect energy use is the conversion of fossil fuels into other products such as fertilizers or pesticides [65].

Intensive farming is concerned above all with productivity and uses a high level of inputs to achieve it. The inputs are usually in the form of chemicals, fertilisers, pesticides and growth-regulators produced by energy intensive industrial processes and additional energy in the form of high levels of mechanisation. Intensive farming does provide large quantities of relatively cheap food—but often at a cost to the environment, nutrient content or animal welfare.

Indirect energy use involves agricultural inputs, such as fertilizers, which have a significant energy component associated with their production. Expenses from indirect energy use increase total energy expenditures to 15 percent of operating costs. Changes in energy prices may have a greater effect on producers of major field crops [85].

According to the OECD, the absolute energy consumption per hectare increased in OECD countries by 39% between 1970 and 1989. On average, some 1734 MJ are consumed per hectare of agricultural land, rising to 46,400 MJ for the highest consumer, Japan [71]. Traditional P fertilizer production is based on chemical processing of insoluble mineral phosphate high-grade ore, which includes an energy intensive treatment with sulphuric acid at high temperature [109]. The manufacture, distribution, delivery and application of different fertilizers and pesticides are energy intensive. The fertilizers and pesticides prices have been continually increasing due to increased energy costs for production, especially natural gas, increased transportation costs and increased demand. The higher energy costs resulting higher fertilizer ans pesticide cost which is reducing the profitability of the crop and vegetables production.

Energy Intensity of Fertilization

Primary production in the agricultural industry is highly dependant upon fossil fuels. Inorganic fertilizers are major consumers of energy in the agricultural sector. In the United States, inorganic fertilization accounts for about a third of total energy input to crop production. In contrast to tractors, irrigation pumps, and other types of equipment, fertilizers are indirect energy consumers. That is, the bulk of energy use associated with fertilizers is not consumed directly at the agricultural site, but indirectly during its production, packaging, and transportation to the site. Additional energy is then used on-site during fertilizer application [29].

Most fertilizer energy use is attributable to the production of nitrogen fertilizers with natural gas. Natural gas is the principal energy resource for creating anhydrous ammonia, a key nitrogen fertilizer. Over 90% of nitrogenous fertilizers contain ammonia and/or other fertilizer elements derived from ammonia (e.g., ammonium nitrate, sodium nitrate, calcium nitrate, ammonium sulphate, ammonium phosphates, and urea). Producing ammonia is a very energy intensive process; it requires about 1090 to 1250 m^3 of natural gas to produce 1 metric ton of anhydrous ammonia (35,000 to 40,000 ft^3 natural gas per short ton). In 1999, over 15 million metric tons (17 million short tons) of ammonia were produced in the United States, with almost 90% of that going to the fertilizer industry. This amount corresponded to about 3% of total US natural gas production [29].

Each of the three primary nutrients in inorganic fertilizers has a different set of energy requirements during its life cycle. However, these requirements can be separated into four main stages: production, packaging, transportation, and application. Table 2 summarizes the world average energy requirements by nutrient type and life cycle stage for inorganic fertilizers. The table clearly shows the relatively high energy intensity of nitrogen production. This corresponds to almost 90% of nitrogen's total energy requirement. In contrast, the production of phosphate and potash account for only about 45% of the total energy requirement for these nutrients. Moreover, the energy requirement for nitrogen fertilizer is 4.5 times that of phosphate fertilizer, and 5.7 times that of potash fertilizer [29]. Table 2 summarizes the energy requirements to produce, package, and apply inorganic fertilizers.

Energy Intensive Production of Major N Fertilizers

Ammoniachich other nitrogen fertilizers are made. It has been manufactured on a commercial scale since 1913, and the process used to produce it, originally developed by Fritz Haber, has changed little since that time. In the Haber process, gaseous hydrogen (H_2) and nitrogen (N_2) are combined under high temperature and pressure conditions to produce ammonia [22].

Table 2. Energy requirements to produce, package, transport, and apply inorganic fertilizers

	Energy requirement (world average) Btu/lb(kj/kg)		
	Nitrogen	**Phosphate**	**Potash**
Produce	29899 (69530)	3313 (7700)	2753 (6400)
Package	1119 (2600)	1119 (2600)	774 (1800)
Transport	1936 (4500)	2452 (5700)	1979 (4600)
Apply	688 (1600)	645 (1500)	430 (1000)
Total	33642 (78230)	7529 (17500)	5936 (13800)

Source: data compiled from [40].

Ammonia production is an energy intensive process, not only because of its temperature requirements, but also because it requires large quantities of purified H_2 that is usually derived from natural gas, or methane (CH_4). Naphtha and coal are also used as feedstocks in some parts of the world. Hydrogen is produced from methane (CH_4) in a multi-step process. First, a "synthesis gas" is produced by "reforming" methane and steam to hydrogen and carbon monoxide (CO) over a catalyst at 1300–1600° F. This synthesis gas is mixed with air in a second reforming reaction over another catalyst, producing a mixture of N_2, H_2, CO, CO_2 and steam. The carbon monoxide and steam then undergo a final "shift conversion", producing carbon dioxide and more hydrogen. This final synthesis gas is then purified to remove CO_2 that would poison the ammonia-synthesis catalyst. The purified N_2-H_2 synthesis gas is compressed to pressures of 2000–5000 psig (depending on process) and passed over an iron catalyst where ammonia is formed [22].

The price of nitrogen fertilizers is directly related to the price of natural gas (methane). Manufacturing 1 ton of anhydrous ammonia fertilizer requires 33,500 cubic feet of natural gas. This cost represents most of the costs associated with manufacturing anhydrous ammonia. Most of the other popular forms of nitrogen fertilizer are made with anhydrous ammonia. Urea is formulated by a reaction between anhydrous ammonia and carbon dioxide at high temperature and pressure. Ammonium nitrate is formulated by combining anhydrous ammonia and nitric acid in a very corrosive manufacturing climate. Solution liquid fertilizers (28 to 32 percent nitrogen) are composed of one-half urea and one-half ammonium nitrate. It's pretty hard to apply a nitrogen fertilizer formulation that doesn't have natural gas in its manufacturing process [28].

Since natural gas is such a critical resource in fertilizer production, natural gas price fluctuations have a dramatic effect on fertilizer costs. As energy costs continue to rise, and the demand for fertilizers increases, this effect is becoming more pronounced [29].

Energy conservation in primary agricultural production must make the most efficient use of non-renewable energy resources while integrating on-farm resources such as biological cycles and controls [35]. At the same time, the conservation measures must strike a balance with the industry-wide goal of maximizing food production and economic returns from a given land base.

6. Nutrient Disturbance in Agricultural and Forest Soil

The degradation of forest ecosystems may be attributed to various natural and anthropogenic factors such as climatical extremes, biotic stresses, selection of tree species, harvesting regimes, litter raking, off-site amelioration measures, former land use, air pollutant deposition and soil acidification, as caused by internal and external processes. An important factor for a loss of tree vitality are nutritional disturbances, eventually leading to declining stand stability and productivity [47].

Both salt-like and lime fertilizers contribute positively to revitalization of nutrient deficient stands. Experience indicates that forest liming may indeed counterbalance the progress of soil acidification as caused by high H deposition [47].

Besides fertilization, forest regeneration with site-adapted tree species may greatly contribute to the rehabilitation of forest ecosystems, since the capacity for storage of C and N and hence, for a closer nutrient cycling, may be improved significantly through this approach [47].

In central Europe the new type forest damages have been observed since the mid 1970s. Various investigations indicate that the declines are frequently associated with nutritional disturbances. Good correlations between the site specific substrate chemistry and the actual nutritional status of the trees/stands were found. To explain the sudden and widespread appearance of the forest declines, adverse anthropogenic impacts mainly due to elevated emissions of air pollutants and their atmospheric derivatives are hypothesized in combination with natural stress factors. Causal mechanisms include soil degradation due to accelerated soil acidification and increased nutrient leaching from the canopy of forest stands. Fertilization and liming experiments have demonstrated that a fast and sustained revitalization and re-stabilization of declining forest ecosystems marked by nutritional disturbances can be achieved [46].

In agriculture the low uptake and utilization of nutrients out of the added fertilizers results in lower production and appearance of nutrient stress symptoms. The poor efficiency of added fertilizers is mainly due to imbalance fertilization or improper management practices. When the soil is deficient in a particular nutrient element and the crop grown on such soil show visible deficiency symptoms, there is urgent need to correct it to improve the crop production. Methods of correcting nutrient deficiencies vary according to agro-climatic regions, the socio-economic situations of the region, the magnitude of disorder, and nutrient element involved [50].

The tolerance in a given plant species or genotype to nutrient stress is closely related to its nutrient use efficiency. Genotype differences in nutrient use efficiency are linked with root nutrient acquisition capacity, or with utilization by the plant, or both. Various plant species have developed morphological and/or physiological mechanisms to improve acquisition and use efficiency of mineral nutrients when grown on poor, infertile soils [79].

Table 3. Dominant soil stresses [3]

Dominant soil stress	Global land area	
	Million km2	**% of total**
Continuous moisture stress	36.5	27.9
Continuous low temperatures	21.8	16.7
Seasonal moisture stress	10.3	7.9
Low nutrient-holding capacity	7.8	6.0
High aluminium	4.1	3.1
Low moisture and nutrient status	3.5	2.7
Low water holding capacity	3.4	2.6
Other stresses	27.2	20.8
Total	130.6	100.0

By such mechanisms plants achieve nutrient deficiency stress tolerance. However, sometimes the ability to adapt to stress is limited, depending on the intensity and duration of stress. For example, some nutrient stress factors, when operating for a long period, can result in inhibited growth and root physiological processes responsible for nutrient uptake with the consequence of an impaired capability for adaptation and stress tolerance. A wide range of morphological, anatomical and physiological plant traits can be responsible for variations in response to nutrient stress within a plant species [50].

The differences in their ability to grow in soils of low phosphorus status might be attributed to several factors including differential influx rates, root length/shoot weight ratios or root system morphology and root hair density. Among the more important factors is the role of root exudates in making available soil P fixed in forms such as iron or aluminium phosphate. [50] Table 3 illustrates the dominant global soil stresses.

7. The Importance of the Nitrogen Mineral Fertilization

Role of Nitrogen in Plants

Plants are surrounded by the nitrogen (N) in our atmosphere. Every acre of the earth's surface is covered by thousands of pounds of this essential nutrient, but because atmospheric gaseous nitrogen is present as almost inert nitrogen (N_2) molecules, this nitrogen is not directly available to the plants that need it to grow, develop and reproduce. Despite nitrogen being one of the most abundant elements on earth, nitrogen deficiency is probably the most common nutritional problem affecting plants worldwide.

Healthy plants often contain 3–4% nitrogen in their above-ground tissues. These are much higher concentrations than those of any other nutrient except carbon, hydrogen and oxygen, nutrients not of soil fertility management concern in most situations. Nitrogen is an important component of many important structural, genetic and metabolic compounds in plant cells. It is a major component of chlorophyll, the compound by which plants use sunlight energy to produce sugars from water and carbon dioxide (i.e., photosynthesis). It is also a major component of amino acids, the building blocks of proteins. It is also a major component of amino acids, the building blocks of proteins. Some proteins act as structural units in plant cells while others act as enzymes, making possible many of the biochemical reactions on which life is based. Nitrogen is a component of energy transfer compounds, such as ATP (adenosine triphosphate) that allow cells to conserve and use the energy released in metabolism. Finally, nitrogen is a significant component of nucleic acids such as DNA (Deoxyribonucleic acid), the genetic material that allows cells (and eventually whole plants) to grow and reproduce [22].

In general, N (nitrogen) is a scarce under natural conditions and organisms quickly use up all that is available at any given time. In other words, under natural conditions, N is a limiting growth factor in nature [42]. Soil nitrogen exists in three general forms: organic nitrogen compounds, ammonium ($NH4^+$) ions, and nitrate ($NO3^-$) ions. At any given time, 95–99% of the potentially available nitrogen in the soil is in organic forms, either in plant and animal residues, in the relatively stable soil organic matter or in living soil organisms, mainly microbes such as bacteria. This nitrogen is not directly available to plants, but some can be converted to available forms by microorganisms. A very small amount of organic nitrogen may exist in soluble organic compounds, such as urea, that may be slightly available to plants [22].

The majority of plant-available nitrogen is in the inorganic (sometimes called mineral nitrogen) NH4+ and NO3- forms. Ammonium ions bind to the soil's negatively-charged cation exchange complex (CEC) and behave much like other cations in the soil. Nitrate ions do not bind to the soil solids because they carry negative charges, but exist dissolved in the soil water, or precipitated as soluble salts under dry conditions. The nitrogen in soil minerals is released as the mineral decomposes. This process is generally quite slow and contributes only slightly to nitrogen nutrition on most soils [22].

Some microorganisms can utilize atmospheric N2 to manufacture nitrogenous compounds for use in their own cells. This process, called biological nitrogen fixation, requires a great deal of energy; therefore, free-living organisms that perform the reaction, such as Azotobacter, generally fix little nitrogen each year (usually less than 20 lb nitrogen/acre), because food energy is usually scarce. Most of this fixed nitrogen is released for use by other organisms upon death of the microorganism. Bacteria such as Rhizobia, that infect (nodulate) the roots of, and receive much food energy from, legume plants can fix much more nitrogen per year (some well over 100 lb nitrogen/acre) [22].

The nitrogen cycle contains several routes by which plant-available nitrogen can be lost from the soil. Nitrate-nitrogen is usually more subject to loss than is ammonium-nitrogen. Significant loss mechanisms include leaching, denitrification, volatilization, and crop removal. The nitrate form of nitrogen is so soluble that it leaches easily when excess water percolates through the soil. This can be a major loss mechanism in coarse-textured soils where water percolates freely, but is less of a problem in finer-textured, more impermeable soils, where percolation is very slow [22]. Crop removal represents a loss because nitrogen in the harvested portions of the crop plant are removed from the field completely. [22]

Most plants take nitrogen from the soil continuously throughout their lives and nitrogen demand usually increases as plant size increases. A plant supplied with adequate nitrogen grows rapidly and produces large amounts of succulent, green foliage [22]. The major mechanisms for nitrogen fertilizer loss are denitrification, leaching, and volatilization. Denitrification and leaching occur under very wet soil conditions, while volatilization is most common when soils are only moist and are drying [22].

8. Environmental Problems of Nitrogen Fertilizers

Commercial fertilizer, especially nitrogen, is easily washed below the level of the plant's root system through the leaching of rain or irrigation. An application which is too heavy or too close to the roots of the plants may cause "burning" (actually a process of desiccation by the chemical salts in the fertilizer). As well, overly heavy applications of commercial fertilizers can build up toxic concentrations of salts in the soil, thus creating chemical imbalances. If organic materials are readily available and cheap, the expense of the commercial fertilizer should also be considered [116].

Nitrogen fertilizer use is currently linked to some water quality problems involving surface and underground drinking water supplies. Current public health regulations in the United States and other countries require water suppliers to notify customers when NO3-nitrogen concentrations in their water exceed 10 micrograms per liter. This is because ingestion of large amounts of NO_3^- by very young infants can lead to development of a potentially fatal, but treatable, condition called methemoglobinemia. Nitrate ingestion is usually not harmful to older children or adults [22]. Nitrate is not absorbed to any appreciable extent by soil particles. Nitrogen is therefore the most likely fertilizer element to be leached out into surface water or groundwater, or to be lost to the atmosphere by denitrification, partly as nitrogen gas and partly as nitrous oxide, which is a greenhouse gas.

9. The Importance of the Phosphorous Mineral Fertilization

Fertilizers are an essential component of industrial agriculture. Plants extract soil nutrients during their growth; harvest, transport, and consumption of crops transfers soil nutrients into animal—including human—bellies. Most agricultural practices remove nutrients faster than they are replenished by natural processes of soil formation; consequently potassium, nitrogen, phosphorus and micronutrient fertilizers are purchased and applied by most farmers [36].

Global food production depends upon availability of phosphate rock, a non-renewable fertilizer ore that is in diminishing global supply [36]. Phosphorus is often a limiting nutrient in natural ecosystems. That is, the supply of available phosphorus limits the size of the population possible in those ecosystems. When the soil is deficient in a particular nutrient element and the crop grown on such soil show visible deficiency symptoms, there is urgent need to correct it to improve the crop production [41].

P is second only to N as the most limiting element for plant growth [108]. Phosphorous (P) plays a series of functions in the plant metabolism and is one of the essential nutrients required for plant growth and development. It has functions of a structural nature in macromolecules such as nucleic acids and of energy transfer in metabolic pathways of biosynthesis and degradation [62]. Next to nitrogen, phosphorous is the most abundant nutrient contained in the microbial tissue, making up as much as 2% of the dry weight [55].

Phosphorous occupies a key place among nutrients because of its relative scarcity and its essential role in all life forms. P is a major constraint on food and fibre production in many parts of the world. Therefore, an economical supply of P is a necessity for a secure production in agriculture and forestry. An inefficient way to satisfy this P demand is by the exclusive use of inorganic fertilizers, which rock phosphate minerals are normally usually containing heavy metal contaminations and which often have relatively low use efficiencies in the field [93]. The elemental state of phosphorus is highly bondable, meaning it readily combines with several other elements, which is why it is predominantly found in phosphate form. As such, phosphorus molecules bond with soil molecules and are not available for plant use until the soil is saturated, or there are no longer any soil particles for the phosphorus to bond with. The result has been a great demand for artificial fertilizers to boost nutrient content in less fertile soil [23].

The ability to obtain the minimum phosphate concentration needed for reasonable growth depends in part on the microbial environment of the root. Under some conditions the rhizosphere microorganisms will contain species that will solubilise phosphate from compounds of very low solubility and so increase the phosphate supply to the plant. Due to the slow rate of diffusion of phosphate, the phosphate supply to a plant depends much more on the size of the root system, the density of its root hairs and the intensity of its ramifications through the soil, than does the supply of the most nutrients [55].

The amount of P in plants ranges from 0.05% to 0.30% of total dry weight. The concentration gradient from the soil solution to the plant cell exceeds 2,000- fold, with an average free P of 1 m in the soil solution [9].

However, although bound P is quite abundant in many soils, it is largely unavailable for uptake [9] [83].

Crop yield on 40% of the world's arable land is limited by P availability. P is unavailable because it rapidly forms insoluble complexes with cations and is incorporated into organic matter by microbes. The acid-weathered soils of the tropics and subtropics are particularly prone to P deficiency and aluminum (Al) toxicity [106] In intensive agriculture, a grain crop yield of 7 metric tons Ha -1 requires the addition of 90 to 120 kg P Ha -1 However, even under adequate P fertilization, only 20% or less of that applied is removed by the 1st year's growth [107].

A low P concentration in the soil solution is usually adequate for normal plant growth. For example, Barber suggested that 0.2 ppm P was adequate for optimum growth. However, for plants to absorb the total amounts of P required to produce good yields, the P concentration of the soil solution in contact with the roots requires continuous renewal during the growth cycle [7].

10. THE PROBLEM WITH THE EFECTIVENESS OF MINERAL FERTILIZERS

The efficiency of fertilizers might be lower than fifty per cent either due to mismanagement or due to reactions of the fertilizers with soil constituents. only a fraction of the fertilizer applied to the soil is taken up by the crop, the rest is either remains in the soil or lost through leaching, physical wash off, fixation by the soil, or release to the atmosphere through chemical and microbiological processes. This low uptake and utilization of nutrients out of the added fertilizers results in lower production and appearance of nutrient stress symptoms. The poor efficiency of added fertilizers is mainly due to imbalance fertilization or improper management practices [41].

There are three general main types of nitrogen fertilizers where the nitrogen in the fertilizer is found in the chemical form of: ammonium, urea and nitrate. Nitrogen in nitrate form fertilizer are soluble and highly mobile in soil. They are prone to various kinds of losses: leaching beyond the root zone in the light textured soils, denitrification in water logged conditions or short time of over irrigation. Ammonium type fertilizers usually transform in the soil within a week or two to nitrate form and the further consideration is the same as for nitrate fertilizers [41].

Phosphate fertilizers, which are highly reactive, are fixed in soil and become immobile.

The assessment of the profitability of P fertilisation is generally based on results of agronomic trials. Fertilizer trials are predominantly carried out under optimum or near-to-optimum management conditions (soil preparation, sowing time, weeding) which do not reflect on-farm conditions in particular when farmers rely on manual labour. For instance, sowing dates are normally distributed over several weeks so that conditions for germination and emergence in many cases are sub-optimal. Therefore, plant establishment and responses to fertilizer are lower on average than those of trial results [93]. Table 4 illustrates the different types of phosphate fertilizers.

Table 4. Different types of phosphate fertilizers

Source	% P2O5	Water soluble %
Normal superphosphate	16	90
Concentrated superphosphate	44-52	95-98
Monoammonium phosphate	48-61	100
Diammonium phosphate	48-53	100
Ammonium polyphosphate	34-37	100
Phosphoric acid	55	100
Super phosphoric acid	76-85	100

P is absorbed mainly during the vegetative growth and, thereafter, most of the absorbed P is re-translocated into fruits and seeds during reproductive stages. P-deficient plants exhibit retarded growth (reduced cell and leaf expansion, respiration and photosynthesis), and often a dark green colour (higher chlorophyll concentration) and reddish coloration (enhanced anthocyanin formation) [62].

The concentrations of inorganic P in soil solutions are low because P is rapidly sorbed to soil colloids [69]. Where a water-soluble Phosphorus (WSP) fertilizer is applied to the soil, it reacts rapidly with the soil compounds. The resulting products are sparingly soluble P compounds and P adsorbed on soil colloidal particles [24] and only up to 45% of Phosphorus fertilizers taken up by crops [89].

Factors Affecting Phosphorous Availability [42]:

1. Amount of Clay: Increased clay content results in greater retention of P by the soil.
2. Type of Clay: Kaolinite and iron oxides retain more P than 2:1 clays
3. Soil pH: Determines the form of P, and the levels of Al, Fe and Ca in soil solution. Also affects pH dependent charge.
4. Soil P Levels: Soils very high in P will eventually saturated binding sites and tend to release higher concentrations of P to soil solution.
5. Temperature: At low and high temperatures, ability of the plant to remove P is decreased
6. Compaction: Limits root penetration and volume of the soil contacted by the root system
7. Aeration: Poorly aerated and poorly drained soils can limit root system development
8. Moisture: Low moisture limits movement of P and affects crop development, excess moisture affects rooting and limits P uptake
9. Time and Method of Application: Increasing the time of contact increases the amount of P retained by the soil in unavailable forms. On soils with high P retention, banded applications may provide more available P to the crop when soil P levels are low. As soil fertility levels increase, the banding advantage decreases., although starter fertilizers frequently result in improved yields due to improved early season development in some crops. Broadcast applications are more rapid and cheaper to apply, can provide larger rates without crop injury, and can result in better mixing within the entire root zone which over time can increase rooting depth.
10. Solubility and/or Particle Size: Retention tends to increases as the solubility increases and as particle size of the applied P decreases.
11. Incorporation vs. Surface Application: Surface applied P is usually the least efficient method of placing P. Incorporation throughout the plow layer leads to improved rooting depth and more effective uptake over time. In no-till situations, incorporation is not possible. Building good P levels prior to initiating a no-till system is strongly recommended. As residues build on the soil surface, natural P cycling in the soil by animals, and movement through macropores, surface applications of can supply the needs of the crop.
12. Other nutrients: Although P availability is little affected by the presence of other essential nutrients within normal ranges, the presence of these nutrients can stimulate P uptake.

Management and research must aim at nutrient and organic matter cycling in combination with sufficient fertilizer use to avoid "nutrient mining" in agriculture and forestry.

Many natural ecosystems and low-input farming systems have adapted to low P supply by recycling P from litter and soil organic materials. Increases in productivity or exports of commodities require external nutrient inputs if they are not to cause a decline in fertility [95].

Phosphorus is a major constraint on food and fibre production in many parts of the world. Therefore, an economical supply of P is a necessity for a secure production in agriculture and forestry. An inefficient way to satisfy this P demand is by the exclusive use of inorganic fertilisers, which normally are expensive imports, and which often have relatively low use efficiencies in the field [95].

Importance of Nutrient Balance

A positive net mass balance indicates the amount of residual nutrient that may remain in the soil or be lost to the air, carried by water runoff into surface-water systems, or carried by percolating water into ground water. A negative net balance indicates that the amount of nutrient removed from the field through the harvested crop exceeds the amount of nutrient applied, with the difference coming from nutrients stored in the soil or available through precipitation. Continued negative balances mine or deplete nutrients in soil, disrupt the soil ecosystem, and can damage soil productivity [39].

In areas with high soil erosion and runoff, the high residual balance of phosphorus could contribute to water quality problems and require improved management. Phosphorus is more stable than nitrogen and more likely to remain in the soil with less loss to the environment unless the soil itself erodes away. Because of this greater stability, and to reduce costs, many farmers apply extra phosphorus one year then skip a year or more [39].

11. Phosphate Rock Is a Finite, Non Renewable Natural Resources

The current world production of food is dependent on large amounts of fossil fuels (fertilizer production) and finite reserves of phosphate rock. Canada has no significant mineable sources of phosphate for inorganic fertilizer production and is entirely dependent on imported sources, mainly the United States.

PR is a finite, non-renewable natural resource. Geological deposits of different origin are found throughout the world. Currently, few PR deposits are mined, and about 90 percent of world PR production is utilized by the fertilizer industry to manufacture P fertilizers, with the remainder being used to manufacture of animal feeds, detergents and chemicals [25].

In fact, if viewed from a global perspective, phosphate rock is not a purchased input, but represents a non-renewable with finite bounds of particular importance to feeding a burgeoning world population [10] and for which there is no substitute [36].

Most of the world's farms do not have or do not receive adequate amounts of phosphate. Feeding the world's increasing population will accelerate the rate of depletion of phosphate reserves. Rocks containing phosphate have been discovered and are being mined at minimal cost. But resources are limited, and phosphate is being dissipated. Future generations ultimately will face problems in obtaining enough to exist [1]. P is an vitally important resource and its future use in agriculture will be impacted by declining availability and increased cost.

The major plant nutrients (nitrogen, phosphorus, and potassium) are critical for maintaining crop yields but have also been associated with the impairment of numerous streams, lakes, and aquifers. Commercial fertilizer is a major agricultural input; farmers typically spend over 10 billion dollars annually on commercial fertilizer, although fertilizer use and prices vary from year to year. However, fertilizer prices, especially for nitrogen, have been volatile and have risen rapidly in recent years. Despite increased fertilizer prices and growing concern about environmental risks from fertilizer use, the use of nutrient management practices on major crops has changed little since the early 1990s.U.S. commercial fertilizer use peaked in 1981 at over 23 million nutrient tons, but has exceeded 22 million tons seldom since then [114] Since 1990, the primary fertilizer products purchased by farmers have exhibited much different nominal price patterns Potash prices, in the form of potassium chloride (KCl), have been stable. A major source of phosphate, diammonium phosphate (DAP), has demonstrated much more price variability—peaking in 1995 at near-record levels, declining steadily through 2002, then rising again in 2003 and 2004. Nitrogen product prices, such as those for anhydrous ammonia, have shown the greatest volatility during the last several years—a spike in 1995, a fall to historically low levels by 1999 and 2000, record-high prices in 2001, followed by a large decline in 2002 and recovery in 2003. During 2001 and 2003, nitrogen fertilizer prices rose dramatically in concert with natural gas prices. In response to higher prices in the United States, nitrogen-based fertilizer imports have increased significantly in recent years [114].

While most economic studies indicate that farmer response to fertilizer price changes is fairly inelastic (i.e., relatively unresponsive to price changes), farmers can make some adjustments in fertilizer use when faced with dramatically higher prices [114].

Table 5. Average U.S farm prices of selected phosphate fertilizers (dollars per ton) [103]

Year	Month	Super-phosphate 20% phosphate	Super-phosphate 44–46% phosphate	Diammonium phosphate (18-46-0
1960	September	38.0	80.1	NA
1970	September	46.9	76.2	95.1
1974	September	104.0	188.0	228.0
1980	May	132	251	298
1990	April	NA	201	219
1995	April	NA	234	263
2000	April	NA	233	240
2001	April	NA	236	244
2002	April	NA	221	227
2003	April	NA	243	250
2004	April	NA	266	276
2005	April	NA	299	303
2006	April	NA	324	337
2007	April	NA	418	442

Table 6. Fertilizer price indexes [103]

Year	Index of prices	
	Paid by farmers for fertilizer	Received by farmers for all crops
1976	74	86
1980	96	105
1984	103	109
1990	97	101
1992	**100**	**100**
1995	105	95
2000	110	96
2001	123	99
2002	108	105
2003	124	111
2004	140	117
2005	162	112
2006	174	120

Table 7. Consumption of selected phosphate fertilizers (metric tons) [103]

Year	Superphosphates Grades 22% and unde	Superphosphates Grades over 22%	Superphosphates Diammonium phosphate (18-46-0) 1	Other nitrogen-phosphate grades 2
1960	510539	399319	20388	NA
1968	401347	1109332	1246953	NA
1970	312032	1204566	1514911	NA
1976	140343	1211060	3232501	NA
1981	112524	1051653	3721845	2510250
1985	68574	751412	3433653	2730920
1990	13717	579841	3580222	3065005
2000	11437	333298	3220179	4114562
2004	21421	213702	3580108	4469690
2005	15635	191061	3392521	4869201
2006	8993	173939	2999062	4925666

Estimates on the remaining amount of phosphorus vary as do projections about how long it will take to deplete the irreplaceable resource entirely. Figures range from 60–130 years [91] and 60–90 years [94] at current market prices with diverse assumptions about the rate of production and demand, but all sources agree that continued phosphorus production will decline in quality and increase in cost. The relatively inexpensive phosphorus we use today will likely cease to exist within 50 years [23]. By some estimates inexpensive rock phosphate reserves could be depleted in as little as 60 to 80 years P fertilizer use increased 4- to 5-fold between 1960 and 2000 and is projected to increase further by 20 Tg year - 1 by 2030 [107]. Between 1950 and 2000, about 1 billion metric tonnes of P has been mined. During this period, about 800 million metric tonnes of fertilizer P were applied to the Earth's croplands. This has increased the standing stock of P in the upper 10 centimetres of soil in the world's croplands to roughly 1,300 million metric tonnes, an increase of 30% [23].

PR sedimentary rock, which accounts for some 85–90% of world production, contains Cd in concentrations ranging from less than 20 to more than 200 mg per kg P_2O_5 [72].

The European fertilizer industry is almost totally dependent on imports of rock phosphate from Africa, the Near East, Russia and the USA.

There is a limit to the phosphate rock reserves that can profitably be recovered at current market prices. High-grade ores with high P_2O_5 concentration and ores of good quality, i.e., containing few contaminants, are being progressively depleted. Consequently, production costs will increase. In the context of declining quality, the levels of certain impurities may pose problems in the processing or in application. The content of heavy metal contaminants are generally higher in sedimentary than igneous rocks. To date, levels have not proved a problem except when producing higher purity phosphoric acid grades. However, certain European countries have applied strict limits to cadmium levels in fertilisers in order to avoid contamination of farmland and crops and cadmium removal could involve further processing costs adding 2-10% to phosphate fertiliser prices [91].

Phosphate content in currently mined rocks can range from over 40% to below 5%. The mined rock is further processed to remove the bulk of the contained impurities and thus upgrade the rock. Consequently, the rock concentrate contains an increased apatite content of an improved quality. Phosphate rock can be beneficiated by many methods, and usually a combination of methods is used. In general, with the lower concentration of phosphate and lower quality deposit, the more waste is generated. Furthermore, more energy and chemicals are required per tonne of useful phosphate produced. Consequently, the cost for recovery and beneficiation of phosphate rock increases significantly in relation to lower grade and lower quality deposit [91].

It imperative that we begin recycling phosphorus and returning it to the soil to decrease the need for mined phosphorus as artificial fertilizer. Within a half century, the severity of this crisis will result in increasing food prices, food shortages and geopolitical rifts [23].

The critical question, when phosphate ore reserves will be exhausted, is hard to answer, however the economic mining of the phosphate ore is estimated to few decades only [93].

There is a growing concern that the world is coming close to its 'limits of growth' as earth's natural resources are limited. P is one of the essential inputs for agricultural production and its growth. Environmental quality comprises two important issues, the maintenance of soil fertility and the avoidance of environmental pollution.

Therefore, it is an urgent need to reduce the loss of nutrients form agriculture, all in order to slow the consumption of finite Phosphate rock deposits. A potential phosphate crisis looms for agriculture in the 21st century. Sustainable management of P in agriculture requires that plant biologists discover mechanisms in plants that enhance P acquisition and exploit these adaptations to make plants more efficient at acquiring P, develop P-efficient germplasm, and advance crop management schemes that increase soil P availability [107].

There are no higher P content (15–18%) materials known other than apatite minerals, which are either in phosphate rock or bone charcoal AGROCARBON formation. The P content of other materials may occur as low as 0.1–0.5 % only, if any at all.

Organic P from manure or sludge should be comparable to P from inorganic fertilizer. Therefore, if a producer has a P recommendation for 33 kg/ha of P_2O_5, applying approximately 73kg of 18-46-0 Diammonium Phosphate fertilizer or 6 tons of 11-6-9 manure (80% available P coefficient) should provide equivalent results [78].

A significant percentage of phosphorus can be recovered by using sustainable agriculture and sanitation. This should be a priority for the global policy agenda.

12. The Inorganic Contamination of Phosphate Rock (PR) Fertilizers

The process to extract phosphorus is particularly damaging, as strip-mining is the most common method of extraction. The by-products of the processing of phosphorus include radioactive gypsum, commonly stored in "stacks" in the United States, and air pollution that includes fluorides. Furthermore, phosphorus deposits are often contaminated by arsenic or cadmium, which can be concentrated during processing and there is a subsequent build-up of these toxic substances in agricultural soils [23].

Once mined, rock phosphorite is crushed and acidified at an enormous scale. 85% of global sulfur is used to make sulfuric acid; 70% of this industrial acid is used to process phosphate rock into usable fertilizer. Several different acidification processes are employed to produce phosphorus-rich liquids and precipitates that are sold and applied to soils as fertilizer and animal feed supplements [36].

PR contains various metals and radionuclides as minor constituents in the ores. Varying amounts of these elements are transferred to phosphate fertilizers in production processes, and later are applied to soils with these fertilizers. Cadmium (Cd) is the heavy metal of most interest because it is potentially harmful to human health, and much attention is being given to its avenues of entry into the human food chain [93].

Cd from phosphate fertilisers poses a potentially serious threat to soil quality and, through the food chain, to human health. The exact size of the problem is hard to determine, due to a lack of sufficient and reliable data. However, it is obvious that in *the long term the continuing input of Cd to agricultural soils through fertilisers could lead to accumulation beyond acceptable levels.* While there are other important sources of soil contamination with Cd as well (e.g., manure, compost, sewage sludge and atmospheric deposition), phosphate fertilisers are generally the major contamination sources and the most important ones as far as farmland is concerned. The European Commission is considering the possibility of using an European Union (EU)-wide charge on Cd in fertilisers so as to improve the competitive position of the 'low-Cd' product *vis-á-vis* the 'high-Cd' one. Such a charge would encourage farmers to use the low-Cd alternatives, and stimulate suppliers to invest in technologies for producing them [72].

Cd occurs naturally as a contaminant in all phosphate rock, but the concentrations vary considerably, depending on the origin of the material. Igneous rock or apatite (found in the former Soviet Union, Finland, South Africa and South America) has low concentrations of Cd (often less than 1 mg per kg P_2O_5). [72] Cadmium from phosphate fertilisers poses a potentially serious threat to soil quality and, through the food chain, to human health. The European Commission is considering the possibility of using an EU-wide charge on cadmium in fertilisers so as to improve the competitive position of the 'low-Cd' product "vis-á-vis" the 'high-Cd' one [72].

Decadmiation is currently too expensive and low-cost technologies are not yet fully developed. In some cases technical restrictions impede the use of either magmatic rock or decadmiation [72].

Obviously, the flow of cadmium to agricultural soils can also be reduced by applying lower amounts of phosphate fertiliser and/or substituting fertilisers by other products containing phosphorus, e.g., manure. While this may be a feasible option in certain cases, it does not take away the need to minimise the amount of cadmium in fertilisers. Moreover, products such as manure, compost and sewage sludge may also contain substantial amounts of cadmium and/or other contaminants [72].

There are still no agreed safe Cd limits for soils, but it appears that concern about Cd build-up in soils may be warranted only where several critical factors combine, i.e., on acid soils with low cation exchange and low P fertility, to which significant P fertilisation is applied as low grade fertilizer or rock phosphate, particularly to Cd accumulators like leafy vegetable crops. In the natural environment, P is supplied through the weathering and dissolution of rocks and minerals with very low solubility. Therefore, P is usually the critical limiting element for plant and animal production. Present legislative limits in the EU Member States variating from 21.5 mg Cd/kg P_2O_5 while the existing Cd concentrations in fertilizers is up to 58 mg Cd/kg P_2O_5. In order to decrease the Cd accumulation in agricultural soils 40 mg upper limit Cd/kg P_2O_5 limit concentration proposed in P fertilizers by 2010 and 20 mg Cd/kg P_2O_5 by 2015. In order to maintaining the status quo, the industry proposes to accept upper limit limit of Cd in phosphate fertilizer of 60 mg Cd/kg P_2O_5. A limit variation for Cd in phosphate fertilizers might also be derived based on a risk assessment approach. However, rock phosphate resources with low Cd content are much less available than expected in the past, meaning that meeting an upper limit for the annual average for each manufacturer might be problematic to achieve [51].

The Cd content of which varies considerably from source to source, revealing as much as a hundred-fold difference when extremes are compared. While the purchase of rock with a low Cd content is clearly desirable, it should be noted that there are insufficient quantities available in the world to meet food production needs. While the 1988 EU directive on Cd emissions to water recognised the desirability of reducing inputs of Cd to soils as part of an overall strategy for reducing environmental contamination by Cd, it also acknowledged that by 2007 there is still no economically viable process to remove Cd from phosphate rock.

A side effect of sustained heavy P fertilizer additions can be the accumulation and introduction into the human food chain of heavy metal contaminants contained in fertilizers. Phosphate fertilizer use has caused small but significant increases soil Cd levels. These inputs are of a magnitude similar to those from the atmosphere in industrialised countries. There are still no agreed safe Cd limits for soils, but it appears that concern about Cd build-up in soils may be warranted only where several critical factors combine, i.e., on acid soils with low cation exchange and low P fertility, to which significant P fertilisation is applied as low grade fertilizer or rock phosphate, particularly to Cd accumulators like leafy vegetable crops, and where the produce is the main source of local food consumption.

Table 8. Average heavy metal concentrations in phosphate rock deposits and estimated input to soil by P fertilizers

	Heavy metal concentration								
PR Deposit	As	Cd	Cr	Cu	Pb	Hg	Ni	V	Zn
	mg kg-1 of PR								
Russia (Kola)	1	0.1	13	30	3	0.01	2	100	19
USA	12	11	109	23	12	0.05	37	82	204
South Africa	6	0.2	1	130	35	0.06	35	3	6
Morocco	11	30	225	22	7	0.04	26	87	261
Other N. Africa	15	60	105	45	6	0.05	33	300	420
Middle East	6	9	129	43	4	0.05	29	122	315
Avg. of 91%	mg kg-1 of PR								
of PR reserves	11	25	188	32	10	0.05	29	88	239
	mg kg-1 of P								
	71	165	1226	209	66	0.29	189	578	1561
Applied with	g ha-1								
20 kg P ha^{-1}	1	3.3	25	4	1	0.01	4	12	31
Tolerable limit	mg kg-1 of soil								
	-	2	100	100	100	2	50	50	300

Availability of fertilizer Cd to plants is related to its solubility in the P fertilizer as well as to soil factors such as pH, cation exchange capacity, and clay and organic matter contents. Plants vary in both their capacities to absorb and to translocate Cd from vegetative tissues to grain [93].

Table 8 presenting the average heavy metal concentrations in phosphate rock deposits and estimated input to soil by P fertilizers [53].

Cd is a toxic heavy metal. Humans exposed to high levels over a number of years may develop chronic Cd poisoning. The most recent concern is the possibility that a growing number of people are being exposed to Cd through its uptake in their diet. This has now brought phosphate fertilizers to the centre of the debate. The major concern is that Cd slowly accumulates in the kidney causing irreversible damage, where the concerns are:

1. Cd do not have an environmental half-life but persists indefinitely in the environment.
2. Cd is removed from the human body only slowly (half-life in the order of 20 years)

3. Cd cause irreversible damage to the target organ the kidney and significantly increases the human health risk.

Maximum levels of Cd in foodstuffs are already set by the Commission's regulation EC 466/2001 of 8 March of 2001. Sedimentary rock, which accounts for some 85–90% of world production, contains Cd in concentrations ranging from less than 20 to more than 200 mg per kg P2O5. De-cadmiation is currently too expensive and low-cost technologies are not yet fully developed. In some cases technical restrictions impede the use of either magmatic rock or de-cadmiation [72].

Concentrations of other heavy metal and radionuclide contaminants in P fertilizers vary considerably, depending on the PR source. Some *PR source metals of possible significance are: arsenic (As), chromium (Cr), lead (Pb), mercury (Hg), nickel (Ni), and vanadium (V). The main radionuclide contaminants in PR are uranium (U), radium (Ra), and thorium (Th)* [93].

Typical levels of gamma radiation from the PR deposits ranged from 5 to 30 kBq kg^{-1} P [53].

The median contents of PR in the USA were reported as 59 mg kg^{-1} of U, 8 mg kg^{-1} of Th, and 18 mg kg^{-1} of Ra. [66] The wide use of phosphate fertilizer will certainly cause high radiological environmental pollutions. Evidence of very high alpha activity in some soil samples may indicate excessive use of phosphate fertilizer in agricultural lands. As phosphate fertilizers are in the form of fine powder, precautionary measures need to be taken to minimize the level of airborne dust during processing, transportation, and application [30].

13. THE ENVIRONMENTAL RISK OF PHOSPHATE FERTILIZER INDUSTRY

Nearly all phosphate fertilizer is produced by treating phosphate rock with sulfuric acid to produce phosphoric acid, which is further processed into various phosphatic fertilizer materials such as superphosphates and ammonium phosphates. The United States has become the world.s largest phosphate fertilizer exporter [39].

Approximately 1.7 tons of phosphate rock and about 1.3 tons of sulfuric acid are required to produce a ton of diammonium phosphate [39].

Fluoride has been one of the largest environmental liabilities of the phosphate industry. The source of the problem lies in the fact that raw phosphate ore contains high concentrations of fluoride, usually between 20,000 to 40,000 parts per million (equivalent to 2 to 4% of the ore). When this ore is processed into water-soluble phosphate (via the addition of sulfuric acid), the fluoride content of the ore is vaporized into the air, forming highly toxic gaseous compounds (hydrogen fluoride and silicon tetrafluoride). In the past, when the industry had little, if any, pollution control, the fluoride gases were frequently emitted in large volumes into surrounding communities, causing serious environmental damage [16].

To make 1 pound of commercial fertilizer, the phosphate industry creates 5 pounds of contaminated phosphogypsum slurry (calcium sulfate). This slurry is piped from the processing facilities up into the acidic wastewater ponds that sit atop the mountainous waste piles known as gypsum stacks. After phosphate rock has been reacted with sulfuric acid to produce phosphoric acid, the precipitated gypsum is removed on filters and is pumped in slurry form to an impoundment where it is allowed to settle. As the gypsum accumulates, a small dragline removes some of the gypsum for raising the height of the dikes. By this process the gypsum settling impoundment, or "stack" as it is often called, increases in elevation. As a stack grows in height (up to 200 feet) the area of the settling impoundment decreases until a point is reached where the pond capacity becomes too small and the pumping height requires too much energy [64].

It is estimated that more than 22 million tonnes of phosphoric acid (as P_2O_5) are produced annually worldwide, generating in excess of 110 million tonnes of gypsum by-product. The free water in the gypsum cake off the filters is highly acidic, having a pH as low as 1.0. While commercial uses, in agriculture and in manufacturing gypsum board and Portland cement, consume less than a few percent of this by-product, the vast majority is disposed of on land in gypsum stacks or is discharged into water bodies [5].

Since the mid-eighties, the annual production rate of phosphogypsum has been in the range of 40 to 47 million metric tons per year. The total amount generated in the United States from 1910 to 1981 was about 7.7 billion metric tons [105]. According to the EPA, 32 million tons of new gypsum waste is created each year by the phosphate industry in Central Florida alone. (Central Florida is the heart of the US phosphate industry). The EPA estimates that the current stockpile of waste in Central Florida's gypsum stacks has reached "nearly 1 billion metric tons." (The average gypsum stack takes up about 135 acres of surface area—equal to about 100 football fields—and can go as high as 200 feet.) [16].

The pH of the phosphogypsum is very acid: pH 2.8. The acidity is due to residual phosphoric and sulfuric acids plus hydrofluoric and fluosilicic acids [64].

Fluoride emissions from phosphogypsum pond water surfaces are primarily HF and have been measured to be in the range of 0.01 to 0.10 kg/hectare/day. Fluoride emissions from dry gypsum surfaces are associated with fugitive gypsum dust. Vegetation very close to a gypsum stack can contain elevated levels of fluorides, which if ingested by cattle for years, can cause fluorosis [5].

Phosphate rock, which is processed to make phosphoric acid, contains concentrations of naturally occurring radioactive elements (radionuclides). Even high grade ores, which contain about 70% calcium phosphate, also contain a large number of impurities, such as calcium fluoride, chlorides, chromium, rare earths, and radionuclides. At the end of the production process, the radionuclides end up in the phosphogypsum [105].

The phosphogypsum, separated from the phosphoric acid, is in the form of a solid/water mixture (slurry) which is stored in open-air storage areas known as stacks. The stacks form as the slurry containing the by-product phosphogypsum is pumped onto a disposal site. Over time the solids in the slurry build up and a stack forms. The stacks are generally built on unused or mined out land on the processing site. The surface area covered by stacks ranges from about 5 to 740 acres. The height ranges from about 10 to 200 feet. In 1989, the total surface area covered by stacks was about 8,500 acres. More than half that acreage is in Florida [105].

Phosphogypsum is virtually useless, and is disposed of in large, aboveground stacks, or piles. A total of 63 phosphogypsum stacks were identified nationwide in 1989, in 12 different states. Two-thirds of these stacks were located in Florida, Texas, Illinois, and Louisiana. In Central Florida, one of the major phosphoric acid producing areas, the industry generates about 32 million metric tons of phosphogypsum each year, which equals the combined weight of approximately 6.4 million elephants [105].

It is sort of a misnomer, however, to call these stacks "gypsum" stacks. Indeed, if the stacks were simply gypsum, they probably wouldn't exist, as gypsum can be readily sold for various purposes (e.g.. as a building material). What can't be readily sold, however, is radioactive gypsum, which is about the only type of gypsum the phosphate industry has to offer. The source of the gypsum's radioactivity is the presence of uranium, and uranium's various decay products (i.e. radium), in raw, phosphate ore [16].

There is a natural and unavoidable connection between phosphate mining and radioactive material. It is because phosphate and uranium were laid down at the same time and in the same place by the same geological processes millions of years ago. They go together. Mine phosphate, you get uranium. While uranium, and its decay-products, naturally occur in phosphate ore, their concentrations in the gypsum waste, after the extraction of soluble phosphate, are up to 60 times greater [16].

The concentrations of uranium and radium-226 in phosphogypsum samples taken in central Florida were about 10 times the background levels in soil for uranium and 60 times the background levels in soil for radium-226.

The radium-226 concentration in phosphogypsum varies significantly at different sampling locations on a single stack and also in phosphogypsum from different stacks within the same geographical area [105].

EPA is currently "weighing whether to classify the gypsum stacks as hazardous waste under federal statutes, which would force the industry to provide strict safeguards" (to nearly 1 billion tons of waste) [16].

Radionuclides that are small particles (i.e., radionuclide dust) can become airborne as wind-blown dust or as dust thrown up into the air by cars and trucks. The radionuclides, uranium and radium-226, are present in the phosphogypsum and can become airborne. Once these radionuclides are in the air, people and animals can breathe them and they can settle out onto ponds and agricultural areas. Radon-222, a decay product of radium-226, is a gas and so may become airborne by diffusing into the air. EPA has determined, however, that the risks associated with stacking phosphogypsum are in line with acceptable risk practices [105].

Phosphogypsum contains some trace metals in concentrations that EPA believes may pose a chemical hazard to human health and the environment. Analysis of samples from various facilities contained arsenic, lead, cadmium, chromium, fluoride, zinc, antimony, and copper at concentrations that may pose significant health risks. The concentrations of these contaminants vary by more than three orders of magnitude among samples taken from various locations. These trace metals may also be leached from phosphogypsum and migrate to nearby surface and groundwater resources [105].

Only two uses (for the gypsum) are permitted: limited agricultural use and research. Other uses may be proposed, but otherwise the phosphogypsum must be returned to mines or stored in stacks [16].

While the presence of uranium decay-products makes gypsum a tough sell for the phosphate industry, the uranium has, at various times, presented the industry with a business opportunity of its own. One of the lesser-known-facts about the phosphate industry is that its processing facilities have produced and sold sizeable quantities of uranium. In 1997, just two phosphate plants in Louisiana produced 950,000 pounds of commercial uranium, which amounted to roughly 16% of the domestically produced uranium in the US. In 1998, the same two plants produced another 950,000 pounds, but due to declining market prices for uranium, both plants have since ceased production. If market prices improve, however, 4 US phosphate plants (2 in Louisiana & 2 in Florida) would have the capacity to produce a combined 2.75 million pounds of uranium per year, according to the Department of Energy (DOE) [16].

Resting atop the phosphate industry's gypsum piles are highly-acidic wastewater ponds, littered with toxic contaminants, including fluoride, arsenic, cadmium, chromium, lead, mercury, and the various decay-products of uranium. This combination of acidity and toxins makes for a poisonous, high-volume, cocktail, which, when leaked into the environment, wreaks havoc to waterways and fish populations [16].

Summary of Problem with RP

4. The economically minable low heavy metal contaminated total RP rock phosphate of the world is sufficient up to 25 year
5. VERY IMPORTANT: there world population and food consumption + energy use level is under acceleration, therefore these calculated numbers are not valid, as the P supply issue is much worse situated than in worst case ever calculated.
6. The Cd, Uranium and other heavy metal issue makes that the accessible LOW CD RPs are exhausted within less than a decade, again calculated as of todays consumption level.
7. If world population and food consumption + energy use level is accelerating faster (for which there are very clear signs in India, China, Indonesia, Philippines and Russia already), than agro business will also accelerate proportionally, which means that LOW CD RP deficit is expected to be a reality already by 2010s.
8. If very significant investments made for RP production and processing than extension of RP access may be prolonged, but the investments to be done are billions and billions, practically almost a new industry has to be built up.
9. There are ongoing technology developments to remove CD from RP by chemical process, but these issues will be extreme high costly and industrial technically not yet available.
10. The transport energy cost through the world is also under explosion, which is additionally accelerating the RP prices, especially the LOW CD RPs imported from long distance.

14. Greenhouse Gas Emission from Intensive Agriculture Activities

Agricultural systems contribute to carbon emissions through several mechanisms: the direct use of fossil fuels in farm operations, the indirect use of embodied energy in inputs that are energy intensive to manufacture (e.g., fertilizers), and the cultivation of soils resulting in the loss of soil organic matter.

Naturally occurring greenhouse gases include water vapour, carbon dioxide (CO_2), methane (CH_4), nitrous oxide (N_2O), and ozone (O_3). Several classes of halogenated substances that contain fluorine, chlorine, or bromine are also greenhouse gases, but they are, for the most part, solely a product of industrial activities.

There are also several gases that do not have a direct global warming effect but indirectly affect terrestrial and/or solar radiation absorption by influencing the formation or destruction of greenhouse gases, including tropospheric and stratospheric ozone. These gases include carbon monoxide (CO), oxides of nitrogen (NOx), and non-CH_4 volatile organic compounds (NMVOCs). Although the direct greenhouse gases CO_2, CH_4, and N_2O occur naturally in the atmosphere, human activities have changed their atmospheric concentrations [104].

While the physics of the greenhouse effect are similar for all gases, each gas differs in its overall effect on the earth's radiation balance, depending on the concentration of the gas, its residence time in the atmosphere, and its physical properties with respect to absorbing and emitting radiant energy. A common measure, termed the global warming potential (GWP), is used to equate the effect of different greenhouse gases on a mass basis. By convention, the effect of carbon dioxide (CO_2) is assigned a value of one (1) and the warming potential of other gases are expressed relative to that standard, i.e., on a CO_2– equivalent basis [73].

The GWP (global warming potentials) of a greenhouse gas is defined as the ratio of the time-integrated radiative forcing from the instantaneous release of 1 kilogram (kg) of a trace substance relative to that of 1 kg of a reference gas (See Table 9) [43].

One tonne of nitrous oxide is deemed to have the same warming effect as 310 tonnes of CO_2, and one tonne of methane is deemed to have the warming effect of 21 tonnes of CO_2.

The reference gas used is CO_2, and therefore GWP-weighted emissions are measured in teragrams of CO_2 equivalent (Tg CO_2 Eq.). One teragram is equal to 1012 grams or one million metric tons [104].

Agricultural systems contribute to carbon emissions through the direct use of fossil fuels in food production, the indirect use of embodied energy in inputs that are energyintensive to manufacture, the cultivation of soils and/or soil erosion resulting in the loss of soil organic matter [6]. In 2006, the agricultural sector was responsible for emissions of 454.1 teragrams of CO_2 equivalent (Tg CO_2 Eq.), or 6 percent of total U.S. greenhouse gas emissions. Methane (CH_4) and nitrous oxide (N_2O) were the primary greenhouse gases emitted by agricultural activities. CH_4 emissions from enteric fermentation and manure management represent about 23 percent and 7 percent of total CH_4 emissions from anthropogenic activities, respectively. Of all domestic animal types, beef and dairy cattle were by far the largest emitters of CH_4. Rice cultivation and field burning of agricultural residues were minor sources of CH_4. *Agricultural soil management activities such as fertilizer application and other cropping practices were the largest source of U.S. N_2O emissions, accounting for 72 percent.* Manure management and field burning of agricultural residues were also small sources of N_2O emissions. [104]

Table 9. Global warming potentials of direct greenhouse gases [44]

Gas	GWP
CO_2	1
CH_4	21
N_2O	310

Nitrous oxide (N_2O) and methane (CH_4) emissions result from both crop and livestock operations and account for approximately 80 percent of U.S. agricultural greenhouse gas emissions on a GWP basis [73].

Nitrogen oxides (NOx), emitted from agricultural soils and through combustion [19]. Increase tropospheric ozone, a component of smog that impacts human health, agricultural crops and natural ecosystems. As much as 35% of cereal crops worldwide are exposed to damaging levels of ozone [14], NOx from agroecosystems can be transported atmospherically over long distances and deposited in terrestrial and aquatic ecosystems. This inadvertent fertilization can cause eutrophication, loss of diversity, dominance by weedy species and increased nitrate leaching or NOx fluxes [110].

Nitrous oxide constitutes the largest agricultural source of GHG emissions in terms of warming potential (48 percent), and almost 70 percent of total U.S. Nitrous oxide emissions are from soils. The best option for reducing these emissions is to use fertilizers more efficiently; adoption of best fertilization practices could reduce agricultural N2O emissions by 30 to 40 percent. [18] Because N_2O emissions are strongly influenced by the availability of nitrogen in soil, some N_2O emissions are an unavoidable consequence of maintaining highly productive crop and pastureland. However, improved control of the amount, timing, and placement of fertilizer can minimize these emissions [73]. Greater than 50 percent of the major cropland area in the United States is rated as having high nitrogen balances, resulting in soils highly susceptible to losses of N_2O to the atmosphere and nitrate (NO_3^-) to water bodies. [102]. Fall application of fertilizer (motivated by lower fertilizer costs in the fall and to save time during spring planting) results in high concentrations of mineral nitrogen remaining in soils over a several-month period with no plant uptake, making that nitrogen vulnerable to losses [73].

In fact, prior to the 1900s, land use was the dominant global source of CO_2 from human activity, and since 1850 an estimated 160 billion metric tons of carbon from biomass and soils have been emitted worldwide as a consequence of land use and land-use changes [45]. Agricultural soils produce the majority of N_2O emissions in the United States. Estimated emissions from this source in 2006 were 265.0 Tg CO_2 Eq. Direct increases occur through a variety of management practices that add, or lead to greater release of, mineral N to the soil, including: fertilization; application of managed livestock manure and other organic materials such as sewage sludge; deposition of manure on soils by domesticated animals in pastures, rangelands, and paddocks (PRP) (i.e., by grazing animals and other animals whose manure is not managed); production of N-fixing crops and forages; retention of crop residues; and drainage and cultivation of organic cropland soils (i.e., soils with a high organic matter content, otherwise known as histosols). Other agricultural soil management activities, including irrigation, drainage, tillage practices, and fallowing of land, can influence N mineralization in soils and thereby affect direct emissions. Mineral N is also made available in soils through decomposition of soil organic matter and plant litter, as well as asymbiotic fixation of N from the atmosphere [104].

Table 10. Selected U.S. greenhouse gas (GHG) emissions from Agriculture (Tg CO_2 Eq.)

Gas/Source	**1990**	**2000**	**2006**
CH_4 (total)	**606.1**	**574.3**	**555.3**
CH_4 (agricultural sector emission)	**165.7**	**171.7**	**174.4**
Percentage of total CH_4 emission (%)	27%	30%	31%
Enteric fermentation	126.9	124.6	126.2
Manure Management	31.0	38.8	41.4
Rice Cultivation	7.1	7.5	5.9
Field Burning of Agricultural residues	0.7	0.8	0.8
N_2O (total)	**383,4**	**385,9**	**367.9**
N_2O (agricultural sector emission)	**281.8**	**276.3**	**279.8**
percentage of total N_2O emission (%)	74%	72%	76%
Agricultural Soil Management	269.4	262.1	265.0
Manure Management	12.1	13.7	14.3
Field Burning of Agricultural Residues	0.4	0.5	0.5
Total CH_4 + N_2O	**989.5**	**960.2**	**923.2**
Total CH_4 + N_2O (agricultural sector emission)	**447.5**	**447.9**	**454.1**
Percentage of total CH_4 + N_2O emission (%)	**45%**	**47%**	**49%**

Source: [104]

Table11. Direct N_2O emissions from agricultural soils by land use and N input (Tg CO_2 Eq.)

Activity	**1990**	**2000**	**2006**
Cropland	**130.9**	**142.0**	**138.9**
Mineral Soils	*128.1*	*139.1*	*136.1*
Synthetic fertilizer	51.3	55.8	53.6
Organic Amendments (a)	9.4	10.2	10.7
Residue N (b)	9.0	10,2	10,1
Other (c)	58.4	62.8	61.7
Organic Soils	*2.8*	*2.9*	*2.9*
Grassland	**87.4**	**74.0**	**75.8**
Total	**218.3**	**216.0**	**214.7**

Source: [104]

(a) Organic amendment inputs include managed manure amendments and other commercial organic fertilizer (i.e., dried blood, dried manure, tankage, compost, and other).
(b) Residue N inputs include unharvested fixed N from legumes as well as crop residue N.
(c) Other N inputs include mineralization from decomposition of soil organic matter as well as asymbiotic fixation of N from the atmosphere. The key to reducing net GHG emissions from agriculture is to promote practices that maintain or increase carbon stocks, while at the same time increasing the efficiency of agricultural inputs (e.g., fertilizer, irrigation, pesticides, animal feed, and animal waste). GHG mitigation policies should address both carbon stocks and emissions of nitrous oxide and methane to achieve the best overall mitigation results [73].

15. Impact of the Kyoto Protocol on Agriculture

The Kyoto Protocol to the United Nations Framework Convention on Climate Change is a proposed treaty that would require industrialized countries to reduce greenhouse gas emissions by sharply restricting their use of coal, oil, and natural gas. Because U.S. agriculture accounts for nearly one-fifth of U.S. greenhouse gas emissions, compliance with the Kyoto Protocol could increase U.S. farm production expenses by 10 billion dollars to 20 billion dollars annually and depress annual farm income by 24 percent to 48 percent. Higher fuel oil, motor oil, fertilizer, and other farm operating costs would also mean higher consumer food prices, greater demand for public assistance with higher costs, a decline in agricultural exports, and a wave of farm consolidations. In short, the Kyoto Protocol represents the single biggest public policy threat to the agricultural community today [27].

U.S. farming is very energy intensive. Fuel and oil costs represent only about 30 percent of the total energy bill farmers pay. The remaining *70 percent is hidden in the prices of manufactured inputs, fertilizer, pesticides, and other chemicals.* Corn and cotton crops use more pesticides and nitrogen fertilizer, products that are quite sensitive to changes in energy prices. Higher energy taxes have the potential for causing an economic downturn in the agricultural sector [27].

16. Global Warming

Rising temperatures due to global warming will likely stimulate soil organic matter decomposition, which may reduce or eliminate the potential to further increase soil carbon stocks [73].

Between 1945 and 1986, the amount of carbon released from UK land tripled. Since 1980, UK soils lost 12–15% carbon of the total carbon in the soil, due mainly to loss from arable lands. This is about 4Million tonnes/yr lost in Greenhouse Gas Emissions. There has been a reduction in GHG emissions from UK agriculture since 1990 of about 14%. A study in the journal Nature looked at the carbon content of soil in England and Wales from 1978–2003 and found that it fell steadily, with some 13 million tonnes of carbon released from British soil each year. Much of it may be entering the atmosphere as the greenhouse gases carbon dioxide and methane, which scientists say has caused global warming. Since the carbon appeared to be released from soil regardless of how the soil was used, they concluded that the main cause must be climate change itself [8].

Soil carbon contents depend on rates of addition from plant growth versus rates of removal in decomposition, leaching and other soil processes, and each of these is sensitive to changes in land use, climate and other variables. The fact that the losses appear to be happening across countries irrespective of land use suggests a link to climate change. Over the survey period, the mean temperature across England and Wales increased by about 0.5 8°C and there were also changes in rainfall distribution. Climate change will affect soil carbon turnover through various processes. Increases in temperature will tend to increase rates of organic matter decomposition by soil microbes, although the magnitude of this effect and differences between soils are uncertain9. The effects of temperature will interact in a complicated way with changes in soil moisture brought about by changing rainfall and evapo-transpiration patterns, and changes in atmospheric CO_2 and nitrogen deposition. In freely drained soils, warmer drier conditions may retard decomposition of organic matter if lack of moisture limits soil microbes. But in wet anoxic soils, increased oxygenation of the soil as evaporation increases the depth to the water table will greatly reinforce the increase in decomposition with increased temperature. Such effects may in part explain the increase in the relative rate of loss with soil carbon content because moreorganic soils tend to be wetter. Also consistent with this, more organic soils necessarily contain a greater proportion of slowly decaying organic matter, and the rate of turnover of this material appears to be more sensitive to temperature changes than that of more-labile organic matter. However the magnitudes of these effects are unknown [8].

17. Importance of SOM and Effects of Losses of Soil Carbon

More than twice as much carbon is held in soils as in vegetation or in the atmosphere, and changes in soil carbon content can have a large effect on the global carbon budget [8]. Soil deterioration from depletion of organic matter is an increasingly serious global problem that contributes to hunger and malnutrition. Often a result of unsustainable farming, overuse of chemical fertilizers and drought, the main weapons to combat the problem—compost, animal manure and crop debris—decompose rapidly [4]. Earth's soil is the largest terrestrial pool of carbon. In other words, most of the earth's carbon is fixed in soil. But if this soil is intensively cultivated by tillage and chemical fertilization, organic matter in soil will be quickly decomposed into carbon dioxide by soil microbes and released into the atmosphere, leaving the soil compacted and nutrient-poor [4]. Most soil fertility experts that higher amounts of soil organic matter are related to increased productivity because of its contribution to water holding capacity, improved soil structure, and supply of nutrients [42]. The lack of good rotations on most crop farms, partially caused by the availability of inexpensive synthetic fertilizers, has resulted in a loss of soil organic matter and a decrease in the diversity of organisms in the soil. This degradation of soil quality allows the growth of large populations of disease organisms and plant parasites that would have been held in check by a diverse community of competing organisms. Also, plants that are unhealthy tend to attract more insect pests than healthier plants.

SOM is defined here as the nonliving component of organic matter in soil. The ultimate source of organic matter in soils is CO_2 fixed by plants, including leaf litter, roots, and root exudates. The activity of soil fauna (especially fungi and microbial communities) metabolizes some of these substrates and transforms others into more resistant organic compounds (collectively referred to as humus). The stabilization and fate of organic matter residues is affected by the quality of the original plant substrate and the physical environment in the soil clay content and mineralogy, pH, O_2 availability), formation and disruption of soil aggregates Carbon is lost from soil mostly as CO_2 produced during decomposition of organic matter, though losses of carbon through leaching or erosion may be important when considering C balance in soils on long time scales [99].

Organic matter in soils plays important roles in determining soil water-holding capacity and soil structure, and provides a long-term store of nutrients needed by plants. Thus changes in SOM will have important feedbacks on hydrology and plant productivity. [99]

There is general agreement that SOM contains at least three identifiable C pools: root exudates and rapidly decomposed components of fresh plant litter (''active'' pool); stabilized organic matter that persists in soils over several thousands of years (''passive'' pool); and a poorly defined "intermediate" or "slow" C pool that has turnover times in the range of years to centuries [99].

The mechanisms through which climate or land-use change may influence soil carbon cycling include a combination of biotic controls that affect the amount and resistance to decay of C added to soils by plants, and physical controls, including the area and chemistry of surfaces for stabilization of organic matter, and the availability of oxidants to decomposers. These mechanisms operate on vastly different time scales [99]. Soil minerals, forming over millennial time scales, control how much of the soil carbon is chemicallyyphysically protected from decomposition. Finally, ecosystem disturbance caused by fire, flooding, tree blow-down, etc. will cause drastic changes in plant production and soil conditions for decomposition that far exceed those associated with interrannual variability [99].

After CO_2 fertilization, the most often-cited mechanism leading to a terrestrial carbon sink is enhanced nitrogen availability to plants due to enhanced nitrate deposition. Nitrogen availability also will be increased if there is net decomposition of SOM, because SOM is a major source of nitrogen for plants. Although nitrogen mineralization is accompanied by CO_2 efflux, the transfer of nitrogen from soil to plants would result in net carbon sequestration because the C:N ratio of plants is roughly 10 times the ratio in SOM [99].

The amount of organic matter in soils is positively correlated with moisture and negatively correlated with precipitation and temperature is a major control of the turnover rate of fast-cycling soil C. Both vegetation inputs and decomposition rates are expected to increase with increases in temperature. Soils become an amplifying feedback if the temperature dependence of decomposition is steeper than that of plant productivity [99].

One consequence of all studies of C turnover as function of temperature is that soils already should be a net source of C to the atmosphere due to increased decomposition rates because of documented temperature increases over the past century. This present net C loss presumably would offset hypothetical increases in soil C stores from increased plant production due to CO_2 fertilization [99]. Climate, vegetation, parent material, and time all affect the processes controlling accumulation and decomposition of organic matter in soils as has been known for decades [99]. Soil organic carbon inventories increase regionally from arid to wet environments, whereas incubation and field decomposition studies show increases in decomposition rates with added moisture in aerobic soils. Decomposition rates for organic matter decrease dramatically under waterlogged conditions. Low C inventories in arid soils may be due to decreased C inputs, faster C turnover, or differences in the quality of plant material. As yet, the relative contributions of these factors are unknown [99].

Soil organic matter consists of a variety of components. In varying proportions and many intermediate stages, these include:

- raw plant residues and microorganisms (1 to 10 per cent)
- active. organic fraction (10 to 40 per cent)
- resistant or stable organic matter (40 to 60 per cent), also referred to as humus [61]

Soil organic matter has the following affects on soil quality [17]

- Stores and supplies plant nutrients (N P K and micronutrients; increases cation exchange capacity. Mitigation of Mineral Deficiency Stress
- Stabilizes and holds soil particles together as aggregates.
- Helps soil to resist compaction, promotes water infiltration, and reduces runoff.
- Aids growth of crops by improving the soil's ability to store and transmit air and water, as measured by improved porosity; water holding capacity, and drought resistance.
- Makes soil more friable and easier to work so that plant roots can penetrate the soil profile better.
- Provides a source of carbon and energy for soil microbes which cycle nutrients and fight plant diseases. A soil rich in organic matter and regularly supplied with different kinds of soil organic matter will support a rich and varied population of soil organisms. Organic matter provides a carbon source for primary producers like cyanobacteria that can convert atmospheric nitrogen to plant available N forms. Organic matter is the principle food source for secondary consumers. A soil populated by a diverse, active microbial population is less likely to support uncontrolled spread of plant pathogens. Enhanced microbial activity also accelerates the breakdown of pesticides.
- Reduces the negative environmental effects of pesticides, heavy metals and other pollutants by binding contaminants. Binds organic pollutants, keeping them out of the soil solution where they would be taken up by plants or leached into ground water. Organic matter also provides sites for microbes to colonize and decompose organic pollutants.
- Soil organic matter has been shown to protect plant roots from potentially harmful levels of aluminum, which occur in some acid soils, and also from certain herbicides that are potentially toxic to plants.
- Soil pH buffering capacity: Organic matter has the ability to moderate major changes in the soil pH. Organic matter buffers the soil against major swings in pH by either taking up or releasing H^+ into the soil solution, making the concentration of soil

solution H^+ more constant. The result is a stable pH close to neutral or suitable for the specific crop to be grown.

Soil organic matter also binds plant micronutrients like iron, aluminum, zinc, copper and manganese by chelation—a chemical association of organic matter and micronutrients that keeps the micronutrients in a form available for plant uptake [17].

When fresh organic matter is added to the soil, soil microbes release long-chain sugars or polysaccharides relatively quickly. These polysaccharides promote formation of large or macro-aggregates. As the organic matter decomposes over the longer term, different sizes of aggregates are formed that are resistant to physical disruption. The number and diversity of stable soil aggregates are what give a soil an excellent physical structure [17].

A soil with a good physical structure will have pores and channels of many sizes, from the tiniest channel to large spaces that will allow rainfall to penetrate without running off and taking soil with it. Soil organic matter helps to form and maintain the air passages and channels, protecting the soil from compaction [17].

The resistant or stable fraction of soil organic matter contributes mainly to nutrient holding capacity (cation exchange capacity) and soil color. This fraction of organic matter decomposes very slowly and, therefore, has less influence on soil fertility than the active organic fraction [61]. Organic matter in soil serves several functions. From a practical agricultural standpoint, it is important for two main reasons. First as a revolving "nutrient bank account" and second, as an agent to improve soil structure, maintain tilth and minimize erosion. The stable organic fraction (humus) adsorbs and holds nutrients in a plant-available form. Organic matter does not add any new plant nutrients, but it does release nutrients in a plant-available form through the process of decomposition. If the rate of addition is less than the rate of decomposition, soil organic matter will decline, and conversely, if the rate of addition is greater than the rate of decomposition, soil organic matter will increase. The term steady state has been used to describe a condition where the rate of addition is equal to the rate of decomposition [61].

Before our soils were cultivated, they had achieved a steady state. In most of our prairie soils, the increased rate of decomposition associated with cultivation, combined with the low rates of crop residue addition associated with crop-fallow rotations, has caused a fairly rapid decline in soil organic matter. The rate of decline decreases with time as the amount of total soil organic matter decreases and particularly as the active. Organic fraction is depleted. Cultivation of soils that are naturally high in organic matter will usually result in a decrease of organic matter [61].

Agricultural soils contain substantial amounts of carbon, typically 20 to 80 tonnes per hectare in the top 20 cm. However, relative to their native ecosystem levels, most agricultural soils are depleted in carbon, having lost 30–50% of their original carbon levels due to changes associated with production agriculture and past management practices.

Historically, agricultural practices often resulted in reduced inputs of carbon through plant residues and increased losses via decomposition and erosion. Lower productivity, particularly prior to the 1950s, and greater removal of crop residues decreased the amount of plant material that could potentially add carbon to the soil [11]. Agricultural soils typically contain 1–4% C on mass basis. The relatively small amounts of C could be considered analogous to a catalyst or to the tip of the iceberg where a small amount visible has a big impact [77]. Carbon loss can also affect soil productivity, soil quality, nutrient cycling, soil fertility, and ultimately crop production. In agricultural production systems, C loss may occur as a result of biomass burning, intensive tillage or cultivation, crop residue removal [87].

Consistent evidence has been found of an organic carbon decline for fertilized soils throughout the world [52].

18. Environmental Benefits of Soil Organic Carbon

Soil organic matter is so valuable for what it does in the soil, it can be referred to as "black gold" because of its vital role in physical, chemical and biological properties and

processes within the soil system [77]. Storing C that will also lead to enhanced air quality, water quality and increased productivity as well as mitigating the greenhouse effect [77]. Carbon sequestration means to "capture and secure or store", in this case soil C [77]. The main goal is to enhance C fixation through photosynthesis and to reduce soil C emissions by securing and storing C that might otherwise be lost to the atmosphere. The soil contains two to three times as much C as the atmosphere. By minimizing the increase in ambient CO_2 concentration through soil C management, we minimize the production of greenhouse gases and minimize potential for climate change [77].

The mineral P fertilizer and equivalent amounts of manure did not result in higher above- or below ground biomass and nutrition in comparison to the un-fertilized control. Cattle manure had a similar effect on plant growth and nutrition as chicken manure. In contrast, a significant growth improvement was found for the Anthrosol and after charcoal amendments in the Ferralsol. Nitrogen uptake of cowpea was significantly ($P < 0.05$) decreased by charcoal additions and in the Anthrosol, which was an effect of poor N nutrition. Phosphorus nutrition and uptake increased when charcoal was added to the Ferralsol and for the Anthrosol. Charcoal amendments improved foliar K nutrition and uptake in contrast to the Anthrosol, whereas K nutrition was even significantly ($P < 0.05$) reduced in comparison to the control. Increasing the amount of charcoal further increased above- and belowground biomass production. Similarly, P, K, Ca, Zn, and Cu uptake by cowpea increased [59]. The Anthrosol contained twice as much soil C compared to the Ferralsol. Charcoal amendments to the Ferralsol, however, resulted in the highest soil C contents which were doubled compared to the Anthrosol [77].

Enhanced soil C storage and management also helps the environment win as improvements in soil, air and water quality are all enhanced with increased amounts of soil C within the soil. Increasing soil C storage can increase infiltration, increase fertility, decrease wind and water erosion, minimize compaction, enhance water quality, decrease C emissions, impede pesticide movement and enhance environmental quality. Accepting the challenges of maintaining food security by incorporating C storage in conservation planning demonstrates concern for our global resources. This concern presents a positive role for conservation agriculture that will have a major impact on global sustainability and our future quality of life [77]. Not only the nutrient contents but also the nutrient retention can be improved with charcoal additions to soil, but not with the addition of coal degradation products such as coal ash or fly ash. This is especially important in highly weathered soils with low ion-retention capacities. Adding charcoal to soil can significantly increase seed germination, plant growth, and crop yields. Low SOM contents may be responsible for the low available water capacity and the weak structure of many agricultural soils. Charcoal may not only change soil chemical properties, but also affect soil physical properties such as soil water retention and aggregation. These effects may enhance the water availability to crops and decrease erosion [32].

The use of charcoal as a soil conditioner for sustainable agriculture in the humid tropics presents the following advantages:

High nutrient contents and nutrient retention capacity lead to improved nutrient supply for plants and reduced nutrient losses by leaching. We assume that two processes are responsible for this: (1) nutrients are physically trapped in the fine pores of amorphous carbonized materials, and (2) slow biological oxidation produces carboxylic units on the edges of the condensed aromatic backbone of the charcoal which increases the CEC.

Transformation of labile plant organic matter into stable C pools can reduce the release of the greenhouse gas CO_2 into the atmosphere during land clearing and can increase C sequestration in the soil. There are strong indications that charcoal is very slowly mineralized in the soil environment. Charcoal from fallow vegetation and/or organic wastes can be easily produced by local farmers and also by those with a low income. Charcoal production is a well-known technique and the required tools and resources are readily available [32].

Interactions of bio-char with soil microorganisms are complex. On the one hand, soil microbial diversity and population size, as well as population composition and activity, may be affected by the amount and type of bio-char present or added to soil. On the other hand, microorganisms are able to change the amount and properties of bio-char in soil. Both effects will have significant influence on nutrient cycles and nutrient availability to plants. A higher retention of microorganisms in bio-char soils may be responsible for greater activity and diversity due to a high surface area as well as surface hydrophobicity of both the microorganisms and bio-char. Activated carbon, which is chemically similar to black carbon or bio-char in soil, has been shown to sorb microorganisms strongly, and this adsorption was seen to increase with higher hydrophobicity. Bio-char additions not only affect microbial populations and activity in soil, but also plant–microbe interactions through their effects on nutrient availability and modification of habitat [56].

Bio-char amendments have so far shown beneficial effects when applied to rice, sorghum, corn, various beans (soybean, common bean, cow pea, moongbean), banana, and vegetables such as carrots. Since bio-char is a valuable commodity, applications can be successfully applied to smaller units of area than entire fields, such as high fertility trenches in contour rows of steep land for vegetables or grain crops, and planting holes for tree crops [56].

The most common system responses attributed to microfloral grazers (protozoa, nematodes, microarthropods) in the literature are increased plant growth, increased N uptake by plants, decreased or increased bacterial populations, increased CO_2 evolution, increased N and P mineralization, and increased substrate utilization. Based on this evidence in the literature, a conceptual model was proposed in which microfloral grazers were considered as separate state variables. To help evaluate the model, the effects of microbivorous nematodes on microbial growth, nutrient cycling, plant growth, and nutrient uptake were examined with reference to activities within and outside of the rhizosphere. Blue grama grass (Bouteloua gracilis) was grown in gnotobiotic microcosms containing sandy loam soil low in inorganic N, with or without chitin amendments as a source of organic N. The soil was inoculated with bacteria (Pseudomonas paucimobilis or P. stutzeri) or fungus (Fusarium oxysporum), with half the bacterial microcosms inoculated with bacterial—feeding nematodes (*Pelodera* sp. or

Acrobeloides sp.) and half the fungal microcosms inoculated with fungal—feeding nematodes (*Aphelenchus avenae*). Similar results were obtained from both the unamended and the chitin—amended experiments. Bacteria, fungi, and both trophic groups of nematodes were more abundant in the rhizosphere than in nonrhizosphere soil. All treatments containing nematodes and bacteria had higher bacterial densities than similar treatments without nematodes. Plants growing in soil with bacteria and bacterial—feeding nematodes grew faster and initially took up more N than plants in soil with only bacteria, because of increased N mineralization by bacteria, NH4+—N excretion by nematodes, and greater initial exploitation of soil by plant roots. Addition of fungal—feeding nematodes did not increase plant growth or N uptake because these nematodes excreted less NH4+—N than did bacterial—feeding nematode populations and because the N mineralized by the fungus alone was sufficient for plant growth. Total shoot P was significantly greater in treatments with fungus or Pelodera sp. than in the sterile plant control or treatments with plants plus Pseudomonas stutzeri until the end of the experiment. The additional mineralization that occurs due to the activities of microbial grazers may be significant for increasing plant growth only when mineralization by microflora alone is insufficient to meet the plants' requirements. However, while the advantage of increased N mineralization by microbial grazers may be short—term, it may occur in many ecosystems in those short periods of ideal conditions when plant growth can occur. Thus, these results support other claims in the literature that microbial grazers may perform important regulatory functions at critical times in the growth of plants [48].

Boreal Forest Wild Fire Investigations

Boreal forest wild fire experiments conducted in each of three contrasting boreal forest sites in northern Sweden [112]. Mesh bags were filled with pure humus collected from the forest, pure charcoal created in the laboratory, or a 50:50 mixture of humus plus charcoal These bags were left in the field and harvested over 10 years. For boreal forest wild fire investigation cases found that, over the 10-year period, loss of mass and C from the bags containing mixtures of charcoal and humus was substantially greater than what was expected on the basis of the components considered separately. Further, nitrogen immobilization was less than expected in the mixture bags. Given that charcoal decomposition rates in soil are extremely slow and that in our study system charcoal persists for thousands of years in the humus layer without evidence of mass loss, most of the enhanced loss of mass and C caused by mixing charcoal and humus must have resulted from charcoal promoting humus loss rather than humus promoting charcoal loss. These results are consistent with charcoal particles serving as foci for adsorption of organic compounds and microbial growth and activity, leading to enhanced decomposition rates and mass loss of associated humus, again in boreal forest cases after wild fire. These Swedish findings indicating that charcoal can promote rapid loss of forest humus and below ground C during the first decade after its formation. Fire often causes substantial losses of ecosystem C, and our results Provide evidence for a previously unreported mechanism that could contribute to these losses. Results show that these effects can be partially offset by its capacity to stimulate loss of native soil C in boreal forests. The effect of charcoal on native soil C needs to be explicitly considered to better understand the potential of black C as an ecosystem C sink and agent of C sequestration [112]. Forest wild fire conditions and effects need to be further investigated.

19. Effects of Biochar Application

Terra Preta

Fifteen hundred years ago, tribes people from the central Amazon basin mixed their soil with charcoal derived from animal bone and tree bark. Today, at the site of this charcoal deposit, scientists have found some of the richest, most fertile soil in the world. Now this ancient, remarkably simple farming technique seems far ahead of the curve, holding promise as a carbon-negative strategy to rein in world hunger as well as greenhouse gases. [4] Tropical forests account for between 20 and 25% of the world terrestrial carbon (C). Soils under tropical forest contain approximately the same amount of C as the lush vegetation above it. The current conversion of Amazonian forest to agricultural land makes disturbance of this C stock important to the global C balance and net greenhouse gas emissions. Changes in land use, particularly by clearing forests, reduce organic C by 20% to 50% in the upper soil layers [92]. Furthermore, this reduction of soil organic matter (SOM) is causing soil degradation. Thus agriculture is not sustainable without nutrient inputs beyond 3 years of cultivation. The efficiency of conventional fertilizers (such as nitrogen (N)) is limited by a low nutrient retention capacity conjoined with strong tropical rains. On the other hand, large amounts of phosphate fertilizers are needed to overcome the soil.s high P-fixation capacity [92]. To overcome these limitations, slash-and-burn agriculture (shifting cultivation) is practiced by about 300 to 500 million people, affecting almost one third of the planet.s 1500 million ha of arable land [92]. The existence of an anthropogenic and C-enriched dark soil in different parts of the world and especially in Amazonia (Amazonian Dark Earths (ADE) or Terra Preta de Índio) proves that the predominant Ferralsols and Acrisols can be transformed into fertile soils. The ADE.s fertility is most likely linked to an anthropogenic accumulation of phosphorus (P), calcium (Ca), and black C as charcoal. Charcoal persists in the environment over centuries and is responsible for the stability of the ADE.s SOM [92]. Charcoal formation and deposition in soils seems to be a promising option to transfer an easily decomposable biomass into refractory SOM pools [92].

Upland soils in the humid tropics such as the Amazon basin are often highly weathered and therefore possess low plant-available nutrient contents. This is a result of both high rainfall and low nutrient retention capacity. In these soils the cation exchange capacity is very low due to the dominance of iron and aluminum oxides and kaolinite as well as generally low soil organic matter contents and pH values [59]. Two basic approaches can be used to reduce nutrient leaching; applying slow-release nutrient forms such as organic fertilizers and increasing adsorption sites thereby retaining applied inorganic nutrients. Additionally, organic fertilizers may mineralize rapidly and therefore have to be applied repeatedly to maintain sufficient available nutrients and exchange capacity. Applications of organic matter may be a feasible way to create sustainably fertile soils similar to the prehistoric Anthrosols. An additional option to manure and composts could be the application of charcoal to soil [59].

When charcoal was applied in un-weathered condition the microbiological parameters (respiration, biomass, population growth, and efficiency) increased linearly and significantly with increasing charcoal concentrations (50, 100 and 150 g kg-1 soil). Application of pyrogenous acid caused a sharp increase in soil respiration, biomass, and reproduction. We suppose that the condensates from smoke contain easily degradable substances which could be utilized by the microbes for their metabolism [92]. In a greenhouse experiment, leaching of N was significantly ($P < 0.05$) reduced if ammonium sulphate was applied with charcoal. Charcoal simultaneously served as K fertilizer and increased the retention of N. Long-lasting soil fertility improvement due to organic fertilization and a synergistic effect if both charcoal and mineral fertilizer were applied was observed in a field experiment [92].

Mineral fertilized soils amended with charcoal and Terra Preta soils had a significantly higher potential for microbial population growth coupled with a low microbial respiration in absence of an easily degradable C source (glucose). The soil respiration before substrate addition correlated positively with the population growth rate on the plots, whereas Terra Preta had a very low soil respiration and very high population growth after substrate additions. *Forest soils had a higher respiration rate but a very low population growth. These results reflect the relatively high biodegradable OM content of primary forest topsoil but low available nutrients (requirement for microbial population growth), in contrast to refractory Terra Preta SOM with high available soil nutrient contents* [92].

The 15N recovery in biomass was significantly higher on compost amended plots due to significantly higher biomass production. The retention in soil was significantly higher in the charcoal-amended plots after the second harvest due to higher N retention and cycling of crop residues which remained on the plots after harvesting. Total N recovery (in soil, crop residues and grains) was significantly ($P < 0.05$) higher on charcoal (18.1%), charcoal plus compost (17.4%), and compost (16.5%) treatments in comparison to only mineral fertilized plots (10.9%) [92]. Charcoal is influencing soil quality in manifold ways, most importantly by reducing available Al and reducing acidity. Furthermore, charcoal adds K to the soil and has the potential to reduce N leaching. Charcoal amendments increased the reproduction rate of the microbial population after substrate addition whether the plot was fertilized or not. The effects of charcoal on soil biological, chemical and physical properties are complex, making it difficult to isolate single significant charcoal effects, but added up they caused significantly increased plant growth and crop production [92].

At the 235th national meeting of the American Chemical Society, scientists report that charcoal derived from heated biomass has an unprecedented ability to improve the fertility of soil—one that surpasses compost, animal manure, and other well-known soil conditioners [4].

They also suggest that this so-called "biochar" profoundly enhances the natural carbon seizing ability of soil. Dubbed "black gold agriculture," scientists say this "revolutionary" farming technique can provide a cheap, straight-forward strategy to reduce greenhouse gases by trapping them in charcoal-laced soil [4].

Soils receiving charcoal produced from organic wastes were much looser, absorbed significantly more water and nutrients and produced higher crop biomass. The authors, with Delaware State University, say "the results demonstrate that charcoal amendment is a revolutionary approach for long-term soil quality improvement" [4].

Applying raw organic materials to soil only provides a temporary solution, since the applied organic matter decomposes quickly. Converting this unutilized raw material into biochar, a non-toxic and stable fertilizer, could keep carbon in the soil and out of the atmosphere [4].

With its porous structure and high nutrient- and water-holding capabilities, biochar could become an extremely attractive option for commercial farmers and home gardeners looking for long-term soil improvement. The biochar-infused soil showed vastly improved germination and growing rates compared to regular soil [4].

Many soils of the lowland humid tropics are thought to be too infertile to support sustainable agriculture. However, there is strong evidence that permanent or semi-permanent agriculture can itself create sustainably fertile soils known as 'Terra Preta' soils. These soils not only contain higher concentrations of nutrients such as nitrogen, phosphorus, potassium and calcium, but also greater amounts of stable soil organic matter. *Frequent findings of charcoal and highly aromatic humic substances suggest that residues of incomplete combustion of organic material (black carbon) are a key factor in the persistence of soil organic matter in these soils.* Our investigations showed that 'Terra Preta' soils contained up to 70 times more black carbon than the surrounding soils. *Due to its polycyclic aromatic structure, black carbon is chemically and microbially stable and persists in the environment over centuries. Oxidation during this time produces carboxylic groups on the edges of the aromatic backbone, which increases its nutrient-holding capacity.* We conclude that black carbon can act as a significant carbon sink and is a key factor for sustainable and fertile soils, especially in the humid tropic [31].

Soil Fertility

There is a long tradition in Japan of using charcoal as a soil improver. The idea that the application of charcoal stimulates indigenous arbuscular mycorrhiza fungi in soil and thus promotes plant growth is relatively well-known in Japan, although the actual application of charcoal is limited due to its high cost" [67].

Soil Water Retention

Hydrophobicity of soils can have severe adverse effects such as reduced plant growth and increased overland flow leading to increased soil erosion. It is therefore important to consider the possibility that biochar applications may introduce hydrophobic compounds into the soil [20].

Water repellency of soils is sometimes observed to increase after fires. In part, this may be due to physical changes in the soil whereby small particles of ash and char block soil pores and reduce water infiltration rates. Of more concern in the context of biochar, is the mechanism whereby hydrophobic organic compounds are produced during combustion and coat soil particles. Organic coatings are a common cause of water repellency in soils. The possibility of such organic compounds or tars residuals occurring in low grade and for energy application produced biochar, therefore there is a cause for appropriate soil application and ecotoxicology concern [115].

The processes involved in distributing volatile organic compounds (VOCs) within the soil will be different in the case of biochar application compared to fires occurring directly on the soil [115].

In the first instance, when char is artificially made, the VOCs will be adhered to the char internal particles and external surface only, so this low grade char may act as bio toxic and water repellent centres. In the second case, the heat of the wild fire will distribute the VOCs throughout the soil structure, while thermally impact top soil organic and inorganic constituents—allowing it to also coat non-char soil particles. Thus, even if the same hydrophobic compounds were present in both cases, we may expect the effect on soil water repellency effects to be different. Nonetheless, the introduction of recalcitrant hydrophobic compounds into the soil with biochar would leave open the possibility of their subsequent redistribution within the soil by physical, chemical or biological processes [115].

Under any circumstances organic tars are biotoxic compounds, which occurrence is *depending on both pyrolysis conditions, applied technology performance and feedstock characteristics.* Biochar organic tar residuals are therefore to be avoided, especially for agricultural soil applications, and its content should be maximized to 1% weight by weight of the biochar [115]. Notwithstanding the possibility of introducing hydrophobic compounds to the soil by biochar application, there is some evidence suggesting that, at least in some instances, biochar may have the opposite effect of increasing soil water retention [115]. It was found that charcoal increased the available moisture in sandy soil, had no effect in loamy soil, and decreased the available moisture in clayey soil. The increase in available moisture observed in sandy soils may make biochar a useful tool in the reversal of desertification [115]. Charcoal, in many soils and sediments, is considered a high affinity sorbent for organic pollutants [81].

Reduction of Nutrient Leaching

It has been suggested that biochar may have the potential to reduce leaching of pollutants from agricultural soils. This possibility is suggested by the strong adsorption affinity of biochar for soluble nutrients such as ammonium, nitrate, phosphate, and other ionic solutes. If this affinity of biochar for ionic solutes can in fact be utilised to reduce run-off in agricultural watersheds, then it will have important benefits in terms of reducing hypoxia of inland and coastal waterways caused by eutrophication. [57] [31] [75].

20. How Much Biochar Can Be Added to Soil?

The amount of biochar that can be added to soils before it ceases to function as a beneficial soil amendment and becomes detrimental will be the limiting factor in the use of biochar as a soil additive. The strongest evidence that high concentrations of black carbon in soil may be beneficial under some conditions comes from the Amazonian Dark Earths (ADEs) such as terra preta and terra mulata—charcoal rich soils which contain approximately three times more soil organic matter, nitrogen and phosphorus than adjacent soils and have twice the productivity [31].

Total amounts of black carbon found in Terra Preta soils were 25±10 Mg ha-1 and 25±9 Mg ha–1 at 0–30 cm and 30–100 cm soil depths, respectively [31].

It is important to note however, that these data come principally from studies on highly weathered tropical soils with very low natural SOC levels. Much less is known about the effect of biochar additions to relatively fertile temperate soils. The lack of research on such soils arises because the potential benefits of raised fertility are unlikely to be as great as in regions with soils of low natural fertility [115].

Versus the Terra Preta tropical applications, *the 3R PROTECTOR product composite dose has been high efficient at temperate soil horticultural farming at such low doses as 400 kg/ha to 1250 kg/ha.*

21. Enchanced Soil Microbiological Activity by Charcoal

It is found that "apparently, some microorganisms were able to live with BC as sole C source". It is also found that BC in soils may enhance the rate of decomposition of labile C compounds [37].

In laboratory condition glucose mineralisation was enhanced over the control by the presence of charred material, i.e., there was an interactive priming of BC and glucose mineralisation. We conclude that BC in soils may promote growth of microorganisms and the decomposition of labile C compounds rather than stabilise them against degradation [37]. Large accumulations of BC in some soils and BC resistance to chemical oxidation, suggests that black carbon is stable over at least such a timescale. Nonetheless, considerable uncertainties remain about just how fast biochar may decompose under different soil conditions [115].

Charcoal seems to assist microbial activity by having a porosity that provides a favourable microhabitat, weak alkalinity and by being a substrate unfavourable for saprophytes. AM fungi easily extend their extraradical hyphae into charcoal buried in the soil and sporulate in the particles encouraging effects of charcoal on arbuscular mycorrhiza (AM) colonisation [80].

The mechanism whereby charcoal stimulates the growth of AMF is described by [67].

Charcoal may stimulate the growth of AMF by the following mechanism: Charcoal particles have a large number of continuous pores with a diameter of more than 100 m. They do not contain any organic nutrients, because of the carbonization process. The large pores in the charcoal may offer a new microhabitat to the AMF, which can obtain organic nutrients through mycelia extended from roots. This may enable the AMF to extend their mycelia far out from the roots, thus collecting a larger amount of available phosphate [67].

The role of mycorrhizal fungi is particularly important in carbon cycling, because they pump the carbon compounds out of the root into a massive network of fine fungal filaments in the soil, where it becomes available to other microbes and also to larger soil organisms like worms, mites and insects. In return, the fungus gathers phosphorus from the soil and delivers it to the plant, helping the plant to grow better. The research confirmed that there were many different fungi in the roots of each plant, but revealed, for the first time, which of these fungi were most active [100].

Arbuscular mycorrhizal fungi, found living on plant roots around the world, appear to be the only producers of glomalin. Wright named glomalin after Glomales, the taxonomic order that arbuscular mycorrhizal fungi belong to. The fungi use carbon from the plant to grow and make glomalin. In return, the fungi's hairlike filaments, called hyphae, extend the reach of plant roots. Hyphae function as pipes to funnel more water and nutrients--particularly phosphorus—to the plants [15]. As carbon gets assigned a dollar value in a carbon commodity market, it may give literal meaning to the expression that good soil is black gold. And glomalin could be viewed as its golden seal [15]. A strong glue, glomalin, is produced by a beneficial fungus that grows on plant roots. The glue comes off of the fungus and is deposited on soil particles. This process leads to build up and stabilization of aggregates [117].

Higher levels of glomalin give greater water infiltration, more permeability to air, better root development, higher microbial activity, resistance to surface sealing (crusts) and erosion (wind/water). Benefit of glomalin: increased aggregate stability which leads to better soil structure which, in turn, leads to better plant production [117]. Soil aggregation is a complex process that is largely dependent upon microorganisms to provide glues that hold soil particles together. These glues are carbon-containing compounds that protect microorganisms from drying out. We are beginning to understand the importance of one group of soil fungi and the glue that is produced in large amounts by these fungi. The fungi are the arbuscular mycorrhizal fungi (AMF) and the glue was named glomalin after Glomales — the taxonomic order of this group of fungi [117].

As a glycoprotein, glomalin stores carbon in both its protein and carbohydrate (glucose or sugar) subunits. Glomalin is causing a complete reexamination of what makes up soil organic matter. That study showed that glomalin accounts for 27 percent of the carbon in soil and is a major component of soil organic matter. Glomalin weighs 2 to 24 times more than humic acid, a product of decaying plants that up to now was thought to be the main contributor to soil carbon. But humic acid contributes only about 8 percent of the carbon. Another team recently used carbon dating to estimate that glomalin lasts 7 to 42 years, depending on conditions. It is glomalin that gives soil its tilth—a subtle texture that enables experienced farmers and gardeners to judge great soil by feeling the smooth granules as they flow through their fingers [15]. As a plant grows, the fungi move down the root and form new hyphae to colonize the growing roots. When hyphae higher up on the roots stop transporting nutrients, their protective glomalin sloughs off into the soil. There it attaches to particles of minerals (sand, silt, and clay) and organic matter, forming clumps. This type of soil structure is stable enough to resist wind and water erosion, but porous enough to let air, water, and roots move through it. It also harbors more beneficial microbes, holds more water, and helps the soil surface resist crusting [15]. These fungi are beneficial to plants because hyphae, hair-like projections of the fungus, explore more soil than plant roots can reach and transport phosphorus and some other nutrients to the plant. In return, plants provide carbon for growth of the fungus [117].

Longer hyphae help plants reach more water and nutrients, which could help plants face drought in a warmer climate. The increase in glomalin production helps soil build defenses against degradation and erosion and boosts its productivity [15].

22. Carbon Sequestration

Although most of the increase in greenhouse gas (GHG) concentrations is due to carbon dioxide (CO_2) emissions from fossil fuels, globally about one-third of the total human-induced warming effect due to GHGs comes from agriculture and land-use change. U.S. agricultural emissions account for approximately 8 percent of total U.S. GHG emissions when weighted by their relative contribution to global warming. The agricultural sector has the potential not only to reduce these emissions but also to significantly reduce net U.S. GHG emissions from other sectors. The sector's contribution to achieving GHG reduction goals will depend on economics as well as available technology and the biological and physical capacity of soils to sequester carbon [73].

Every tonne of carbon added to, and stored in, plants or soils removes 3.6 tonnes of CO_2 from the atmosphere [73]. Carbon stocks in agricultural soils are currently increasing by 12 million metric tonnes (MMT) of carbon annually. If farmers widely adopt the best management techniques now available, an estimated 70 to 220 MMT of carbon could be stored in U.S. agricultural soils annually. Together with attainable nitrous oxide and methane reductions, these mitigation options represent 5 to 14 percent of total U.S. GHG emissions [73].

Increasing the organic matter content of soils (which accompanies soil carbon storage) improves soil quality and fertility, increases water retention, and reduces erosion [73]. Due to past management practices, most agricultural soils have relatively depleted stocks of soil carbon, compared with native ecosystems [60]. On farmland, intensive cultivation practices, along with comparatively low productivity, harvesting and removal of residues, and soil erosion depleted the carbon stocks of many agricultural soils by 30 to 50 percent or more compared to their condition under native vegetation. A cycle of soil exhaustion and land abandonment, in part, fueled the westward expansion of agriculture in North America [73]. U.S. agricultural soils have lost, on average, about one-third of the carbon they contained before wide-scale cultivation began in the 1800s. Soil science studies suggest that changes in land use and land management practices could increase the carbon content of crop and grazing land soils by 104-318 million metric tons per year [60].

Increasing the quantity of carbon sequestered—or stored—in soils and biomass is an alternative to reducing emissions of carbon and other greenhouse gases (GHG) in an overall strategy to mitigate global climate change and its negative economic and environmental effects. Conceptual discussions in the climate change literature often acknowledge that a unit of carbon emissions reduction and a unit of carbon sequestration may have very different contributions to net GHG mitigation over time. First, to be equivalent with emissions reductions, carbon sequestration must be maintained for a period equal to the time emitted carbon remains in the atmosphere, which is referred to as "permanence". Second, after undergoing a change from one management practice to another (e.g., from conventional tillage to conservation tillage), terrestrial systems tend to move to new equilibrium carbon levels over time. Terrestrial systems then accumulate additional carbon from activity changes for a finite period, until they reach a new "C-stock equilibrium."

Generally, for a unit of carbon sequestered in soil or biomass in year t to have the same climate change mitigation effect as a similar unit of emissions reduction in year t (and to be of equal mitigation value), the unit sequestered must remain in the soil or biomass for a period of about 100 years [60].

Organic carbon is maintained in soils through a dynamic process. Plants convert carbon dioxide into tissue during photosynthesis; after a plant dies, a portion of the stored carbon is left behind in the soil by decomposing plant residue and a larger portion is emitted back into the atmosphere. Decomposition rates tend to be proportional to the amount of organic matter in the soil. Hence, over time and under relatively constant environmental and management conditions, rates of carbon additions and emissions tend to equilibrate and the amount of organic carbon in soils stabilizes at a constant, or steady-state, level (i.e., the C-stock equilibrium). If the relationship between additions and losses changes due to a change in soil management, the soil will gradually move to a new C-stock equilibrium [60].

On organic farms, increasing soil organic matter and microbial biomass is a fundamental principle to support agro-ecosystem stability.

Agricultural soils play a key role in C cycling and can act as a source or sink of C and thereby impact greenhouse gas emissions [87]. As TOC declines due to cultivation, the more resistant charcoal fraction increases as a portion of the total C [87].

23. BIOCHAR PRODUCTION

Competition between Biochar and Biomass Energy Production

When biochar is added to soil, we are essentially choosing to forgo a renewable energy source – the energy that could be released by the combustion of the char. Thus, even though it may be possible with some feedstocks to obtain some energy co-production along with biochar, this will always be less than the amount of energy that might be obtained by complete energy conversion of the original biomass [115].

However, in the practice and in many cases, the prime objective of the solid fuel based renewable energy production is the conversion of solid fuels to electric power, such as with the efficient turbine techniques; for which efficient electric power production concentrated energy is required.

Therefore, the question arises as to whether it is more efficient in terms of avoided carbon dioxide emissions to use biomass as a source of energy to displace fossil fuels or whether it would be better to sequester a fraction of the carbon in the biomass as biochar and meet energy demand from other sources [115]. Co-production of biochar and energy is possible [115]. However, *there is an economical conflict between maximising energy versus biochar production, unless additional, short termed and attractive economical values are added to the biochar use, such as fertilization, plant growth promotion and other simultaneously achieved multiple benefits. If the additional and short termed economical values are higher than the present energy value of the biochar, than the market will automatically select the higher economic value valorization, e.g., the short term available higher profit.*

From industrial technology point of view, there is a significant technical difference if the char production primarily objective is made for energy or biochar, which two applications require different product quality requirements, such as following *Table 12.*

Fowles (2007) used a simplified high level model to compare these two options by considering just two variables—the energy obtained per unit of CO_2 emission for different fuels, and the percentage of carbon in biomass that can be sequestered by pyrolysis. From this analysis it emerged that in almost all cases the reduction in carbon dioxide emissions is greater if biomass is used to produce biochar for sequestration than if the same biomass had been used to produce energy to displace fossil fuels. Perhaps the weakest aspect of Fowles' analysis is that he neglects the possibility of combining CCS with biomass energy. Where this option is economically and technically feasible, it has the potential for a greater reduction in atmospheric CO_2 than biochar production as it may simultaneously sequester a greater proportion of the biomass carbon than biochar whilst also offsetting a greater amount of fossil fuel use [115]. For all industrial installations and operations, there are direct and indirect economical and environmental costs, which life cycles to be analyses case by case.

Table 12. Application based specific 3R AGRICARBON char characteristics, application concepts and impacts

Items	ENERGY CHARS characteristics	3R AGRICARBON BIO-CHAR characteristics
Orano toxic residuals	Medium level (5–15% w/w) of organic (tar) residuals, related to the needed ignition characteristics and ignition performance	Very low organic (tar) residuals <1% w/w
Heavy metals	Present and allowable EU/US regulation provides up to maximum limits applied for heavy metal content	Very low total and high ecotox targeted heavy metal content. In practical terms the limit should be far below the present and allowable EU/US regulation limits to make the system long termed sustainable.
Halogenic residuals	Present and allowable EU/US regulation limits applied for halogen residual content	Low halogen residual content, far below the present and allowable EU/US regulation limits
Pore size structures	Wide range of micro, meso and few macro porosity	Macroporosity bio optimized
Pore distribution	As of combustor performance need	Size distribution bio optimized
Surface	Surface characteristic not relevant	Surface characteristic modified and bio optimized
pH	PH is not relevant	Bio optimized pH
Climate impacts	Negative direct climate impacts in air	Positive climate impact effects directly and indirectly in air, biosphere, soil and subsurface water
Soil impacts	Soil impacts not applicable	The application targets restoration of natural balance in an complex ecological system □the soil”, with long list of biotic and abiotic elements
Science and technology	The energy char application management requires understanding and high science disciplines from thermal treatment and environmental sciences	The biochar management requires complex understanding and high science disciplines from thermal treatment, agricultural biotechnology, soil and environmental sciences
Authority permits	Traditional industrial permitting procedures for thermal units	Combined Authority permitting: carbonization industrial production and ecotox tests needed for product applications

Table 13. The economical and environmental costs and impact factors of 3R AGROCARBON production

Economical costs	Environmental costs and impacts.
Equipment manufacturing and installation	Low energy
Plant operations	Total recycling and repeated reuse of all output streams for zero emission solution
Input energy use	High energy efficiency utilization
Input feed material	Regionally available agro, forest, food industrial byproduct and waste utilization
Input feed and output product transport	Decentralized installations
Output product application adaptation	Organic and low input farming

How quickly this greenhouse gas payback occurs will depend upon the rate at which the biomass would have released greenhouse gases were it not pyrolysed.

The repeated recycle and reuse of all elements of the feed stream, is one of the most important factor for sustainable 3R AGROCARBON industrialization as zero emission solution.

24. POTENTIAL AGROCARBON: BIOCHAR FOODSTOCK CHARACTERIZATION

The Characteristics of the Food and Meat Industrial Animal Bone-Based Charcoals

Organic wastes from human food chain are products directly derived from food industrial processing, which may fit for human consumption in accordance with Community legislation, but are not intended for human consumption for commercial or other reasons. Despite the high qualities of the food chain inputs for human consumptions, the technical by-products will be microbiologically affected during and after the processing (cross-contamination), while there is no economical cover to more properly handle these low value by-products. As a result these by-products will rapidly turn into negative valued problem substances, especially from the meat industry. The conventional uses of the Meat and Bone Meals (MBMs) for animal protein consumption have been banned because of microbiological safety issues. As MBMs may contain harmful food borne pathogens direct use as fertilizer in the agriculture as the conventional thermal treatment inactivation at 133°C/20min/3 bars heat/pressure conditions may not be sufficient to guarantee the microbiological sterility of the product. Even if the Technical By-product properly inactivated, the organic waste will still be optimal nutrient substratum for pathogenic microbes. As the food chain processing environment is optimal environment for these pathogens as well, thus the risk of cross contamination is very high. These bacterial infections of public health concern are: Salmonella, Shigella sp., Campylobacter, Escherichia coli, Mycobacterium bovis and others. Therefore, *850°C material core temperature thermal treatment would be required for safe inactivation* (EC 1774/2000). Most of the traditional solutions, because of safety reason are not viable anymore, nor the disposal and nor the incineration. (If high P content material is incinerated, such as the bone meal is, than the P will be highly corrosive in oxidative environment, resulting costly maintenance on the boiler steel components though the system).

More than half of the animal by-products are not suitable for normal consumption, because of their unsuitable physical, chemical and microbiological characteristics. More than 1 million tonnes/year bone by-products are generated from the slaughtering of bovine animals in the EU countries and more than 7 million tonnes/year total in the world. The bone by-products need to be considered as potentially microbiological risk materials with more than 70% organic matter content. The technical by-products are safe materials at the point of their genesis, however, if no rapid transformation is implemented, they may turn into risk materials. The total revenue from the sale of animal technical by-products is estimated in the order of 1.5 bn €. The cost of disposing all animal by-products, largely through incineration, is estimated in the order of 3 bn € and there are also important environmental implications from this disposal route. Therefore it is important that the farmers will not lose the value of the by-products, but rather have them transformed into added value product. In practice, there is no financial frame to make proper treatment of food waste with conventional technology, as available techniques resulting low value added end products only.

Animal bone charcoals containing high mineral content and low carbon levels. Plant based charcoals containing high carbon content and low mineral levels. In both cases the carbonization technology performance, thermal conditions and the input material characteristics are the critically important factors for the output charcoal quality performance, especially when applied as biocarbon in relation to diverse living environment, such as the soil. Micro-organisms are highly sensitive for ecologically toxic effects from organic and/or inorganic residuals, therefore advanced carbonization technology is required for production of bio specific chars.

25. The Solution: Integrated Protector Technology for Recycling and Added Value Carbon Economy Upgrading of Organic and Inorganic Agro By-products

The overall objective of the PROTECTOR technology is to realize integrated thermal inactivation (carbonization) and biotechnological recycling of high P containing organic waste of non Specified Risk Material (SRM) bone meal and upgrade it into a high added value and safe biotechnological crop protection and nutrition product for environmentally friendly vegetable cultivation. The biological control effect targets primarily combat against crown rot of tomato and damping off plant pathogens and improvement of plant natural resistance as well. The risk of cross contamination at food chain organic waste streams requires new technological solutions. The developed technology utilizes animal bone char for microbiological carrier and food industrial by-products as nutrients during the solid substrate fermentation and formulation process.

Natural soil borne microbiological strains are selected for entrapment into the internal sphere of the solid carrier and formulated in a way that long term storage of the substance will be viable on ambient temperature. The method and product provides enhanced survival of the protected microorganism during introduction into the soil.

The 3R Environmental Technology Ltd Group subsidiary Terra Humana Ltd has been developed an innovative thermal recycling treatment system for high temperature 850°C carbonization of the Phosphorous rich bone meal and integrated to innovative Solid Substrate Fermentation Technology. *The added value upgrade of the waste to plant available mineral nutrient product is archived by integrated thermal and biotechnological means. The production and use of ABC mineral for biotechnological solid carrier is a highly innovative carbonization method and product*

Figure 1. Carbonization plant.

PROTECTOR is a conscious design, maintenance and harmonious integration of agriculturally productive ecosystems which have the diversity, stability, and resilience of natural ecosystems. The soil biological diversity is critically important for variety of life. It reflects the number, variety and variability of living organisms, change over time, providing diversity within species (genetic diversity), between species (species diversity), and between ecosystems (ecosystem diversity).

The objective driven scientific and technical objectives of the integrated PROPTECTOR recycling and upgrading technology is consisting of the following main elements:

1. Providing a *strategy for thermal recycling of the phosphorus rich bone meal and/or other plant based organic and/or inorganic agricultural by-products* into safe and sterile char and mineral, which is suitable for specific microbiological carrier.
2. Providing objective driven *selection of soil microbial strains* compatible for P and other mineral mobilization for plant uptake and biological control effect.
3. Including *scale up optimization of innovative SSFF technology.*
4. Providing design, implementation and tests of *integrated ABC and SSFF "product-like" field demonstration production plant.*

5. Providing assessment and mathematical descriptions of the PROTECTOR *P dissolution kinetics* and the impacts of PROTECTOR on P and organic matter cycles in soil.
6. Providing detailed *risk assessment* of all components of the input materials, the recycling process and the final product, including the applied microorganisms, in order to establish conclusively that there is no risk to the environment and human health.
7. *Field demonstrating the PROTECTOR effectiveness* using tomato as the test plant against at least two economically important plant pathogens. Field demonstrations are executed under different climatic, soil and agro-industrial production conditions in Hungary, Israel, Italy, Germany and The Netherlands. The independent and accredited Authority tests are executed in Hungary as of EU permit procedures for plant growth promoter with biological control by-effect.
8. Providing justification for socio-economic and technical viability by executed *Cost-Benefit-Analysis, Strength-Weakness-Threats-Opportunities* and socio-economic assessment of impacts, including the evaluation of its acceptance by consumers and retailers.
9. Providing completed *industrial scale-up engineering design* and detailed technical implementation plan.
10. Providing a clear product and technology-oriented *dissemination and marketing strategy,* and a plan for increasing the PROTECTOR - technology based innovation, rapid take-up and commercialization.

In this context the *prevention* (prevent exposure of soil borne plant diseases by biological control effect), *protection* (enhance plant natural resistance and promote plant growth) and *preservation* (supply natural, clean and plant high efficient available fertilizer), are important drivers.

The patented 3R carbonization technology is the only known technology which is capable to carbonize organic material up to 850°C material temperature, which *end product is specifically manufactured for biotechnological adaptation purposes.*

Also *the 3R plant is the only known technology system, where thermal process treatment for biotech specific solid carrier manufacturing and biotechnological systems are integrated and built together into one industrial unit.*

The ABC mineral solid carrier is characterized by having grain size is around <5 mm, macro porous structured, specific surface internal area is < $120m^2/g$,and *does not contain heavy metal or organic/inorganic contamination* which can inhibit the microbial activity. The main characteristics of the carrier material are variable according to different strain specific microbiological substance need within wide bounds by the special carbonization manufacturing process.

Natural soil microorganisms (bacteria and fungus) with P mobilizing and biocontrol activity have been successfully selected and specific fermentation process developed. The different microorganisms having *strain and char solid carrier specific requirements, resulting in different strain specific optimizations within the process.*

Innovative two stage SSFF (Solid Substrate Fermentation and Formulation) technology and pilot plant has been developed for effective formulation of the microbiological substances on high phosphorous based ABC mineral carrier. SSFF field demonstration unit have been developed, scientific modelled, engineering designed and basic frame equipments installed in Hungary. The developed SSFF technology and pilot plant have the advantages of superior productivity both for fungi and/or bacterial agents, advanced automated and totally closed technique, reduced energy requirement, zero waste water output and improved product recovery. Different alternative types of food industrial by-products used as a C/N source for the liquid phase deep tank fermentation process. The inoculum product from liquid phase fermentation has been adopted to bone char mineral solid fermentation. All the fermentation process has been optimized economically towards industrial production scenarios in the European Union and in the USA.

An economically effective nutrient and two stage fermentation and formulation strategy has been developed for effective production of the PROTECTOR substance. The total economics has played key role in the fermentation medium development.

The different microorganisms produced different levels of colony-forming units (*cfu*) per gram ABC mineral, ranging from $3x10^7$ to $5x10^8$. Improved growth during optimization of the SSFF process have been measured, as well as high survival rate after the drying. High percentage of effective and germinable propagules has been achieved with long shelf life.

Figure 2. Fermentor.

The most important experience we have achieved is that: there is a very significant difference of applicable results from micro scale in-vitro laboratory versus large scale "product-like" optimization conditions, both for thermal and biotechnology processes.

Only sufficient "product-like" scale results providing progressive and application oriented results, while theoretical and micro scale laboratory tests results may lead to incorrect conclusions towards wrong development direction. Our results confirming that only large scale laboratory fermentation and formulation conditions with industrial type nutrient and field application strategy can provide true value results that leads to successful scale up.

The results of the SSFF tests clearly confirmed our assumption that a properly produced ABC is advantageous support material for microbial colonization. Enumeration of colony-forming units within the PROTECTOR products and microscopic follow up was found to be an easy and valuable method for quality control during and after the production of ABC with micro-organisms. Improved growth during optimization of the SSFF process could be measured, as well as survival during the drying process.

The following PROTECTOR scale up "product-like" conditions have been applied:

- Micro scale laboratory fermentation optimization: LIQUID:<5 litre SOLID: <5kg (SSFF semi sterile)
- Small scale laboratory fermentation optimization: LIQUID: <12 litre, SOLID <15kg (SSFF semi sterile)
- Medium scale laboratory fermentation optimization: LIQUID: <160 litre, SOLID <150kg (SSFF semi sterile)
- Large scale lab LIQUID < 500 litre (SSFF semi sterile)"product like" test SOLID <500kg

Based on the advanced scientific RTD and product like tests, industrial scale ABC carbonization system and integrated SSFF has been detailed industrial designed by Edward Someus, for production of PROTECTOR products with output capacity of 20,000 m^3/year. The PROTECTOR has been successfully tested with formulations of individual stains, formulations of strain consortium and co-composting with ABC mineral/charcoal containing abundant microorganisms.

Table 14 presents the inorganic analyses of the ABC mineral. Table 15 presents the inorganic element analysis of ABC mineral material with X-ray fluorescens spectrophotometer after digestion with LiBO2(mg/kg) indicating change of inorganic content during the different carbonisation temperature.

The innovative PROTECTOR method and technology is international intellectual property right "IPR" protected and patented original solution, invented and designed by the Swedish environmental engineer Edward Someus.

Table 14. Inorganic element analysis of ABC mineral material with X-ray fluorescens spectrophotometer after digestion with $LiBO_2$ (mg/kg)

Elements	Sample number		
	019 SV	**022 SV**	**SS20**
P	122000	129000	130000
Ca	320000	333000	330000
Mg	6000	6350	6880
K	1670	1920	917
S	<200	<200	<200
Mn	5	3	2
Fe	85.7	71.4	143
Zn	173	185	170
Ba	<10	<10	<10
Pb	<2	<2	<2
Cu	3	3	4
Cr	<10	<10	<10
As	<2	<2	<2
Ni	<5	<5	<5
Cd	<2	<2	<2
Co	<10	<10	<10
Sr	138	151	136

Table 15. Inorganic element analysis of ABC mineral material with X-ray fluorescens spectrophotometer after digestion with $LiBO_2$ indicating change of inorganic content during the different carbonisation temperature (mg/kg)

Elements	Sample number Carbonization temperature			
	019 (500°C)	**020 (550°C)**	**021 (600°C)**	**022 (700°C)**
P	127000	121000	134000	132000
Ca	327000	308000	341000	333000
Mg	5770	5470	6590	5940
K	1830	2330	2500	2330
S	<200	<200	<200	<200
Mn	6	5	5	3
Fe	92.9	107	100	85.7
Zn	184	182	199	186
Sr	138	121	148	136
Ba	<10	<10	<10	<10
Pb	<2	2	<2	<2
Cu	3	2	4	3

Table 15. Continued

Elements	Sample number Carbonization temperature			
	019 (500°C)	020 (550°C)	021 (600°C)	022 (700°C)
V	<10	<10	<10	<10
Cr	<10	<10	<10	<10
Co	<10	<10	<10	<10
Ni	<5	<5	<5	<5
As	<2	<2	<2	<2
Mo	<2	<2	<2	<2
Cd	<2	<2	<2	<2
Hg	<2	<2	<2	<2

26. Protectorapplication Targets, Concepts and Priorities

Humus is that fraction of the soil medium which aids in making minerals available to plant rootlets. The principle components of humus are a group of humic acids which have very unique properties, acting as "oscillating catalysts." With higher levels of humus in the soil it takes less nutrients to produce. Humus is produced in the soil by the proper decay of animal and plant matter. The principle component of humus is carbon. In order to build humus levels in soil the proper ratio of water, air, residue, fungi, bacteria, protozoa, algae and yeasts are needed. In this context *PROTECTOR high efficiently promotes the restoration and maintenance of this complex soil system, promotes humus building and mineral mobilization towards plant availability, and naturally balance the soil life with positive effect to all the interacting soil components.* Versus this natural process, the use of acid and salt based fertilizers resulting lowering and by time total destruct the amount of humus in the soil.

Natural soil ecological systems may loose carbon. Once pastureland converted to human agricultural cultivation land, soil carbon levels may decline. In healthy soils, carbon exists as long, sticky string-like molecules. Ploughing and burning off - two farming practices which reduce soil carbon levels. These strings twist around individual soil particles and literally bind them together. Soil micro-organisms tend not to bother consuming these large, unpalatable molecules, preferring fresh or rotted plant matter – the stems, roots and other plant parts which over time become incorporated into the soil. But if the soil loses this plant content (because the stubble is burnt or removed), the soil microorganisms have no choice but to make a meal of the carbon molecules. Once the carbon is gone, the structure of the soil breaks down making it difficult to retain water and nutrients. Trees versus crops. Carbon levels in forest soils are usually much higher than those under agriculture.

PROTECTOR CARBON ECNOMY: the two major application concepts in low input and/or organic farming are as follows:

1. *NATURAL RESTORATION:* restoration of soil natural balance by enhancing the symbiosis of the soil elements by directed introduction of selected mineral

mobilization micro-organism immobilized on specific surface modified mineral solid carriers. In this context *LOW DOSE of C content PROTECTOR is introduced,* 400 kg/ha to 1250 kg/ha (0,1% v/v), or even more. The PROTECTOR natural restoration soils develop into organic communities that are capable of growing biotic culture towards truly living earth, renew the soil and sequester carbon. The combined effects are: plant growth promotion with biological control by-effect against soil borne plant pathogens and natural fertilization of soil. In case soils are rich in P but in bound form unavailable to plants, the P mobilization selected and formulated PROTECTOR will enhance the mobilization of the in situ minerals as well for available to plants. The additional effect of the PROTECTOR application is the stimulation of the natural defence mechanisms of the plants.

2. *RE-FERTILIZATION–PROTECTOR fertile soil additive*: re-fertilization of low grade unfertile soil by PROTECTOR COMPOST (carbon-mineral adapted microbiological substance with specific compost blend) to anthropogenically (man made) modifiy and rebuild the soil structure of the soil towards high C content organic matter in its top horizon approx. 20 cm depth, but without hydromorphic (water saturation) characteristics. In this context, for specific cases *HIGH DOSE of C content plant based PROTECTOR may also be introduced,* 10 tons/ha to 25 tons/ha (2% v/v), or even more, subject to economy versus achieved results considerations. The combined effects are: re-fertilization of soil, biological control against soil borne plant pathogens, plant growth promotion.

27. Development of Alternative Protector Strategies for Different Climatic, Environmental and Soil Conditions

The strategy of the re-fertilization is the production of specific *PROTECTOR FERTILE SOIL* additive products by controlled co-composting of agricultural and food industrial organic waste with addition of PROTECTOR microbiological substance adapted on 3R specific high Phosphorous ABC mineral and wood charcoal carrier. The introduced microorganisms are supporting the decomposition and mineralization of the organic waste material, whereas the produced different organic compounds are enhancing the solubilization of the P from the apatite structure of the ABC mineral by the following mechanisms.

During the decomposition of organic waste materials (from agriculture and food industry) the PROTECTOR introduced microbial consortium can be produce a large number of organic acids which enhancing the solubilization of the apatite content of the ABC mineral resulting a specific compost material with rich in available P and other key minerals and organic content.

The enhancement of P release from apatite seems to be a function of the acidification of the P mineral by organic acids and more importantly their chelating ability on calcium (Ca), iron (Fe) and aluminium (Al) [78]. The greater ability of organic acids, compared with mineral acids of comparable strength, to release P from mineral and the direct evidence of their chelating ability have been documented. [49] Another important factor in the release of P is the participation of the OH groups in the organic acids. For example, it has been shown that citric acid with three carboxyl (COOH) groups and an OH group was able to dissolve more P from mineral than cis-aconitic acid with three carboxyl groups but without the OH group [54] Fulvic acid is the most reactive of the humic substances in adsorbing significant amounts of Ca^{2+} and releasing H^+ ions, thereby enhancing bon char mineral dissolution. Humic acid may form complexes with P and Ca, and create a sink for further dissolution of the mineral [86].

The application of humic substances to soil also makes more P available to plants by competing for, and by forming a protective coating over, soil phosphate-sorption sites. An additional benefit that accrues from the application of phospho-compost is the movement of dissolved P to a greater soil depth, which provides a larger soil volume for P uptake by plants [25]. Increasing soil acidity, high cation exchange capacity, low levels of calcium (Ca) and phosphate in solution and high organic matter content favour apatite dissolution [25].

Through PROTECTOR composting, naturally occurring decay mechanisms are controlled and encouraged to break down organic matter into compost, a relatively stable organic product that improves soil structure, aeration, and water and mineral nutrient retention.

Humus and partly decomposed soil organic matter can greatly increase the water- and nutrient-holding capacities of sandy soils [21]. For different climatic, environmental and soil conditions alternative PROTECTOR strategy can be developed, including the enhancing of locally selected specific microbiological fungus and bacteria strain cultivations. Table 16 demonstrates the different PROTECTOR solutions and priority issues.

Table 16. PROTECTOR solutions and priority issues

Protector priorities	Protector objectives
Nutrient management and sustainable use of natural resources	Increase production with regard for the environment. Increasing nutrient use efficiency and agriculture and food industrial by-products and waste recycling through integrated nutrient sources management in cropping systems. The ABC mineral provides not only P for plant growth but that they may also supply secondary organic and inorganic nutrients, including calcium, magnesium and micro-nutrients in combination with organic compost and soil microbiological substances.
Transformation of waste into high added value products	The PROTECTOR transformation of food chain waste into a high added value healthy and safe product is technically, environmentally and economically very effective. PROTECTOR performs integrated carbonization and biotechnological recycling of mineral and/or carbon biotechnological carriers from organic and inorganic agricultural by-products.
Natural and environmentally friendly Phosphorous supply for plant	Continuously maintaining adequate P level for plants by application of slow-release natural and environmental friendly P fertilizers. Balanced and adequate supply of phosphorous mineral nutrient.
Carbon sequestration	As important result of the PROTECTOR technologies will turn carbons in waste streams into useful products, in many cases used as soil improvement products, which in fact is enforcing the carbon sink for log term. The Kyoto Protocol and the US Biochar Provisions S.1884 Salazar Act highlights that soil is a major carbon store which must be protected

	and increased where possible.
Preventing and reversing soil degradation and increasing soil fertility	Re-fertilization of low grade - unfertile soil by PROTECTOR COMPOST (carbon-mineral adapted microbiological substance with specific compost blend) Providing integrated solution for biological control, plant growth promotion and natural fertilization, while supporting the restoration of soil natural balance.
Reduced fossil consumption and sustainable greenhouse gas emission management	Energy saving integrated thermal and biological utilization and recycling organic waste and recovery of energy from it can preserve virgin materials and reduce the use of fossil fuels (so reducing greenhouse gas emissions). By increasing the amount of organic waste that is recycled, composted or has energy recovered, there is considerable scope for reducing greenhouse gas emissions from the waste produced in the agricultural, aquaculture and food industry sector.
Minimizing or substitution of the external agriculture chemicals and artificial fertilizer inputs	In order to maintain sustainable agriculture and production of safe/healthy food, the high input of chemicals and artificial fertilizers need to minimized or even substituted by natural products towards low input and organic farming.

28. Sustainability of Protector Technology

- ENERGY SUSTAINABLITY = uses minimum energy for agro product recycling and processing, all energy is recycled and reused
- ECONOMICAL SUSTAINABLITY = high cost efficiency and economical value creation for farmers
- MARKET AND CONSUMERS SUSTAINABLITY= "WIN-WIN" for all stake holders from producers to Consumers, where the enduser is main winner, all in order to receive very high quality and purity nutritious crop food products on the table for less cost.
- ENVIRONMENTAL SUSTAINABILITY = recycles and reuses waste materials to added value products by an *advanced ZERO EMISSION carbonization process solution.* Environmentally compatible agriculture practices with maintaining the nutrient balance of the soil.
- TECHNICAL SUSTAINABILITY = industrialized production system with controlled output quality.
- LEGAL SUSTAINABIOILITY = The 3R technology operations and product applications are designed to meet the EU, USA, Canadian, Australian and Japanese norms, standards and Authority permit requirements.
- *CLIMATE SUSTAINABILITY = carbon negative process*

Protector Benefits

- *OVERALL BENEFITS*
  - *Mitigation of mineral and nutrient deficiency stress of the food crops, n*atural balance and functionality restoration of degraded continental agro soils in temperate climatic regions, with controlled microbiological activity and precision farming nutrient supply for sustainable, improved, economical and ecological food crop production.
  - Prevention of nutrient loss, biochar may help to reduce nutrient run-off from soils and the associated problems of eutrophication.
  - Supporting beneficial soil life forms towards balanced living soil
  - Soil conditioning:
    - The increase in available moisture observed in sandy soils make biochar a useful tool in the reversal of desertification.
    - Restoration of degraded lands by increasing soil carbon.
    - Increasing soil carbon storage
    - Improved soil moisture capture and retention
- *ENVIRONMENTAL BENEFITS*
  - Efficient use of nutrient inputs for efficient GHG mitigation
  - Providing more conservation-oriented management practices
  - Carbon sequestration of degraded and eroded sites and areas.
  - Supporting environmentally-orientated farming, development of environmentally compatible agriculture practices with maintaining the nutrient balance of the soil.
  - Prevention of groundwater pollution
  - Less N2O (very powerful GHG), even for same rate of fertilizer application -
  - Reducing direct and indirect energy use to avoid carbon emissions.
  - Increasing renewable energy production to avoid carbon emissions. Biomass energy can be used to avoid greenhouse gas emission by providing equivalent energy for heat and electricity generation, and for transportation fuel. If biomass is harvested and burned, and the same area replanted or regenerated, there are no net carbon emissions over the harvest cycle.
- *ECONOMICAL BENEFITS*
  - Improved farm income per farmer from increased tons/hectare, which is leading to rural social improvements and gender equality.
  - Reduced per hectare expenditures on fertilizers, therefore less pressure on increasingly limited oil and natural gas supplies—leading to lower fuel costs.
  - Improved rural employment opportunities
  - Combined Heat and Power harvested from bioenergy as a byproduct with charcoal production from biomass
  - More available farm and forest residue to support more biofuel utilization, helping on both CO_2 emission reduction and peak oil/peak gas supply constraints
  - Lowered disaster insurance costs (and similar for areas with reduced or increased rainfall, etc.)

- *APPLICATION BENEFITS*
  - Improved protection of subsurface water resources from less fertilizer runoff
  - Nutrient balance and slow release fertilizers: Using nitrogen more efficiently means better matching its availability to plant needs.Remaining nitrogen is released by the charcoal slowly into the soil for the plants to absorb.
  - Biological and chemical methods for manipulating soil microbial processes to increase efficiency of nutrient uptake, suppress N2O emissions, and reduce leaching.
  - Promotion of increased plant growth
  - Less malnutrition and better health conditions from supplying more food crops per hectare.
  - Increased use of all the non-dispatchable renewables (wind, PV, biomass, etc), as biochar facilities provide much of the required backup (heaviest char and electrical output during peak electrical demand periods - in West it is July - August for air-conditioning).
  - Buildup of soil organic matter that accompanies carbon sequestration provides many environmental, social, and economic co-benefits
  - *The charcoal stimulates indigenous arbuscular mycorrhiza fungi in soil and thus promotes plant growth*
  - *Weed control* and suppression by development of stronger and larger root system of the host plants.
  - The addition of fine charcoal in compost preparation improves the process by improving: aeration; water retention and probably by providing microorganisms refuge and adsorption of metabolites; metabolite from microorganisms entrapped in the charcoal structure.
  - Charcoal and its cellular structure may provide habitat for this beneficial bacterium.
  - Porous substance with high water and air holding capacity; Suitable habitat for some microbes and plant growth, good material for soil amendment, absorption of chemicals and humidity control
  - High alkalinity; Neutralization of acidic soil and improvement of chemical components of soil and selection of microorganisms
  - Non organic matter; exclusion of saprophytes and propagation of autotrophic and symbiotic microorganisms, free living nitrogen fixing bacteria, root nodule bacteria, Frankia and some AM.
  - Charcoal as media for immobilizing useful microorganisms (spores of bacteria, root nodule bacteria, mycorrhizal fungi, Frankia and nitrogen fixing bacteria.
  - Charcoal seems to assist microbial activity by having a porosity that provides a favourable microhabitat, weak alkalinity and by being a substrate unfavourable for saprophytes. AM fungi easily extend their extraradical hyphae into charcoal buried in the soil and sporulate in the particles encouraging effects of charcoal on AM colonisation. The role of mycorrhizal fungi is particularly important in carbon cycling, because they pump the carbon compounds out of the root into a massive network of fine fungal filaments in the soil, where it becomes available to other microbes and also to larger soil organisms like worms, mites and insects.

In return, the fungus gathers phosphorus from the soil and delivers it to the plant, helping the plant to grow better. Arbuscular mycorrhizal fungi, found living on plant roots around the world, appear to be the only producers of glomalin. The fungi use carbon from the plant to grow and make glomalin. In return, the fungi's hairlike filaments, called hyphae, extend the reach of plant roots. Hyphae function as pipes to funnel more water and nutrients--particularly phosphorus--to the plants. Higher levels of glomalin give greater water infiltration, more permeability to air, better root development, higher microbial activity, resistance to surface sealing (crusts) and erosion (wind/water). Benefit of glomalin: increased aggregate stability which leads to better soil structure which, in turn, leads to better plant production. As a glycoprotein, glomalin stores carbon in both its protein and carbohydrate (glucose or sugar) subunits. Glomalin is causing a complete re-examination of what makes up soil organic matter.

- Important by-product generated in the process is fine charcoal that, in some cases represents up to 15% of the produced charcoal.addition of fine charcoal in compost preparation improves the process by improving: aeration; water retention and probably by providing microorganisms refuge and adsorption of metabolites.

29. Protecting Soil Resources and Combating Climate Change

There is an urgent need to make effective measures to control GHG emissions from human activities, including the agricultural production, with special concern to the carbon dioxide and methane emissions from anaerobic decomposition of livestock. The atmospheric CO_2 has increased from 280 ppm in 1750 to 367 ppm in 1999 and today's CO_2 concentrations have not been exceeded during the past 420,000 years. The industrialised countries continue to struggle with how to reduce greenhouse gas emissions by 60 per cent by the year 2050, the target recommended by the Kyoto Protocol. Soil organic carbon is an important pool of carbon in the global biogeochemical cycle. The total amount of organic carbon in soils is estimated to be 2011 Gt C, which constitutes about 82% of the global organic carbon in terrestrial ecosystems [113].

The new U.S. legislation July 26, 2007 act S.1884 (by Senator Salazar) biochar provisions promotes the commercial development of technologies that will simultaneously create clean, renewable energy from agricultural and forestry biomass products, while protecting and restoring soil resources and helping to address global climate change. Unlike most carbon-neutral biomass energy systems, aerobic biochar technology, such as the PROTECTOR, is carbon-negative: it removes net carbon dioxide from the atmosphere and stores it in the form of stable soil carbon 'sinks', improving soil fertility, water retention, productivity and crop yields. More and more research shows that soils amended with the char have very beneficial effects on crop growth. The enhanced nutrient retention capacity of biochar-amended soil not only reduces the total fertilizer requirements but also the climate and environmental impact of croplands. Char-amended soils have shown reductions in nitrous oxide emissions and reduced runoff of P into surface waters and leaching of nitrogen into groundwater. As a soil amendment, biochar significantly increases the efficiency of and reduces the need for traditional chemical fertilizers, while greatly enhancing crop yields. The method for sequestering carbon in the soil involves a gentler way of farming the land, and the potential for sequestering carbon is huge. The PROTECTOR is reducing impacts on the climate, by reducing own carbon footprint in the form of carbon offsetting.

30. Product Permit EU Authority and Commercial Field Cultivation Tests for Validation and Demonstration of Protector Agronomic and Nutrition Effectiveness

During the course of the applied scientific RTD and application oriented optimization the PROTECTOR products have been successfully *multiple tested for several years under different climatic, soil and agro-industrial production conditions* in Hungary, Israel, Italy, Germany and The Netherlands. This validation test case study below is executed by the independent and accredited *Agricultural Office, Plant Protection & Soil Conservation Directorate,* the Authority in Hungary, as part of the EU permit procedures for plant growth promoter with biological control by-effect [2]. Both our own Europe-wide past years tests results and the Authority tests are clearly indicating positive PROTECTOR product efficiency under different field conditions, which true tests have been executed as of agricultural industrial practice. It is estimated that by the end of 2008 the Authority permit tests will be completed and product permit granted.

Authority 2007 field test conditions:

Test plant:

Test Program I.

- Tomato: Lycopersicon lycopersicum
- Sweet pepper: Capsicum annuum

Test Program II.

- Broccoli: Brassica oleracea Italica

Test soil type:

The test location and soil type was the same during the Test program I. and II.

Before selecting of the experimental plots soil samples were taken for determination of soil nutrient content. The results indicated reach macro- and micro element content of the test soil *(see Table 17).*

Arrangement of test plants and repetitions in Test Program I. -II.: Random block arrangement were applied with 4 repetitions.

Application method of PROTECTOR products and doses:

Before planting Protector product and artificial fertilizers were dispersed and worked it immediately with rotational hoe.

The following doses were applied during the test program I.-II. (*see Tables 18 and 19*):

Test Program I.

ABC mineral without microbial adaptation: 600kg/hectare

Microbial (3R-T10) adapted ABC mineral: 400kg/ha, 600kg/ha and 1000kg/hectare.

Test Program II.

ABC mineral without microbial adaptation: 600kg/hectare
Microbial (3R-T10) adapted ABC mineral: 400kg/ha, 600kg/ha and 1000kg/hectare.
Microbial (3R-T10) adapted ABC mineral (3 month stored): 600 kg/hectare

Table 17. Measurement of nutrient condition of the test soil (soil type: chernozem with lime deposit) (PROTECTOR Authority test program I.–II.)

Measured soil parameter	Values
pH_{KCl}	7.49
Salt content (%)	<0.02
$CaCO_3$ (%)	9
Humus content (mass %)	3.50
$NO_2 + NO_3$-N (mg/kg)	18.4
P_2O_5 (mg/kg)	195
K_2O (mg/kg)	228
Na (mg/kg)	98.6
Mg (mg/kg)	181
Cu (mg/kg)	2.88
Zn (mg/kg)	1.86
Mn (mg/kg)	44.4
SO_4-S (mg/kg)	11.9

Table 18. Data of the different PROTECTOR treatments (PROTECTOR Authority test program I)

N^o	Treatment	Dose kg/ha	Time	Fenophase	Mode of treatment
1	Untreated Control	-	-	-	-
2	ABC mineral	600	2007.06.13	Before planting	Manual spread to soil
3	ABC mineral+ Trichoderma sp.	400	2007.06.13	Before planting	Manual spread to soil
4	ABC mineral+ Trichoderma sp.	600	2007.06.13	Before planting	Manual spread to soil
5	ABC mineral+ Trichoderma sp.	1000	2007.06.13	Before planting	Manual spread to soil

Fertilization practice: Only mineral fertilizations were applied. No adjunctive organic matter based fertilizers were adopted. To every parcel, exception of the Control treatment, 20 kgs of N active ingredient in form of ammonium-nitrate and 200 kgs of K_2O ingredient in form of potassium-sulphate nutrients per ton were added.

Irrigation condtition: Because of extraordinary warm and droughty weather during the Test Program I.-II. season regular artifical irrigation control was required for providing the sufficient water demand of the plants.

Table 19. Data of the different PROTECTOR treatments (PROTECTOR Authority test program I)

No	Treatment	Dose kg/ha	Time	Fenophase	Mode of treatment
1	Untreated Control	—	—	—	—
	ABC mineral	600	2007. 08. 07.	Before planting	Manual spread to soil
	ABC mineral +Trichoderma sp. (3 month stored)	600	2007. 08. 07.	Before planting	Manual spread to soil
	ABC mineral +Trichoderma sp.	400	2007. 08. 07.	Before planting	Manual spread to soil
	ABC mineral +Trichoderma sp.	600	2007. 08. 07.	Before planting	Manual spread to soil
	ABC mineral +Trichoderma sp.	100	2007. 08. 07.	Before planting	Manual spread to soil

Assessment of the PROTECTOR qualitative and quantitative effects

During the test program I-II. regular physiological and phenological observations were performed. The tomatoes were cultivated until the first, sweet peepers until the third harvest period while broccoli until the total bearing and investigated for the following important parameters: measurement of total weight of fruits, soil and plant leaves nutrient content changing after the PROTECTOR treatments.

Results

Sweet Pepper

At the first harvesting the size and yields of the 3 types PROTECTOR treated plants were 3 to 5 percent higher compared to the untreated control, but differences between the treatments were not significant *(see Table 20)*.

At the second harvesting the tendency was the same as after the first period, but the differences were not significant because of the high variance. However it can be seen that the largest quantity of fruits were measured in the 1000 kg/ha dose of PROTECTOR treatment *(see Table 21)*.

Table 20. Measurement of fruit weight in sweet pepper at first harvest (PROTECTOR Authority test program I)

No	Treatment	Dose (kg/ha)	Repetitions				Average kg/parcel	Control %
			I.	II.	III.	IV.		
1	Untreated control	—	3.308	3.487	3.985	3.258	3.510	100
2	ABC	600	3.747	3.033	3.987	3.700	3.617	103.1
3	ABC+ Trichoderma sp.	400	3.300	3.790	3.986	3.517	3.648	104.0
4	ABC+ Trichoderma sp.	600	4.711	3.433	3.114	3.496	3.689	105.1
5	ABC+ Trichoderma sp.	1000	3.114	3.192	3.789	4.591	3.672	104.6

The results of the last measurement *(see Table 22)* were remarkable from different viewpoints. It can be clearly seen that the positive tendency was the same as received after the first and second harvesting period. The effect of the blank ABC and low dose PROTECTOR treatment were similar, while in the 600 and 1000 kg PROTECTOR treatments the yields were significantly exceed the untreated control.

Table 21. Measurement of fruit weight in sweet pepper at second harvest (PROTECTOR Authority test program I)

No	Treatment	Dose (kg/ha)	Repetitions				Average kg/parcel	Control %
			I.	II.	III.	IV.		
1	Untreated control	—	6.224	5.791	8.991	8.670	7.419	100
2	ABC	600	8.135	7.258	7.525	7.958	7.719	104
3	ABC+ Trichoderma sp.	400	7.606	8.777	6.809	7.795	7.747	104.4
4	ABC+ Trichoderma sp.	600	7.370	8.005	7.736	8.180	7.823	105.4
5	ABC+ Trichoderma sp.	1000	8.269	7.716	7.724	8.523	8.058	108.6

Table 22. Measurement of fruit weight of sweet pepper at third harvest (PROTECTOR Authority test program I)

No	Treatment	Dose (kg/ha)	Repetitions				Average kg/parcel	Control %
			I.	II.	III.	IV.		
1	Untreated control	-	4.020	3.330	3.540	3.580	3.618	100
2	ABC	600	3.120	3.820	4.040	3.920	3.725	103.0
3	ABC+ Trichoderma sp.	400	3.420	4.120	3.970	3.540	3.763	104.0
4	ABC+ Trichoderma sp.	600	3.740	4.380	3.980	3.680	3.945	109.1
5	ABC+ Trichoderma sp.	1000	4.000	4.800	4.650	4.300	4.438	122.7

These results indicated the effective and clear phosphorous and nutrient mobilization and plant growth promotion effects of the PROTECTOR treatments. The results also indicated that the effective mobilization required minimum 3 months.

The application of the specific ABC mineral increase the yields by 1 tonne/ha. The formulated ABC mineral carrier by Trichoderma harzianum sp. 3R-T10 further increased the yields in the field test with 2 additional tonnes/has. The largest yields have received with the 1000 kg/ha PROTECTOR treatment dose, where production yield increased by about 3 tonne/ha, which was significant result (*see Table 23* and *Table 24).*

Table 23. Final results of sweet pepper test yield values (PROTECTOR Authority test program I)

No	Treatment	Dose (kg/ha)	Repetitions				Average kg/parcel	Control %
			I.	II.	III.	IV.		
1	Untreated control	—	13.55	12.61	16.52	15.51	14.546	100
2	ABC	600	15.00	14.11	15.55	15.58	15.061	103.5
3	ABC+ Trichoderma sp.	400	14.33	16.69	14.77	14.85	15.158	104.2
4	ABC+ Trichoderma sp.	600	15.82	15.82	14.83	15.36	15.456	106.3
5	ABC+ Trichoderma sp.	1000	15.38	15.71	16.16	17.41	16.167	111.1

Table 24. Sweet pepper yield values at different treatment per 1 hectare (PROTECTOR Authority test program I)

No	Treatment	Dose (kg/ha)	Repetitions				Average t/hectare	Control %
			I.	II.	III.	IV.		
1	Untreated control	—	25.80	14.01	31.45	29.53	27.696	100
2	ABC	600	28.56	26.87	29.61	29.66	28.676	103.5
3	ABC+ Trichoderma sp.	400	27.28	31.77	28.11	28.28	28.860	104.2
4	ABC+ Trichoderma sp.	600	30.12	30.12	28.24	29.24	29.429	106.3
5	ABC+ Trichoderma sp.	1000	29.29	29.91	30.77	33.16	30.782	111.1

The results of the soil nutrient analysis clearly indicated that (*see Table 25*) the plant available Phosphorous content significantly increased. The results also indicated the effective phosphorous mobilization activity of the selected *Trichoderma* sp.

Table 25. Variation of the soil available P_2O_5 content after the different treatments in sweet pepper test (PROTECTOR Authority test program I)

No	Treatment	Dose (kg/ha)	Repetitions				Average mg/kg	Control %
			I.	II.	III.	IV.		
1	Untreated control	—	190	178	180	179	181.750	100
2	ABC	600	160	180	196	209	186.250	102.5
3	ABC+ Trichoderma sp.	400	199	188	194	195	194.000	106.7
4	ABC+ Trichoderma sp.	600	209	190	184	208	197.750	108.8
5	ABC+ Trichoderma sp.	1000	251	223	193	197	216.000	118.8

The increasement of plant available nitrogen and potassium content is resulted from the addition of mineral fertilizers. Significantly higher (10 percent) available soil potassium concentration (*see Table 26*) has been measured in all treatetment compared to the non treated control.

Table 26. Variation of soil available K_2O content after the different treatments in sweet pepper (PROTECTOR Authority test program I)

No	Treatment	Dose (kg/ha)	Repetitions / mg/kg				Average mg/kg	Control %
			I.	II.	III.	IV.		
1	Untreated control	—	223	173	263	232	22,75	100,0
2	ABC	600	261	339	226	360	296,50	133,1
3	ABC+ Trichoderma sp.	400	300	323	263	376	315,50	141,6
4	ABC+ Trichoderma sp.	600	336	284	349	292	315,25	141,5
5	ABC+ Trichoderma sp.	1000	342	270	404	279	323,75	145,3

Table 27. Variation of soil available NO_3 and NO_2 content after the different treatments in sweet pepper test (PROTECTOR Authority test program I)

No	Treatment	Dose (kg/ha)	Repetitions				Average mg/kg	Control %
			I.	II.	III.	IV.		
1	Untreated control	—	2,27	2,41	4,43	4,44	3,377	100,0
2	ABC	600	3,72	4,94	3,14	4,67	4,118	121,5
3	ABC+ Trichoderma sp.	400	4,07	4,21	3,95	3,79	4,005	118,2
4	ABC+ Trichoderma sp.	600	4,9	4,08	3,28	4,31	4,143	122,3
5	ABC+ Trichoderma sp.	1000	4,16	3,94	3,55	4,87	4,130	121,9

The results of the plant analysis indicated that PROTECTOR treatments increased the generally stable P content in the plant leaves *(See Table 28)*. The most impressive results have been archived by 1000 kg/ha PROTECTOR treatment.

Table 28. Variance of the Phosphorous content in sweet pepper leaves after different treatments (PROTECTOR Authority test program I)

No	Treatment	Dose (kg/ha)	Repetitions				Average m/m %	Control %
			I.	II.	III.	IV.		
1	Untreated control	—	0.244	0.230	0.261	0.254	0.247	100
2	ABC	600	0.251	0.249	0.247	0.254	0.250	101.2
3	ABC+ Trichoderma sp.	400	0.261	0.248	0.235	0.263	0.252	101.8
4	ABC+ Trichoderma sp.	600	0.276	0.244	0.260	0.265	0.261	105.7
5	ABC+ Trichoderma sp.	1000	0.275	0.269	0.254	0.254	0.263	106.4

Non significantly higher Nitrogen and Potassium concentration (*see Table 29 and Table 30*) has been measured in all treated sweet pepper leaves compared to the non treated control.

Table 29. Variation of Nitrogen content in sweet pepper leaves after different treatments (PROTECTOR Authority test program I)

No	Treatment	Dose (kg/ha)	Repetitions				Average m/m %	Control %
			I.	II.	III.	IV.		
1	Untreated control	—	3.19	2.96	3.45	2.89	3.123	100.0
2	ABC	600	3.09	2.86	3.41	3.19	3.138	100.5
3	ABC+ *Trichoderma* sp.	400	3.58	3.56	2.61	3.55	3.325	106.5
4	ABC+ *Trichoderma* sp.	600	3.58	3.41	3.01	3.29	3.323	106.4
5	ABC+ *Trichoderma* sp.	1000	2.97	3.41	3.49	3.63	3.375	108.1

Table 30. Variation of potassium content in sweet pepper leaves after different treatments (PROTECTOR Authority test program I)

No	Treatment	Dose (kg/ha)	Repetitions				Average m/m %	Control %
			I.	II.	III.	IV.		
1	Untreated control	—	2.460	1.930	2.650	1.977	2.254	100.0
2	ABC	600	2.620	2.630	2.787	2.560	2.649	117.5
3	ABC+ *Trichoderma* sp.	400	3.100	2.541	2.400	2.810	2.713	120.3
4	ABC+ *Trichoderma* sp.	600	2.785	3.150	2.450	2.500	2.721	120.7
5	ABC+ *Trichoderma* sp.	1000	2.660	3.110	2.660	2.687	2.779	123.3

Tomato

The tomato test results indicated a similar effects than we have received in the sweet pepper field test. The tomato yield received with the the largest dose (1000 kg/ha) PROTECTOR treatment was 7.5 percent higher as compared to a non treated control, which is equivalent with 5.2 tonnes per hectare surplus. But the lower dose (600 kg/ha) of PROTECTOR application was also still resulted 2.7 tonnes per hectare surplus as compared to a non treated Control (see *Table 31* and *Table 32*).

Table 31. Total tomato yield values after different PROTECTOR treatments (PROTECTOR Authority test program I)

No	Treatment	Dose (kg/ha)	Repetitions				Average kg/parcel	Control %
			I.	II.	III.	IV.		
1	Untreated control	—	12.64	11.60	11.20	13.10	12.135	100
2	ABC	600	10.80	12.86	13.10	12.34	12.275	101.2
3	ABC+ *Trichoderma* sp.	400	13.10	11.80	11.98	12.90	12.445	102.6
4	ABC+ *Trichoderma* sp.	600	12.80	12.58	12.20	12.88	12.615	104
5	ABC+ *Trichoderma* sp.	1000	13.0	13.50	11.90	13.79	13.048	107.5

Table 32. Results of tomato yields per hectare (PROTECTOR Authority test program I)

No	Treatment	Dose (kg/ha)	Repetitions				Average kg/ha %	Control %
			I.	II.	III.	IV.		
1	Untreated control	-	72.0	66.1	63.8	74.7	69.170	100
2	ABC	600	61.6	73.3	74.7	70.3	69.968	101.2
3	ABC+ *Trichoderma* sp.	400	74.7	67.3	68.3	73.5	70.937	102.6
4	ABC+ *Trichoderma* sp.	600	73.0	71.7	69.5	73.4	71.906	104.0
5	ABC+ *Trichoderma* sp.	1000	74.1	77.0	67.8	78.6	74.371	107.5

The results of the soil nutrient analysis clearly indicated that *(see Table 33)* the plant available Phosphorous content significantly increased in the tomato test.

Table 34 demonstrates the increasing P content in the tomato leaves after different treatments.

Table 33. Variation of soil available P_2O_5 content after the different treatments in tomato test (PROTECTOR Authority test program I)

No	Treatment	Dose (kg/ha)	Repetitions				Average mg/kg	Control %
			I.	II.	III.	IV.		
1	Untreated control	-	202	189	195	191	194.250	100
2	ABC	600	253	202	180	190	206.250	106.2
3	ABC+ *Trichoderma* sp.	400	253	211	192	207	225.000	115.8
4	ABC+ *Trichoderma* sp.	600	240	238	215	207	225.000	115.8
5	ABC+ *Trichoderma* sp.	1000	313	208	210	209	235.000	121.0

Table 34. Variation of Phosphorous content in tomato leaves after different treatments (PROTECTOR Authority test program I)

No	Treatment	Dose (kg/ha)	Repetitions				Average m/m %	Control %
			I.	II.	III.	IV.		
1	Untreated control	-	0.279	0.331	0.280	0.288	0.295	100
2	ABC	600	0.312	0.282	0.301	0.285	0.295	100.2
3	ABC+ *Trichoderma* sp.	400	0.321	0.274	0.287	0.307	0.297	100.9
4	ABC+ *Trichoderma* sp.	600	0.325	0.295	0.318	0.277	0.304	103.1
5	ABC+ *Trichoderma* sp.	1000	0.289	0.314	0.327	0.292	0.306	103.7

Broccoli

Yield measurement has been executed in each experimental plot after the total maturation.

No any fitotoxic effect, development anomaly and disease development have been observed during the test.

Application of ABC mineral alone has not caused significant yield increasement. The largest yield has been received in the 600 kg/ha dose PROTECTOR treatment, where the 12% increasement was significant result. The higher doses (1000 kg/ha) has not caused further yield increasement (*see Table 35*). The stored Protector product has smaller effect than received in the fresh product.

The results of the soil nutrient analysis (*see Tables 36-37-38)* is clearly demonstrated the positive effect of all treatments. The increasement of plant available Nitrogen and Potassium content is resulted from the addition of mineral fertilizers (*see Tables 37 and 38*). The increasing available Phosphorous content (*see Table 36*) is evidently resulted from the P content of the ABC mineral caused by mobilisation activity of Trichoderma fungi. The plant available Phosphorous content significantly increased in the broccoli test. The higher amount received in the higher application doses.

Non significantly lower Nitrogen concentration (*see Table 39*) has been measured in all treated broccoli leaves compared to the non treated control. The Phosphorous (*see Table 40*) and Potassium (*see Table 41*) concentration in the broccoli leaves have been significantly increased by application of both 600kg/ha and 1000 kg/ha Protector doses.

Table 35. Measurement of broccoli yield after the test (PROTECTOR Authority test program II)

No	Treatment	Dose (kg/ha)	Repetitions				Average kg/parcel	Control %
			I.	II.	III.	IV.		
1	Untreated control	—	24.74	22.90	23.81	21.29	23.338	100.0
2	ABC	600	24.76	23.24	23.62	23.20	23.705	101.6
3	ABC+ *Trichoderma* sp. (stored)	600	21.90	23.62	28.65	24.76	24.733	106.0
3	ABC+ *Trichoderma* sp.	400	24.69	22.97	24.80	27.96	25.105	107.6
4	ABC+ *Trichoderma* sp.	600	26.86	24.95	27.35	25.30	26.114	111.9
5	ABC+ *Trichoderma* sp.	1000	24.95	29.03	24.30	21.98	25.067	107.4

Table 36. Variation of soil available P_2O_5 content after the different treatments in broccoli test (PROTECTOR Authority test program II)

No	Treatment	Dose (kg/ha)	Repetitions / P_2O_5 mg/kg				Average P_2O_5 mg/kg	Control %
			I.	II.	III.	IV.		
1	Untreated control	—	251.0	262.0	254.0	248.0	253.750	100.0
2	ABC	600	249.0	242.0	285.0	311.0	271.750	107.1
3	ABC+ *Trichoderma* sp. (stored)	600	241.0	280.0	300.0	269.0	272.500	107.4
4	ABC+ *Trichoderma* sp.	400	260.0	241.0	251.0	254.0	251.500	99.1
5	ABC+ *Trichoderma* sp.	600	252.0	261.0	270.0	261.0	261.000	102.9
6	*ABC+ Trichoderma sp.*	*1000*	*310.0*	*252.0*	*307.0*	*256.0*	*281.250*	*110.8*

Table 37. Variation of soil available NO_3 and NO_2 content after the different treatments in broccoli test (PROTECTOR Authority test program II)

No	Treatment	Dose (kg/ha)	Repetitions				Average mg/kg	Control %
			I.	II.	III.	IV.		
1	Untreated control	—	5.16	4.02	5.63	5.12	4.983	100.0
2	ABC	600	6.10	5.25	7.44	7.95	6.685	134.2
3	ABC+ *Trichoderma* sp. (stored)	600	5.26	6.14	8.74	7.20	6.835	137.2
4	ABC+ *Trichoderma* sp.	400	7.13	5.30	7.14	5.79	6.340	127.2
5	ABC+ *Trichoderma* sp.	600	5.32	7.27	7.21	5.25	6.263	125.7
6	ABC+ *Trichoderma* sp.	1000	6.20	5.41	6.23	5.54	5.845	117.3

Table 38. Variation of soil available K_2O content after the different treatments in broccoli test (PROTECTOR Authority test program II)

No	Treatment	Dose (kg/ha)	Repetitions / mg/kg				Average mg/kg	Control %
			I.	II.	III.	IV.		
1	Untreated control	—	219.0	273.0	220.0	226.0	234.500	100.0
2	ABC	600	250.0	242.0	234.0	274.0	250.000	106.6
3	ABC+ *Trichoderma* sp. (stored)	600	216.0	232.0	316.0	278.0	260.500	111.1
4	ABC+ *Trichoderma* sp.	400	244.0	235.0	237.0	268.0	246.000	104.9
5	ABC+ *Trichoderma* sp.	600	232.0	223.0	267.0	277.0	249.750	106.5
6	ABC+ *Trichoderma* sp.	1000	235.0	246.0	251.0	296.0	257.000	109.6

Table 39. Variation of Nitrogen content in broccoli leaves after different treatments (PROTECTOR Authority test program II)

No	Treatment	Dose (kg/ha)	Repetitions				Average m/m %	Control %
			I.	II.	III.	IV.		
1	Untreated control	—	4.36	3.78	4.46	4.39	4.248	100.0
2	ABC	600	2.78	4.19	4.05	3.57	3.648	85.9
3	ABC+ *Trichoderma* sp. (stored)	600	4.46	3.87	3.74	4.71	4.195	98.8
4	ABC+ *Trichoderma* sp.	400	3.28	4.32	3.75	5.29	4.160	97.9
5	ABC+ *Trichoderma* sp.	600	4.82	4.14	4.40	3.35	4.178	98.4
6	ABC+ *Trichoderma* sp.	1000	4.67	3.35	4.09	3.95	4.015	94.5

Table 40 demonstrates the increasing P content in the broccoli leaves after different treatments.

Table 40. Variation of Phosphorous content in broccoli leaves after different treatments (PROTECTOR Authority test program II)

No	Treatment	Dose (kg/ha)	Repetitions				Average m/m %	Control %
			I.	II.	III.	IV.		
1	Untreated control	—	0.229	0.335	0.306	0.306	0.294	100.0
2	ABC	600	0.317	0.316	0.355	0.347	0.334	113.5
3	ABC+ *Trichoderma* sp. (stored)	600	0.359	0.323	0.309	0.390	0.345	117.4
4	ABC+ *Trichoderma* sp.	400	0.268	0.296	0.306	0.426	0.324	110.2
5	ABC+ *Trichoderma* sp.	600	0.371	0.359	0.331	0.346	0.352	119.6
6	ABC+ *Trichoderma* sp.	1000	0.400	0.296	0.372	0.371	0.360	122.4

Table 41. Variation of Potassium content in broccoli leaves after different treatments (PROTECTOR Authority test program II)

No	Treatment	Dose (kg/ha)	Repetitions				Average m/m %	Control %
			I.	II.	III.	IV.		
1	Untreated control	—	0.898	1.020	0.797	0.983	0.925	100.0
2	ABC	600	0.813	0.997	1.340	1.090	1.060	114.7
3	ABC+ *Trichoderma* sp. (stored)	600	1.140	1.180	0.742	1.260	1.081	116.9
4	ABC+ *Trichoderma* sp.	400	0.809	1.200	0.952	0.990	0.988	106.8
5	ABC+ *Trichoderma* sp.	600	1.080	1.260	0.809	1.710	1.215	131.4
6	ABC+ *Trichoderma* sp.	1000	1.130	1.130	1.280	1.280	1.205	130.3

31. Conclusion

The test program results clearly indicated the biological effectiveness of the Microbial (3R-T10) adapted ABC mineral product. The application of PROTECTOR substances are resulting balanced plant available mineral content of soil and balanced nutrient uptake for plant. Significant yield increase and fruit quality improvement has also been demonstrated in tomato, sweet pepper and broccoli.

Summarizing the results of the tomato and sweet pepper trials, it is clearly indicated that the developed PROTECTOR product is effective plant natural nutrient, plant growth promoter and plant health improver with biological control effects against soil bore plant diseases. The biological efficiency of the ABC mineral carrier colonized by Trichoderma sp is evident. Other fungus and bacteria strains have also been successfully tested. After the PROTECTOR treatment the soil plant available Phosphorous content increased, which resulted 11 percent higher plant yield compared to the control in average. Both non treated tomato and sweet pepper control were produced a good average yield, which is indicating that under more disadvantageous soil and climate conditions the different PROTECTOR treatments would have been even higher surplus effects on the yield. The tendency and positive effect of the PROTECTOR treatments to the soil available Phosphorous content and plant yields were similar to observed during the sweet pepper test. After 600–1000 kg/ha PROTECTOR treatments the plant available Phosphorous content of the soil versus control is significantly increased, in average 20%.

Further Authority tests with other horticultural test plants on sandy loam soil with low available nutrient content are also in progress for completion of the permit procedure. The results on low quality soils so far are more than favourable and indicate a similar positive effect tendency with higher efficiency than test under different climatic and soil conditions. After the 600 kg/ha PROTECTOR treatment, the soil plant available phosphorous content remarkably increased, resulting in 25% higher plant yield compared to the control on average.

32. Future Trends

By identification of future trends and driving forces, which will be the framework for the agricultural economy on the horizon of 2020, the economic development, agricultural and environmental policy have been considered as major impacting factors [82]. There are *food supply uncertainties* that we face in the 21st century, for which there are several reasons:

(a) *increasing use of crops for bio-energy, resulting increased agricultural product prices*
(b) *increasing demand for meat and milk products in the developing world*
(c) *poor harvests around the world following droughts and floods*

Consumer awareness for higher food quality is rising, which is also supported by the easy information access via the Internet, resulting in the following:

(d) explosive *demand for low input farming food products, wherein fewer and milder chemicals are used, and heavy metals are removed* from the cultivation inputs
(e) explosive *demand for organic farming food products, wherein all chemicals and heavy metals are removed* from the cultivation inputs, as already specified for organic farming
(f) *demand for safer, more nutritious and more tasty food products, for affordable and reasonably low prices.*

The *agricultural, environmental and ecological issues* are further expanded, and these are considered more than ever critically important for sustainable development:

(g) comprehensive measures taken to *minimize the agricultural environmental and ecological impacts towards protection and preservation of biological diversity*
(h) healthy soil is recognized as one of the most *important natural resources*
(i) *The biochar carbon-negative* [101] *technology* (that is, the PROTECTOR) for protecting and restoring soil resources might be an *important element to address global climate change and at the same time increase high-quality food crop production in a sustainable way in the low input and organic cultivation sectors.*

The predicted future trend is that while both soil and water of good quality and sufficient quantity are becoming rare resources [82], advanced, but economical, technological solutions are needed to combat these negative trends, including reducing the levels of biodiversity. The existing social demand for aesthetically pleasing, "traditional" landscapes is further increased. A further trend is that the consumer requirement for choice of food applies not only to the variety but also to the origin and mode of production of agricultural commodities. The evolution of trade and, consequently, of markets, determines whether these commodities are sourced locally or by suppliers outside borders. The combination of agricultural technology and markets frame the production possibilities, and evidently directly determine how the land is used, and—by extension—if it will come out of production. Finally, both employment opportunities and life-style choices are jointly having a profound effect on the migratory movements that are occurring, with the added dimension that communication and transport technology gives additional scope for the individual freedom to choose where to live.

In all of these trends, the PROTECTOR innovative technology offers sustainable and economical solutions, where the natural preservation concept from the ancient *Lithica 400 BC is considered: "For all of the pests that out of earth arise, the earth itself the antidote supplies."*

33. Sources of Information and Advice

The 3R Environmental Technologies Ltd and the Terra Humana Ltd are continuously developing its web site http://www.terrenum.net *to provide information and consultation for public. Further reference can be found in table 16.1.* as sources of further information and advice.

References

[1] Abelson, P. H. A Potential Phosphate Crisis. *Science*, 26 March 1999: Vol. 283. no. 5410, p.2015)

[2] Agricultural Office, Plant Protection & Soil Conservation Directorate, Directorate Fejer County, Hungary. Report on evaluation of PROTECTOR plant growth promotion material, October 15 2007,

[3] Almaraz, R., van den Berg, E., Reich P. (1997). An Assessment of the Soil Resources of Africa in Relation to Productivity. *Geoderma,* 77, 1-18.

[4] American Chemical Society (2008). Ancient Method, 'Black Gold Agriculture' May Revolutionize Farming, Curb Global Warming. *ScienceDaily*. Retrieved May 9, 2008. Available from URL: http://www.sciencedaily.com/releases/2008/04/080410153658.htm

[5] Anwar, E. Z., Wissa, M.A. Phosphogypsum Disposal and The Environment. Available from URL: http://www.fipr.state.fl.us/pondwatercd/phosphogypsum_disposal.htm

[6] Ball, A.S., Pretty, J.N. Agricultural influences on carbon emissions and sequestration. Powell et al. (ed), *UK Organic Research 2002: Proceedings of the COR Conference*, 26-28th, March 2002, Aberystwyth, pp. 247-249.

[7] Barber, S.A. (1995). Soil nutrient bioavailability. A mechanistic approach. New York, USA, John Wiley & Sons.

[8] Bellamy, P. H., Loveland, P. J., Brabley, R. I., Lark, R. M., Kirk, G. J. D. Carbon losses from all soils across England and Wales 1978-2003. *Nature,* 2005. vol. 437 pp. 245-248.

[9] Bieleski, R.L. (1973). Phosphate pools, phosphate transport, and phosphate availability. *Annu Rev Plant Physiol,* Volume 24: 225–252.

[10] Brandt-Williams, S., Pillet, G. Fertilizer co-producrs as agricultural emternalities. Quantifying Environmental Services Used in Production of Food, Biennial Emergy Analysis Conference., publ. 2003

[11] Brenner, J., Paustian, K., Bluhm, G., Cipra, J., Easter, M., Elliott, T., Kautza, T., Killian, K., Schuler, J., and Williams, S. Quantifying the change in greenhouse gas emissions due to natural resource conservation practice application in Iowa. Final Report to the Iowa Conservation Partnership. March, 2001, USDA Natural Resources Conservation Service

[12] Bumb, B.L., Baanante, C.A. (1996). The Role of Fertilizer in Sustaining Food Security and Protecting the Environment. Food, Agriculture and the Environment Discussion Paper, 17. International Food Policy Research Institute, Washington, DC.

[13] Bundesministerium für Ernährung, Landwirtschaft und Verbraucherschutz. 2006. Die EU-Agrarreform–Umsetzung in Deutschland 2006. Available from URL: www.bmelv.de

[14] Chameides, W. L., Kasibhatla, P. S., Yienger, J. & Levy, H. (1994). Growth of continental-scale metro-agroplexes, regional ozone pollution, and world food production. *Science,* 264. 74–77.

[15] Comis, D. Glomalin hiding place for a third of thc world's stored soil carbon. Agricultural Research. Sept 2002. *FindArticles.com.,* 09. May. 2008. Available from URL: http://findarticles.com/p/articles/mi_m3741/is_9_50/ai_92589768

[16] Connett, M. (2003). The Phosphate Fertilizer Industry: An Environmental Overview, Fluoride Action Network, Available from URL: http://www.fluoridealert.org/phosphate/overview.htm

[17] Cooperband, L. (2002). Building Soil Organic Matter with Organic Amendments, *published by the Center for Integrated Agricultural Systems (CIAS)*, College of Agricultural and Life Sciences, University of Wisconsin-Madison.

[18] Council for Agricultural Science and Technology. (2004). Climate Change and Greenhouse Gas Mitigation: Challenges and Opportunities for Agriculture. Ames, IA: CAST, 120 p.

[19] Delmas, R., Serca, D. & Jambert, C. (1197). Global inventory of NOx sources. *Nutr. Cycl. Agroecosyst.*, 48, 51–60.

[20] Doerr, S. H., R. A. Shakesby, and R. P. D. Walsh. (2000). Soil water repellency: its causes, characteristics and hydro-geomorphological significance. *Earth-Science Reviews* 51, no. 1-4 (August): 33-65.

[21] Dunn, R.A. (1994). Soil Organic Matter, Green Manures and Cover Crops For Nematode Management. SS-ENY-918, Entomology and Nematology Department, Florida Cooperative Extension Service, Institute of Food and Agricultural Sciences, University of Florida.

[22] Eckert, D.,Efficient Fertilizer Use Mannual - Nitrogen, Available from URL: http://64.51.188.251/Assets/pdffiles/efu/Nitrogen.pdf

[23] EcosSanres Fact Sheet 4, Closing the Loop on Phosphorus. Available from URL: http://www.ecosanres.org/pdf_files/Fact_sheets/ESR4lowres.pdf

[24] FAO. (1984). *Fertilizer and plant nutrition guide.* FAO Fertilizer and Plant Nutrition Bulletin No. 9. Fertilizer and Plant Nutrituion Service, Land and Water Development Division.

[25] FAO. (2004). Use of phosphate rocks for sustainable agriculture. *FAO Fertilizer and plant nutrition bulletin* 13., Technical editor: E. Zapata.

[26] Firbank, L. Sowing the seeds of farming's future. BBC NEWS, Science, November 13, 2007.

[27] Francl, T., Nadler, R., and Bast, J. (1998). Impact of the Kyoto Protocol on Agriculture. American Council for Capital Formation, policy conference presentation sponsored by the ACCF Center for Policy Research.

[28] Funderburg,E., Why are nitrogen prices so high, Available from URL: http://www.noble.org/Ag/Soils/NitrogenPrices/Index.htm

[29] Gellings, C.W., Parmenter, K.E.. Energy Efficiency in Fertilizer Production andUse. *Encyclopedia of Life Support Systems*(EOLSS), Available from URL: http://www.eolss.net/ebooks/Sample%20Chapters/C08/E3-18-04-03.pdf

[30] Ghosh, D., Deb, A., Bera, S., Sengupta, R., Patra, K.K. (2007). Measurement of natural radioactivity in chemical fertilizer and agricultural soil: evidence of high alpha activity. *Environ Geochem Health,* DOI 10.1007/s10653-007-9114-0.

[31] Glaser, B. (2007). Prehistorically modified soils of central Amazonia: a model for sustainable agriculture in the twenty-first century. *Philosophical Transactions of the Royal Society – Biological Sciences*, 362: 1478.

[32] Glaser, B., Lehmann, J., Zech,W. (2002). Ameliorating physical and chemical properties of highly weathered soils in the tropics with charcoal – a review, *Biol Fertil Soils,* 35:219–230.

[33] Glaser, B., Haumaier, L., Guggenberger and G., Zech, W. (2001). The 'Terra Preta' phenomenon: a model for sustainable agriculture in the humid tropics, *Naturwissenschaften* 88: 1.

[34] Glaser, B., Haumaier,L., Guggenberger,G., Zech,W., The 'Terra Preta' phenomenon: a model for sustainable agriculture in the humid tropics, *Naturwissenschaften,* Volume 88, Number 1 / February, 2001, pages 37-41.

[35] Gold, M.V. (1994). Sustainable agriculture: definitions and terms. SRB 94-05. Beltsville, MD: U.S. Department of Agriculture.

[36] Golley F, Baudry J, Berry RJ, Bornkamm R, Dahlberg K, Jansson A-M, King J, Lee J, Lenz R, Sharitz R et al. (1992). What is the road to sustainability? *INTECOL Bull* 20: 15–20

[36] Grimm, K. Phosphorites feed people: finite fertlizer ores impact canadian and global food security. Available from URL:: http://www.eos.ubc.ca/personal/grimm/phosphorites.html

[37] Hamer, U., Marschner,B., Brodowski,S., Amelung, W., (2004). Interactive priming of black carbon and glucose mineralisation. *Organic Geochemistry,* 35, no. 7 (July): 823-830.

[38] Heerink, N., Kuyvenhoven, A. & Van Wijk, S.M. (2001). Economic policy reforms and sustainable land use in LDCs: issues and approaches. In N. Heerink, H. Van Keulen & M. Kuiper, eds. Economic policy reforms and sustainable land use in LDCs. Recent advances in quantitative analysis, pp. 1–20. Heidelberg, Germany, Physica Verlag.

[39] Heimlich, R. (2003). Agricultural Resources and Environmental Indicators. US Department of Agriculture, Agriculture Handbook No. (AH722)

[40] Helsel Z.R. (1992). Energy and alternatives for fertilizer and pesticide use. *Energy in Farm Production*, Volume 6 (ed. R.C. Fluck), pp. 177–201. New York: Elsevier.

[41] Hera, C. (1996). The role of inorganic fertilizers and their management practices. In: C. Rodriguez-Barrueco (Ed.), *Fertilizer and Environment*, Kluwer Academic Publishers, Dordrecht, 131-149.

[42] Hodges, S.C. SOIL FERTILITY BASICS. Soil Science Extension North Carolina State University, Available from URL: http://www.soil.ncsu.edu

[43]Houghton, J.T., Ding, Y., Griggs, D.J., Noguer, M., van der Linden, P.J., Dai, X., Johnson,C.A., and Maskell, K. (eds.)., IPCC (2001) *Climate Change 2001: The Scientific Basis,* Intergovernmental Panel on Climate Change, Cambridge University Press. Cambridge, United Kingdom.

[44] Houghton, J.T., Meira Filho, L.G., Callander, B.A., Harris, N., Kattenberg, A. and Maskell,K. (eds.). *IPCC (1996) Climate Change 1995*: The Science of Climate Change. Intergovernmental Panel on Climate Change, Cambridge University Press. Cambridge, United Kingdom.

[45] Houghton, R.A. (2003). Revised estimates of the annual net flux of carbon to the atmosphere from changes in land use and land management 1850–2000. Tellus 55B: 378–390.

[46] Huettl,R.F., Liming and fertilization as mitigation tools in declining forest ecosystems, Water, Air and Soil Pollution, Volume 44, Numbers 1-2 / March, 1989, Pages: 98-118.

[47] Hüttl, R.F., Schneider, B.U., Forest ecosystem degradation and rehabilitation. *Ecological Engineering*, Volume 10, Issue 1, February 1998, Pages 19-31.

[48] Ingham, R.E, J. A. Trofymow, Elaine R. Ingham, and David C. Coleman Interactions of Bacteria, Fungi, and their Nematode Grazers: Effects on Nutrient Cycling and Plant Growth. *Ecological Monographs*, 1985 March.Volume 55, Issue 1.

[49] Johnston, H.W. 1954. The solubilization of "insoluble" phosphate. II – A quantitative and comparative study of the action of selected aliphatic acids on tricalcium phosphate. *N. Z. J. Sci. Tech. B.*, 36: 49–55.

[50] Kant, S., Kafkafi,U. Mitigation by Crop Management. Available from URL: http://www.plantstress.com/articles/min_deficiency_m/mitigation.htm

[51] Kauwenbergh, S. J. Van. Natural and agricultural (fertilizer) inputs of Cadmium. International Fertilizer Development Center

[52] Khan, S.S, Mulvaney,R.L., Ellsworth, T.R., Boast, C.W., The Myth of Nitrogen Fertilization for Soil Carbon Sequestration, *J Environ Qual.,*2007; 36: 1821-1832

[53] Kongshaug, G., Bockman, O.C., Kaarstad, O. and Morka, H. (1992). Inputs of trace elements to soils and plants. Proc. Chemical Climatology and Geomedical Problems, Norsk Hydro, Oslo, Norway.

[54] Kpomblekou, K. & Tabatabai, M.A. 1994. Effect of organic acids on release of phosphorus from phosphate rocks. *J Soil Sci*., 158: 442–453.

[55] Lampkin, N. (1990). Organic Farming. *Published by Farming Press Books*, United Kingdom

[56] Lehmann, J., Rondon,M. Bio-Char Soil Management on Highly Weathered Soils in the Humid Tropics. A vailable from URL: http://www.css.cornell.edu/faculty/lehmann/

[57] Lehmann, J., Gaunt, J. and Rondon, M. (2006). "Bio-char sequestration in terrestrial ecosystems—a review". *Mitigation and Adaptation Strategies for Global Change,* 11, 403–427.

[58] Lehmann, J., da Silva Jr, J.P., Rondon, M., Cravo, M.S., Greenwood, J., Nehls, T., Steiner, C. and Glaser, B. (2002). 'Slash-and-char – a feasible alternative for soil fertility management in the central Amazon?', *Proceedings of the 17th World Congress of Soil Science*, (pp. 1–12) Bangkok, Thailand

[59] Lehmann, J., da Silva Jr, J.P., Steiner, C., Nehls, T., Zech, W. & Glaser, B. (2003). Nutrient availability and leaching in an archaeological Anthrosol and a Ferralsol of the Central Amazon basin: fertilizer, manure and charcoal amendments. *Plant and Soil,* 249: 343–357.

[60] Lewandrowski, J., Peters,M., Jones,C., House,R., Sperow,M., Eve,M., Paustian,K., (2004). Economics of Sequestering Carbon in the U.S. Agricultural Sector, US Department of Agriculture, Economic research service, Technical Bulletin Number 1909.

[61] Lickacz, J., Penny,D. Soil Organic Matter. Agri Facts. Practical Information for Alberta's Agriculture Industry, Agdex 536-1.

[62] Marschner, H. (1993). Mineral nutrition of higher plants. London, Academic Press Ltd., Harcourt Brace & Co. Publishers.

[63] Matson, P. A., Naylor, R. & Ortiz-Monasterio, I. (1998). Integration of environmental, agronomic, and economic aspects of fertilizer management. *Science*, 280: 112–115.

[64] McFarlin, R. F., Florida institute of phosphate Research (FIPR) (1995). Establishing Vegetation Cover on phosphogypsum in Florida. Publication number: 01-086-116.

[65] Mclaughlin,N.B., Hiba,A., Wall, G.J., King,D.J. Comparison of energy inputs for inorganic fertilizer and manure based corn production, *Canadian Agricultural Engineering,* Vol. 42, No. 1 January/February/March 2000.

[66] Menzel, R.G. (1968). Uranium, radium, and thorium content in phosphate rocks and their possible radiation hazard. *J. Agr. Food Chem.*, 16, 231-284.

[67] Michinori, N. Microbial Fertilizers in Japan. Available from URL: http:// www.agnet.org/library/eb/430/,accessed November 19, 2007

[68] Mizuta, K., Matsumoto, T., Hatate, Y., Nishihara, K. and Naka.nishi T. (2004). Removal of nitratenitrogenfrom drinking water using bamboo powder charcoa', *Bioresource Technology* 95, 255–257.

[69] Miller, B.W. (2006).The long-term effects of a single phosphorus fertilizer application on phosphorus availability in forest soils., 18th World Congress of Soil Science July 9-15, 2006 - Philadelphia, Pennsylvania, USA)

[70] Naylor, R. (1996). Energy and resource constraints on intensive agricultural production. *Annu. Rev. Energy Environ.*, 21: 99–123.

[71] OECD (1993) World Energy Outlook. OECD, Paris.

[72] Oosterhuis, F.H., Brouwer, F.M., Wijnants, H.J. (2000). A possible EU wide charge on cadmium in phosphate fertilisers: Economic and environmental implications. Report number E-00/02, commissioned by the European Commission.

[73] Paustian, K., Antle,J. M., Sheehan,J., Paul, E.A., (2006). Agriculture's role in greenhouse gas mitigation. Prepared for the Pew Center on Global Climate Change.

[74] Pohlman, A.A. & McColl, G.J. 1986. Kinetics of metal dissolution from forest soils by organic acids.,*J. Env. Qual.*, 15: 86–92.

[75] Radovic, L.R., Moreno-Castilla, C. and Rivera-Utrilla, J. (2001).Carbon materials as adsorbents inaqueous solutions', in L.R. Radovic (ed.). *Chemistry and Physics of Carbon,* (pp. 227–405) New York, Marcel Dekker

[76] Ragothama. K.G. (1999). Phosphate acquisition. *Annu Rev Plant Physiol Plant Mol Biol.,* 50: 665–693).

[77] Reicosky, D.C., Environmental Benefits of Soil Carbon Sequestration, Available from URL: http://www.depweb.state.pa.us/watersupply/lib/watersupply/InnovTechForum-IIE-Reicosky.pdf

[78] Rehm, G. Schmitt, M., Lamb, J. Randall, G. Busman, L. 1998 Understanding phosphorous fertilizers. Communication and Educational Technology Services, University of Minnesota Extension. Available from URL: Http://www.extension.umn.edu

[79] Romheld, V. (1998). The soil-root interface (rhizosphere): Its relationship to nutrient availability and plant nutrition. In: *International workshop on role of environmental and biological factors in acquisition of toxic and essential elements by plants*. Research Institute of Pomology and Floriculture, Skierniewice. pp. 41-58.

[80] Saito, M. and Marumoto,T. Inoculation with arbuscular mycorrhizal fungi: the status quo in Japan and the future prospects, *Plant and Soil*, IssueVolume 244, Numbers 1-2 / July, 2002, 273-279.

[81] Sander, M., Pignatello, J. J. Characterization of charcoal adsorption sites for aromatic compounds: insights drawn from single-solute and bi-solute competitive experiments. *Environmental Science & Technology*, 2005 (Vol. 39) (No. 6) 1606-1615

[82] SCENAR 2020 (2006). Scenario study on agriculture and the rural world. European Commission, Directorate-General Agriculture and Rural Development, Directorate G. Economic analysis and evaluation, G.4 Evaluation of measures applicable to agriculture (Contract No. 30 – CE – 0040087/00-08).

[83] Schactman DP, Reid RJ, Ayling SM (1998). Phosphorus uptake by plants: from soil to cell. Plant Physiol 116: 447–453.

[84] Schimel, D. S., Braswell, B. H., Holland, E., McKeown, R., Ojima, D. S., Painter, T. H., Parton, W. J. & Townsend, A. R. (1994). *Global Biogeochem. Cycles,* 8, 279–294.

[85] Shoemaker; R., McGranahan, D., McBride, W. (2006). Agriculture and Rural Communities Are Resilient to High Energy Costs. Rising energy prices may prompt farmers and rural residents to make tradeoffs in their production practices and daily lives. Economic Research Service/USDA, WWW.ERS.USDA.GOV/AMBERWAVES

[86] Singh, C.P. & Amberger, A. (1990). Humic substances in straw compost with rock phosphate. *Bio. Wastes*, 31: 165–174.)

[87] Skjemstad, J.O, Reicosky,D.C, Wilts, A.R, McGowan, J.A. (2002). Charcoal Carbon in U.S. Agricultural Soils, *Published in Soil Sci. Soc. Am. J.,* 66:1249–1255.

[88] Skjemstad, J.O., R.C. Dalal, L.J. Janik, and J.A. McGowan. (2001). Changes in chemical nature of soil organic carbon in Vertisols, under wheat in southeastern Queensland. *Aust. J. Soil Res*., 39:343–359.

[89] Smil, V. 2000. Phosphorus in the environment: natural flows and human interferences. *Annu. Rev. Energy Environ.*, 25: 53–88.

[90] Spaargaren, O. (2006). Mineral soils influenced by human activity: Anthrosols (new reference group: Technosols), In: 3rd European summer school on soil survey / Michéli, E., Panagos, P., Jones, A., Montanarella, L., - Luxembourg : Office for Official Publications of the European Communities, (EUR 22193 EN) - p. 16-1 - 16-5.

[91] Steen, P. (1998). Phosphorus Availability in the 21st Century: Management of a Nonrenewable Resource. Phosphorus and Potassium 217. Available from URL: www.nhm.ac.uk/mineralogy/phos/p&k217/steen.htm
[92] Steiner, C. (2006). Slash and Char as Alternative to Slash and Burn - Soil charcoal amendments maintain soil fertility and establish a carbon sink, Dissertation, Faculty of Biology, Chemistry and Geosciences, University of Bayreuth, Germany.
[93] Tiessen, H, ed. (1995). Phosphorus in the Global Environment - Transfers, Cycles and Management, SCOPE/UNEP International Phosphorus Project, Published by John and Sons Ltd.
[94] Tiessen, H. ed. 1995. Scope 54:Phosphorus in the Global Environment: Transfers, Cycles and Management. Wiley, 1995. Available from URL: www. icsu-scope.org
[95] Tiessen, H. SCOPE 54 Phosphorus in the Global Environment - Transfers, Cycles and Management, H. Tiessen ed., 1995, 480 pp
[96] Tilman, D. et al. (2001). Forecasting agriculturally driven global environmental change. *Science*, 292: 281–284.
[97] Tilman, D., Cassman, K. G., Matson, P.A., Naylor, R., & Polasky, S. 2002.Agricultural sustainability and intensive production practices. *Nature,* 418: 671-677.
[98] Tilman D, Fargione J, Wolff B, D'Antonio C, Dobson A, Howarth R, Schindler D, Schlesinger WH, Simberloff D, Swackhamer D (2001). Forecasting agriculturally driven global environmental change. *Science,* 292: 281–284.
[99] Trumbore S.E., Potential responses of soil organic carbon to global environmental change. *Proc. Natl. Acad. Sci.,*August 1997. Vol. 94, pp. 8284–8291,
[100] University of York (2007, October 21). Hungry Microbes Share Out The Carbon In The Roots Of Plants. *ScienceDaily.*, Retrieved May 9, 2008, Available from URL: http://www.sciencedaily.com /releases/2007/10/071018123523.htm
[101] US Biochar Provisions S. (1884). Salazar Act (July 26, 2007).
[102] U.S. Department of Agriculture (USDA). (2003). Agricultural resources and environmental indicators, 2003. Agriculture Handbook no. 722 (AH222), Washington DC: Economic Research Service. Available from URL: www.ers.usda.gov/ publications/arei/ah722/.
[103] US Department of Agriculture, U.S. Fertilizer Use and Price Data Sheets. Available from URL: http://www.ers.usda.gov/Data/FertilizerUse/.
[104] U.S. EPA – Inventory of U.S. Greenhouse Gas Emissions and sinks: 1990 – 2006 APRIL 15, 2008, U.S. Environmental Protection Agency, EPA 430-R-08-005
[105] U.S. Epa Environmental protection Agency. About phosphogypsum. Available from URL: http://epa.gov/radiation/neshaps/subpartr/about.html#general.
[106] von Uexku ll, H.R., Mutert, E. (1995) Global extent, development and economic impact of acid soils. *Plant Soil,* 171: 1–15.
[107] Vance, C. P., Symbiotic Nitrogen Fixation and Phosphorus Acquisition. Plant Nutrition in a World of Declining Renewable Resources, *Plant Physiology*, October 2001, Vol. 127, pp. 390–397.
[108] Vance CP, Graham PH, Allan DL (2000). Biological nitrogen fixation: phosphorus Ba critical future need? In FO Pederosa, M Hungria, MG Yates, WE Newton, eds, Nitrogen Fixation from Molecules to Crop Productivity. *Kluwer Academic Publishers*, Dordrecht, The Netherlands, pp 509–518.

[109] Vassilev, N., Vassileva, M. 2003. Biotechnological solubilization of rock phosphate on media containing agro-industrial wastes. *Applied Microbiology and Biotechnology.* Publisher Springer Berlin / Heidelberg ISSN0175-7598. Volume 61, Numbers 5-6.

[110] Vitousek, P. M., Mooney, H. A., Lubchenco, J. & Melillo, J. M. 1997. Human domination of earth's ecosystems. *Science*, 277: 494–499.

[111] Vitousek, P. M. et al. Human alteration of the global nitrogen cycle: sources and consequences. *Ecol. Applic.,* 7, 737–750 (1997).

[112] Wardle, D.A., Nilsson, M.C., Olle Zackrisson, Fire-Derived Charcoal Causes Loss of Forest Humus, *Science*, VOL 320 2 MAY 2008

[113] Watson, R.A. and Osborne, P.L. (1979). An algal pigment ratio as an indicator of the nitrogen supply to phytoplankton in three Norfolk broads. *Freshwater Biol.*, 9, 585-594.

[114] Wiebe, K., Gollehon, N., editors, Agricultural Resources and Environmental Indicators, 2006 Edition, United State Department of Agrculture, Economic information Bulletin 16.

[115] Woolf, D. (2008). Biochar as a soil amendment: A review of the environmental implications. Available from URL: http://www.orgprints.org/13268/01/Biochar_as_a_soil_amendment_-_a_review.pdf

[116] Williams,S. Fertilizer: Application (Organic Vs Inorganic.) Available from URL: http://gardenline.usask.ca/misc/fertili2.html

[117] Wright, S. Glomalin, A Manageable Soil Glue. Available from URL: www.ars.usda.gov/sp2UserFiles/Place/12650400/glomalin/brochure.pdf.

[118] Worrell, E., Phylipsen, D., Einstein, D., Martin, N., (2000) Energy Use and Energy Intensity of the U.S. Chemical Industry, Energy Analysis Department Environmental Energy Technologies Division Ernest Orlando Lawrence Berkeley National Laboratory, LBNL-44314, Available from URL: http://www.energystar.gov/ia/business/industry/industrial_LBNL-44314.pdf

In: Environmental Chemistry Research Progress
Editors: P. Robinson and R. Gallo

ISBN 978-1-60741-055-3

Chapter 2

PHOTOCHEMISTRY OF ALKYL NITRITES, NITRO COMPOUNDS AND CARBOXYLIC ACID: DYNAMICS OF OH FORMATION

Qiang Wei and Ke-Li Han*
State Key Laboratory of Molecular Reaction Dynamics, Dalian Institute of Chemical Physics, Chinese Academy of Sciences, Dalian 116023, P. R. China

ABSTRACT

In recent years, the photodissociation dynamics of alkyl nitrites, nitro compounds and carboxylic acid has been a subject of intensive studies, which is closely related to the atmospheric chemistry. Here we present a review on the photochemistry of alkyl nitrites, nitro compounds and carboxylic acid, with emphasis on the recent progress in photodissociation dynamics at 266 and 355 nm by using laser induced fluorescence (LIF) technique. The OH photofragments were observed in these photolysis processes. Internal state distributions of the nascent OH from the photodissociation were measured. The OH spin-orbit state and Λ-doublet state were observed. The *ab initio* calculations have also been employed to investigate those photodissociation processes. Based on the experimental and theoretical results, the photodissociation dynamics of these molecules are revealed.

Keywords: Photodissociation dynamics; laser induced fluorescence; OH spectroscopy, nascent rotational distribution; H transfer; *Ab initio* calculation

* Tel: +86-411-84379293; Fax: +86-411-84675584;E-mail Address: klhan@dicp.ac.cn; Group Homepage: http://beam.dicp.ac.cn

INTRODUCTION

The study of the photodissociation reactions is important for both practical and fundamental reasons. On the practical side, photodissociaton reactions play a vital role in initiating many chemical reactions in the atmosphere. Knowledge of the dissociation products and the detailed energy distribution in those products is necessary in order to understand the reactions in which the products may participate in any ensuing collisions. In addition, such information is useful in such diverse applications as laser assisted synthesis and even laser design itself [1-4]. From the fundamental viewpoint, the molecules of interest collide with photons having well defined momenta, the theories are much simpler than those or reactive scattering events and applicable to even moderately large molecules. The photodissociation processes are governed by an initially prepared state and shape of potential energy surfaces along the reaction coordinate. Information about the excited state of the parent molecule, i.e. the geometrical structure of the excited state, and the photodissociation dynamics can be obtained from the fragments formed in the dissociation process. By examining the dynamics of the process, the shape of the potential energy surfaces can be studied in detail [5-10].

The photodissociation of alkyl nitrites has been widely studied over the last 20 years [11-48]. Alkyl nitrites, generally denoted as RONO where R represents an alkyl moiety (e.g., methyl, ethyl, isopropyl, etc.), are excellent candidates for photodissociation investigations since their dissociation is fast (< rotational period) and product quantum states carry memory of the initial excitation [11]. The UV absorption spectra of RONO species are very similar and consist of two excitation bands. The first weak absorption band is the S_1 ($n\pi^*$) $\leftarrow$ S_0 transition in the range of 300-400 nm corresponding to the excitation of one electron of the lone pair of oxygen to the π^* antibonding orbital. The second strong absorption band is the S_2 ($\pi\pi^*$) $\leftarrow$ S_0 transition centered on about 220 nm [12]. In both absorption bands, the primary photodissociation process is cleavage of the weak RO-NO bond (~ 41 kcal/mol) [13] to yield alkoxy radical and nitric oxide,

$$\text{RONO}(S_0) + h\nu \rightarrow \text{RO(A or X)} + \text{NO}(X^2\Pi) \qquad (1)$$

Thus, most studies are focused on the detection of NO products in the photodissociation process. For example, Lahmani et al. investigated the photodissociation of CH_3ONO between 364 and 318 nm wavelengths by probing the NO product state distributions with the two-photon LIF technique [14-16]. Schwartz-Lavi et al. studied the photodissociation of $(CH_3)_3CONO$ at 365.8 and 351.8 nm. They measured the rotational alignment, μ, v, and J vector correlations, energy disposals, and internal state distributions (vibration, rotation, Λ-doublet, and spin-orbit components) of the nascent NO fragments by single photon LIF combined with polarization and sub-Doppler spectroscopy [37-38]. In our present experiments, not only the NO but also a new photoproduct OH were measured [45-48].

Nitroalkanes and nitroaromatic compounds are of interests as energetic and functional molecules that can have applications as explosive materials [49-52]. Nitroalkanes and nitroaromatic compounds may also have a possible role in atmospheric chemistry and photochemistry [53-58]. There have been extensive experimental and theoretical studies of photodissociation reactions of nitroalkanes and nitroaromatics[59-107]. Experimentally [59-93], the photodissociation dynamics of nitro compounds have widely been studied in the gas phase by vacuum ultraviolet (VUV) photoionization mass spectrometry technique, by the ultra electron diffraction (UED) technique, by the multimass ion imaging technique, by the LIF technique, by photofragment translational spectroscopy(PTS), by resonance enhanced multiphoton ionization(REMPI) technique, and so on. Upon UV irradiation, the nitro compounds can dissociate through different pathways including simple bond cleavage, concerted molecular dissociation, and rearrangement or elimination through cyclic transition states, which may compete with each other. Theoretically, quantum chemical calculations have been performed for these nitro compounds [94-104]. All the studies mentioned above predict that photodissociation mechanism of nitro compounds is very complicated because multiple dynamical channels are involved. In our experiment, the photodissociation of nitromethane, nitroethane, nitrobenzene, nitrotoluenes and nitrophenol was investigated using LIF technique [105-107]. We have focused on the OH production from the photodissociation processes.

Photodissociation of carboxylic acids, which play important roles in atmospheric, combustion and interstellar chemistry, has been investigated extensively [108-126]. These acids are end-products or important intermediates in the oxidation of hydrocarbons. Some carboxylic acids, such as formic acid, have been identified, and some, such as propiolic acid, are expected to be present in interstellar space. Also, photodissociation of carboxylic acids generates the important OH radical in atmosphere. In our laboratory, the benzoic acid monomer was studied by using LIF technique. From the experimental results, we can not only identify the dissociative state but also verify the energy threshold for OH fragment produced from photodissociation of benzoic acid [125,126].

The outline of the present review is as follows. In Section 2, we present the experimental method in our studies. In Section 3, some recent results about photochemistry of gaseous nitro compounds and carboxylic acid are discussed. The main conclusions are summarized in Section 4.

2. Experimental Method

The apparatus used in the studies is laser induced fluorescence machine, which has been applied to study the internal distribution of photodissociation products. Briefly, the second-harmonic output (532 nm) of an Nd:YAG laser (Spectra- Physics, GCR-170, 10Hz) was used to pump a tunable dye laser (Lumonics HD-500) operating with DCM dye. The output of the dye laser was frequency-doubled again through a Harmonic Generator (Lumonics, HT-1000) and used for the probe laser beam. The residual part of the 1064 nm output of the Nd:YAG laser was converted into 266 nm (4~5 ns duration) or 355 nm (5~6 ns duration) by KD*P crystals and provided for the photolysis laser beam. The two laser beams were collinearly counterpropagated and softly focused on the center of the photolysis cell by two 70 cm focal length lens, respectively. The photolysis laser pulse energy used in the experiments was reduced to less than 1.2 mJ to ensure that the OH formation from multiphoton process was negligible. The probe laser pulse energy was reduced to ~0.1 mJ. The probe laser beam was optically delayed with respect to the photolysis beam by ~15 ns. This delay period was sufficient to separate the two laser beams, and was short enough to make collision effects negligible under the pressure (typically 70 mTorr) used in the experiments.

The resulting fluorescence was collected by a lens with a focal length of 50 mm (diameter=50 mm), and detected by a photomultiplier tube (PMT, Hamamatsu CR161). A suitable band pass filter was placed in front of the PMT to cut off scattering from the photolysis laser. The signal from the PMT was then gate-integrated by a Boxcar (SRS, SR250), A/D converted by a homemade interface, and stored in a personal computer using a data-taking program. In the experiments, the linearity of the LIF intensity with respect to variations in the power of each one of the lasers and the pressure of the photolysis cell was carefully checked to avoid the saturation effects. Each LIF spectra was recorded at least three times to ensure a proper signal. All experiments were conducted on "slow flows" so that photolysis products do not build up for successive 10 Hz pulses. All the reagents except Methyl nitrite are obtained commercially without further purification. The sample/He mixture was continuously expanded into the center of the photolysis cell through a 0.4 mm inside diameter tube. This gas mixture overlapped with the focused photolysis and probe laser beams at the center of the photolysis cell. The pressure of the photolysis cell was monitored by a capacitance manometer and maintained at ~70 mTorr in the data acquisition.

3. Results and Discussions

3.1. Methyl Nitrite [46, 47]

Methyl nitrite (CH_3ONO) is one of the very few prototypes of medium-sized polyatomic molecules. Its near-UV absorption spectra consist of two different band systems [36]: one is a structured band in the range of 300-380 nm (S_1, n $\rightarrow$ π^*) and the other is a broad and unstructured band centering near 210 nm (S_2, $\pi \rightarrow \pi^*$). Lahmani et al. investigated the photodissociation of CH_3ONO between 364 and 318 nm wavelengths by probing the NO product state distributions with the two-photon LIF technique [14-16]. Pfab and co-workers reported on the photolysis of jet-cooled CH_3ONO at 387, 380, and 355 nm [17-19]. Mestdagh et al. determined the dissociation time (about 125 fs) of the S_1 state of methyl nitrite, n-, and t-butyl nitrite and i-amyl nitrite with the time resolved LIF technique [39]. Huber et al. [20, 36] have measured the translational energy distributions at 248 and 193 nm with a molecular beam time-of-flight apparatus. Winniczek et al. [32] studied the photodissociation of methyl nitrite in the range of 223-227 nm of the S_2 state by resonance-enhanced multiphoton ionization (REMPI) photoelectron spectroscopy. Farmamara et al. [12] have determined the lifetime of the S_2 state of the methyl nitrite molecule within (25 ± 15) fs using the pump-probe technique with 125 femtosecond laser pulses at wavelengths near 200 nm. Yin et al. have reported the internal energy state distributions and the rotational alignments of the NO fragments from the photodissociation of the S_2 state of CH_3ONO by employing the polarized single-photon laser-induced fluorescence (LIF) technique [33]. In our present experiment, a new photoproduct OH was found at the photodissociation of CH_3ONO [45-48].

The LIF spectra of the OH fragments from CH_3ONO photolysis at 266 nm were recorded by scanning the OH $X^2\Pi$ (υ'' =0) $\rightarrow$ $A^2\Sigma^+$ (υ' =0) transitions in the 306-315 nm range. In order to precisely assign the J'' levels of the OH populated in the dissociation process, the experimental spectra ware compared with the simulated ones calculated from the known spectroscopic constants [127]. The intensities of the fluorescence signal (I_F) are related to the populations of a ground rovibrational state N (J'', υ'') by the expression:

$$I_F \propto \frac{N(J'',\upsilon'')q_{\upsilon'\upsilon''}S_{J'J''}f(\nu,\nu_0)}{(2J''+1)} \tag{2}$$

Where $q_{\upsilon'\upsilon''}$ are the known Franck-Condon factors for the OH *A—X* transition [127]; $S_{J'J''}$ are the line strengths for the OH *A—X* one-photon rotational transitions [128]; and $f(\nu, \nu_0)=\rho(\nu_0)\exp(-a(\nu-\nu_0)^2)$ is the laser intensity function. A portion of the typical LIF spectra of the OH fragment is shown in Figure 1, together with the corresponding simulated one. The rotational lines were monitored up to $J''=15.5$ and peaked at $J''=7.5$.

In order to compare the measurement of each rotational level, intensities of the observed LIF rotational lines were normalized with respect to the pressure of the sample in the photolysis cell, the probe-laser and the photolysis-laser energies. Based on the normalized experimental data, we found that the rotational state distribution could be characterized by a Boltzmann temperature. The equation of the Boltzmann population distribution is expressed as:

$$\ln\frac{N(J'')}{(2J''+1)} = \frac{-\varepsilon(J'')hc}{kT_R} + \text{constant} \tag{3}$$

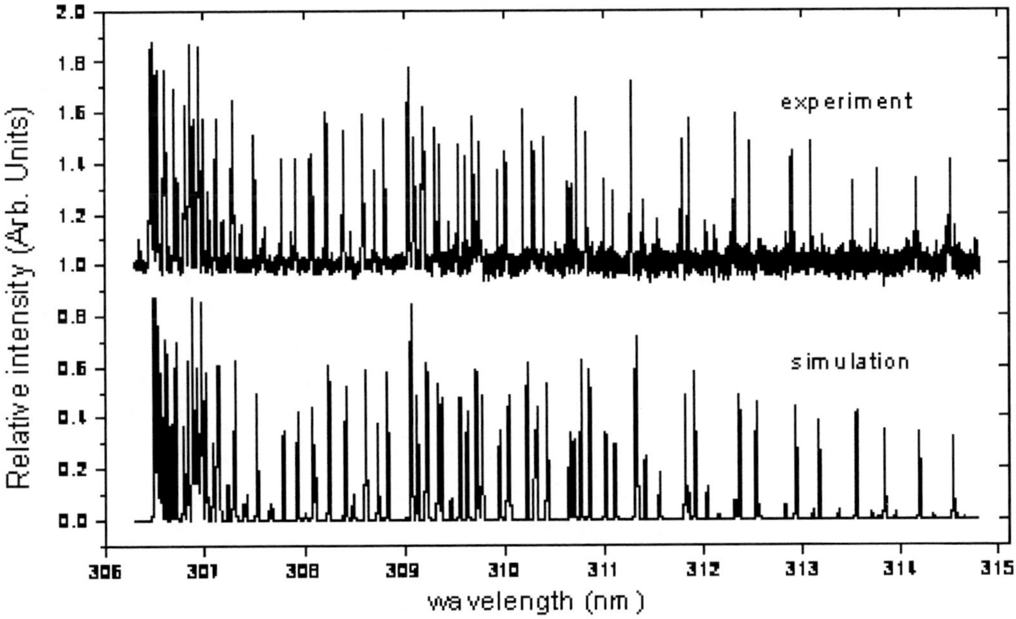

Figure 1. A portion of the $A^2\Sigma^+$ (v′=0) ←$X^2\Pi$ (v″=0) of the one-photon LIF spectrum of the nascent OH fragment from the photodissociation of CH_3ONO at 266 nm. Upper spectrum: experiment; lower spectrum: simulation.

The best linear fit to the experimental data is given in Figure 2. The rotational temperature was calculated from the line slope, $T_R = 2200 \pm 150$ K. The OH fragment was found to be vibrationally cold and no LIF signals of the OH ($X^2\Pi$, v″=1) were detected in our experiments.

The OH transitions were labeled following Hund's case (a) [129]. The P, Q, or R branches are for the cases of ΔJ = -1, 0, or 1, respectively. Because of spin-orbit coupling, the ground state ($X^2\Pi$) of the OH radical splits into two different states: $^2\Pi_{3/2}$ and $^2\Pi_{1/2}$. The ratios of the populations of the $^2\Pi_{3/2}$ and $^2\Pi_{1/2}$ states are close to unity for all values of J''. Thus, we can conclude that the distribution of the OH population between the two spin-orbit states is statistical, and there is no preferential population for either of these two states. The Λ-doublets arise from the orientation of pπ lobes of OH with respect to the plane of rotation. In the Π^+ (A′) state, the pπ lobe lies in the plane of rotation, while in the Π^- (A″) state the pπ lobe is perpendicular to the plane of rotation. According to the parity selection rule (+ ↔ −), the Q branches are for the detecting of the OH fragment in the Π^- state, while the R (and P) branches are due to the Π^+ state [130]. The R/Q (or P/Q) intensity ratios represent the ratios of the Λ−doublets, which give information about the relative population of the OH fragment in the Π^+ and Π^- states [131]. The distribution of the Π^- state is rather preferential than that of the Π^+ state in both $\Pi_{3/2}$ and $\Pi_{1/2}$ states, which suggests that the pπ electronic orbital of the OH radical is predominantly perpendicular to its plane of rotation.

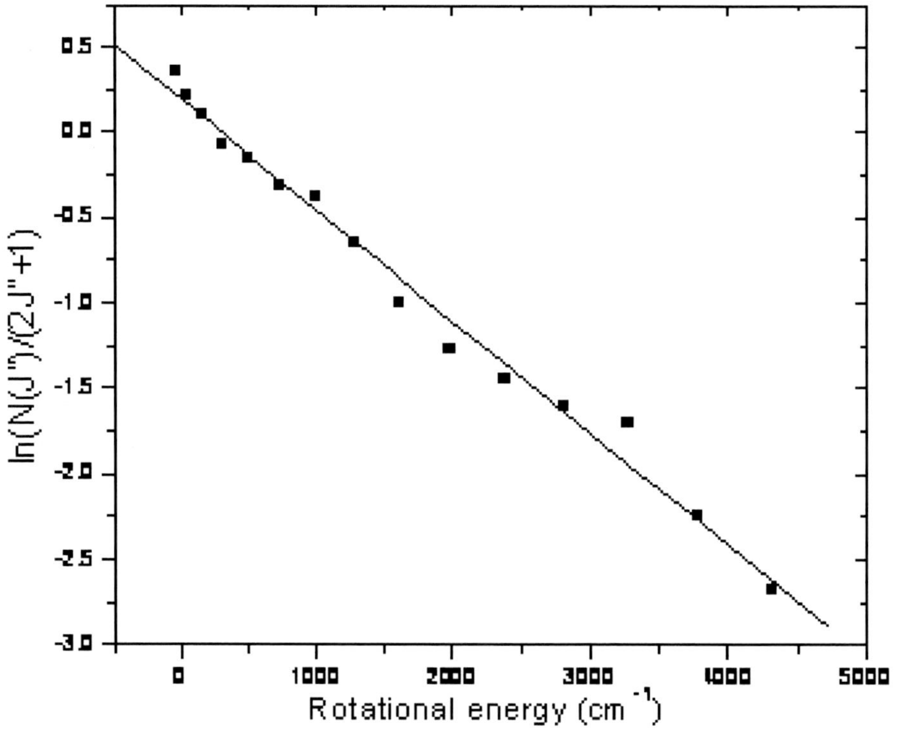

Figure 2. Boltzmann plot of the OH rotational state population for the $^2\Pi_{3/2}$ state (closed rectangles) after photodissociation of CH_3ONO at 266 nm. The best linear fit (solid line) was obtained at a Boltzmann temperature of T_{rot}=2200K.

As is well known that alkyl nitrites occur as cis and trans (syn and anti) conformers in the gas phase around room temperature. The intramolecule hydrogen bonds widely exist in the cis conformers, which generally leads the cis conformers to be more stable [132-137]. In the case of the simplest alkyl nitrite, CH_3ONO, Günthard and co-workers [135-136,138] investigated the stabilization of both cis and trans conformers using the microwave and matrix IR experimental techniques. Felder and Günthard [138] found that the cis conformer was stabilized by 0.63 kcal/mol with respect to the trans conformer due to the intramolecule hydrogen bonds between the terminal oxygen atom and the methyl hydrogen atoms. Very recently, Southern et al. [139] and Sobolewski et al. [140] studied the intramolecule hydrogen bond of anthranilic acid and salicylic acid in the S_0 and S_1 excited states. All these authors reached an identical conclusion that the intramolecule hydrogen bond was strengthened when the parent molecule was excited into the excited states [139-140]. Thus, a reasonable photodissociation mechanism was proposed for the OH product pathway for methyl nitrite: after absorbing a 266 nm photon, the CH_3ONO molecule was excited into the S_2 state, and then the intramolecular hydrogen bonds between the terminal oxygen atom and the methyl hydrogen atoms were strengthened, followed by the intramolecule hydrogen atom transfer from the methyl group to the nitrite group via a five-membered ring transition state. Subsequently the N-OH bond broke to release OH fragment. As an aid to understand the proposed dissociation mechanism above, we carried out the DFT calculations on the PES of CH_3ONO corresponding to the OH photoproduct channel in the ground electronically excited state with the GAUSSIAN 03 program[141] at the B3LYP/6-311+G(d, p) level. Figure 4 displays the relative energies after the zero-point energy corrections and optimized geometries of methyl nitrite, intramolecule hydrogen atom transfer isomer, transition state (TS), and products.

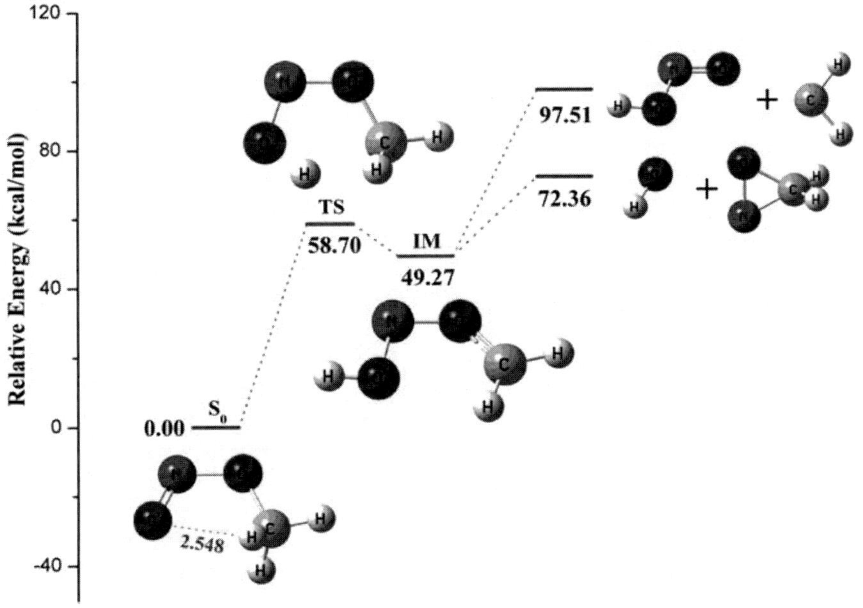

Figure 3. Schematic energy diagram for the hydrogen atom transfer and dissociation reaction of CH_3ONO calculated at the B3LYP/6-311+G(d,p) level, where energy units are in kcal/mol.

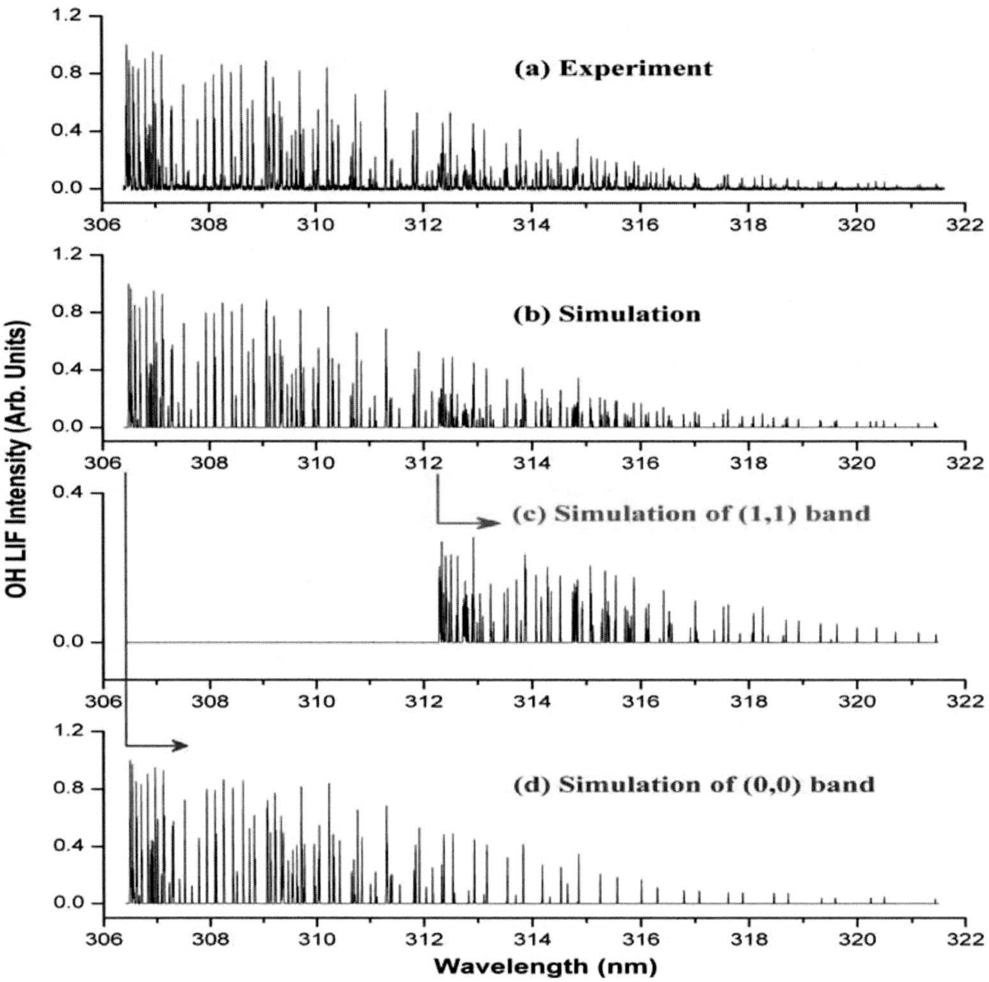

Figure 4. A portion of the $A^2\Sigma^+$ (v′=0) ←$X^2\Pi$ (v″=0) and $A^2\Sigma^+$ (v′=1) ←$X^2\Pi$ (v″=1) bands of the one-photon excited LIF spectra of the nascent OH from the photolysis of $CH_3(CH_2)_3ONO$ at 266 nm. (a) experiment; (b) simulation; (c) (1, 1) band simulation; (d) (0,0) band simulation.

As shown in Figure 3, the parent molecular CH_3ONO firstly isomerizes to CH_2ONOH via a five-membered ring transition state with a 50.70 kcal/mol barrier, and then dissociates to H_2CON+OH by breaking the N−OH bond (72.36 kcal/mol). On the other hand, the intermediate CH_2ONOH can generate the HONO photoproduct through the cleavage of the C−O bond. Once the nascent HONO product contains internal energy in excess of 48 kcal/mol, it decomposes into HO+NO [142]. Under the present experimental conditions, the internal energy (E_0) of the trans conformer of CH_3ONO had been evaluated about 2.14 kcal/mol [33]. The internal energy (E_0) of the cis conformer of CH_3ONO was 0.63 kcal/mol lower than that of the trans conformer, with E_0 = 1.51 kcal/mol. The energy of one 266 nm photon is 107.49 kcal/mol (hv). It needs 97.51 kcal/mol (D_0) to generate HONO photoproduct via breaking the C−O bond. Thus, the available energy corresponding to the HONO fragment channel (Figure 4) from the 266 nm photolysis of CH_3ONO was $E_{avl} = hv + E_0\,(CH_3ONO) - D_0(CH_2\text{−}ONOH) \approx 11.49$ kcal/mol, which was lower than 48 kcal/mol. So this available energy was not enough to allow the HONO photoproduct to further dissociate to OH+NO. For the OH product pathway depicted in Figure 3, the energy barrier was 72.36 kcal/mol, which was 25.15 kcal/mol lower than that for the HONO product pathway. Therefore, it was much easier to produce OH than to produce HONO when CH_3ONO was photolyzed.

3.2. n-Butyl Nitrite [47, 48]

In contrast to photochemistry studies of CH_3ONO, there exists less literature about photodissociatio of $CH_3(CH_2)_3ONO$. We have studied the $CH_3(CH_2)_3ONO$ photodissociation at 266 and 355 nm.

For $CH_3(CH_2)_3ONO$, vibrationally excited state (v"=1) of OH(X, $^2\Pi$) was found in the LIF spectra of the nascent OH($X^2\Pi$) fragments at 266 nm. Figure 4 depicts the experimental and simulated spectra of the nascent OH (0, 0) and (1, 1) bands from the photodissociation of n-butyl nitrite at 266 nm. The LIF intensities of (0, 0) band is much stronger than that of the (1, 1) band. The (1, 1) rotational state distribution conformed to a Boltzmann behavior, and the rotational Boltzmann temperature was calculated to be 1462 ± 120 K. But, the rotational state distribution of OH v"=0 state from the photolysis of $CH_3(CH_2)_3ONO$ at 266 nm is non-single Boltzmann behavior, and thereby it can not be described by a simple rotational temperature. The measured relative population of v"=0:1 was 0.78:0.22.

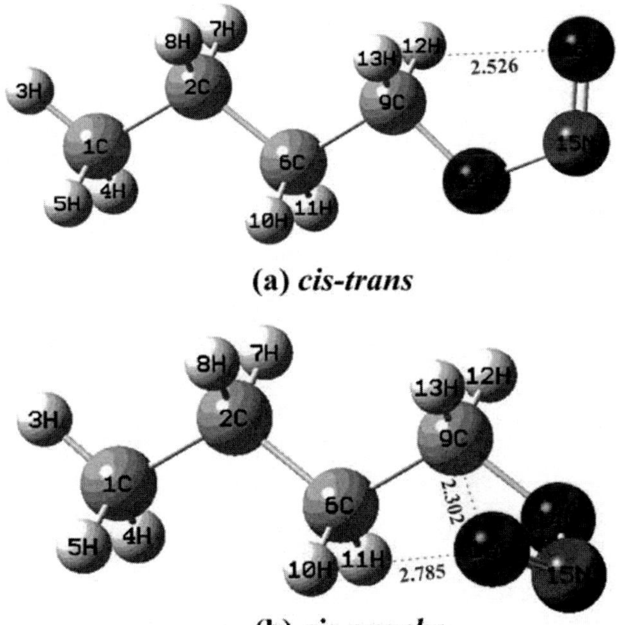

(a) *cis-trans*

(b) *cis-gauche*

Figure 5. Equilibrium geometries of (a) the cis-trans and (b) the cis-gauche conformers of $CH_3(CH_2)_3ONO$, optimized at B3LYP/6-311+G(d,p) level. Distances are given in Å.

In the photolysis of $CH_3(CH_2)_3ONO$ at 355 nm, we have only found the rotational state distribution of OH v"=0 state and it is also non-Boltzmann. The maximum rotational population was monitored up to J"=13.5, which was smaller than that (J"=20.5) of OH v"=0 state from the 266 nm photolysis of $CH_3(CH_2)_3ONO$.

Four different equilibrium geometries (cis-trans, cis-gauche, trans-gauche, and trans-trans) of the ground electronic state of $CH_3(CH_2)_3ONO$ have been determined in a very recent study by McLaughlin et al [143]. They found that the cis-trans (cis with respect to the C-O-N=O dihedral angle and trans with respect to the C-C-O-N dihedral) and cis-gauche geometries were the two most stable conformers due to their ability to form the hydrogen bond between the nitrite group and its α-carbon hydrogen atom [143]. To better understand the photodissociation mechanism, we optimized these two geometries with the GAUSSIAN 03 program [141] at the B3LYP/6-311+G(d,p) level. Figure 5 displays the cis-trans and cis-gauche conformers, respectively. Similar to the photolysis mechanism of CH_3ONO mentioned above, the photodissociation mechanism of the cis-trans conformer of $CH_3(CH_2)_3ONO$ was proposed for the OH product channel: after absorbing a 266 or 355 nm photon, the parent molecule $CH_3(CH_2)_3ONO$ was excited into the S_2 or S_1 excited state, and then the intramolecule hydrogen bond between the terminal oxygen atom O_{16} and its α-carbon hydrogen atom H_{13} was strengthened, followed by the intramolecule hydrogen atom transfer from the hydrogen atom H_{13} to the terminal oxygen atom O_{16} via a five-membered ring transition state. Subsequently the N-OH bond broke to release OH fragment. On the other hand, the photodissociation mechanism was proposed for the OH product pathway for the cis-gauche conformer: after absorbing a 266 or 355 nm photon, the parent molecule $CH_3(CH_2)_3ONO$ was excited into the S_2 or S_1 state, and then the hydrogen bond of terminal oxygen atom O_{16} with its α-carbon hydrogen atom H_{16} and the pseudo hydrogen bond of terminal oxygen atom O_{16} with its β-carbon hydrogen atom H_{10} were strengthened, followed by the intramolecule hydrogen atom transfer from the α-carbon hydrogen atom to the nitrite group via a five-membered ring transition state and from the β-carbon hydrogen atom to the nitrite group via a six-membered ring transition state. Subsequently the N-OH bond broke to release OH fragment.

According to the photodissociation mechanisms proposed above, three dissociation channels for the OH fragments are involved in the photolysis of n-butyl nitrite: the first one is from the photolysis of the cis-trans conformer via a five-membered ring transition state, the second one is from the photolysis of the cis-gauche conformer via a five-membered ring transition state, and the last one is from the photolysis of the cis-gauche conformer via a six-membered ring transition state. In the present experiments, we detected the total OH product from the photolysis of n-butyl nitrite. Each of these three OH product channels could not be discriminated, which would lead to the complicated internal state distributions of the OH from the photolysis of n-butyl nitrite. This inference was in good agreement with the experimental findings: non-Boltzmann rotational state distributions of OH v"=0 state from the photolysis of n-butyl nitrite at 355 and 266 nm; Boltzmann rotational state distribution of OH v"=1 state from the photolysis of n-butyl nitrite at 266 nm.

3.2. Nitromethane and Nitroethane [105]

Nitromethane (CH_3NO_2), as the simplest member of the nitro-compound family, is one of the most prototypical systems for mechanistic studies due to its rich chemistry exhibited both on the ground and excited electronic states contrast to its simple structure. The ultraviolet absorption spectrum of CH_3NO_2 consists of two bands: a weak band centered at 270 nm [59] and a much stronger band centered at 198 nm [60], which were assigned as $\pi^*\leftarrow n$ and $\pi^*\leftarrow\pi$ transitions, respectively. For nitroethane, the UV absorption spectrum consists of two smooth, broad bands centered roughly at 280 nm and 200 nm [72]. The band at longer wavelength has been assigned to a $\pi^*\leftarrow n$ transition that is located on the NO_2 functional group, while the feature at 200 nm is assigned to a $\pi^*\leftarrow\pi$ transition also located on the NO_2 chromophore [60].

In our experiment, we have measured the OH photoproducts from the photodissociation of nitromethane and nitroethane at 266 nm. This was unambiguously identified by their characteristic LIF spectrum in the 306-316 nm region due to the $X^2\Pi$ (υ'' =0) $\rightarrow$ $A^2\Sigma^+$ (υ' =0) transition. It is good agreement with the results of Zabarnick *et al.* [69], which observed a small yield of OH radicals with 266 nm photolysis nitromethane. But, this result is contradictory with the experiments of Greenblatt et al. [70], which observed no OH production from nitromethane photolysis at 282 nm. According to our observation and the different conclusions of the above two groups, we proposed that there possibly exit a threshold to produce the OH radical from the photolysis of CH_3NO_2 between 266 nm and 282 nm. On both photodissociation processes of CH_3NO_2 and $CH_3CH_2NO_2$, no OH (υ'' =1) could be observed.

From the normalized signals, we have found that the rotational state distributions of the OH fragments (υ'' =0) are peaked at J''=4.5 and has a long tail up to the highest J''=15.5 for CH_3NO_2. The rotational temperatures are 2045 ± 150 and 1923 ± 150 K for the $X^2\Pi_{3/2}$ and $X^2\Pi_{1/2}$ states, respectively. The rotational energy of the ground state $X^2\Pi$ (v''=0) was calculated to be 3.43 ± 0.2 kcal/mol. The rotational distribution of the nascent OH products from photolysis of $CH_3CH_2NO_2$ at 266 nm is rotationally much "colder" than that produced from CH_3NO_2. Its highest distribution only attains to J''=9.5. Complicatedly relative to nitromethane, the rotational state distributions cannot be characterized by a simple Boltzmann temperature. This result is in consistent with Greenblatt's conclusion [57], who studied the photodissociation of 2-nitropropane using one-photon LIF technique at 222, 249 and 355 nm, respectively. In their experiment, the OH and HONO were detected at the same time, and the production of HONO increased with the longer wavelength of the photolysis laser. A number of the nascent HONO will dissociate to NO + OH when their internal energy exceeds 48 kcal/mol [142]. The relative complicated rotational state distribution give an evidence that the origin of the nascent OH production should come from both the direct and indirect pathway from the photolysis of nitroethane at 266 nm. The rotational energy of ground vibrationally OH product is 1.19 ± 0.2 kcal/mol.

Comparing the different rotational distributions of the nascent OH product from photodissociation of nitromethane and nitroethane at 266 nm, we can see that different dissociation mechanisms existed in both systems although they are simple two nitroalkanes. Greenblatt *et al.* [70] have suggested a five-membered ring transition state existed in producing OH product using isotopic substitution studies for nitroethane. Similar dissociation mechanism for nitromethane in producing OH production also has been tentatively proposed here that it will form a four-ring transition state in the excited state to dissociate just like the ground state transition state calculated by He's group [97].

3.3. Nitrobenzene and o-, m- and p-Nitrotoluene [106]

The gas phase UV absorption spectrum of nitrobenzene and nitrotoluene between 190 and 350 nm was given by Abbott *et al.* [80] Nagakura *et al.* assigned this absorption spectrum of nitrobenzene to three different bands [144]. The first absorption band centered at 280 nm was assigned to a mixture of an asymmetric charge transfer state and another state that is equivalent to the B_{2u} state in benzene. The second absorption band centered at 240 nm was assigned to a symmetric charge transfer state involving electron transfer from the phenyl ring toward the nitro group. The last absorption band centered about 195 nm was assigned to the $B_{1u} \leftarrow A_{1g}$ transition in benzene. The UV absorption spectra of m-, o- and p-nitrotoluene are similar to that of nitrobenzene. Due to the methyl substitution effect, the maximum absorption positions of three absorption bands of nitrotoluene are slightly red-shifts relative to nitrobenzene. In this work, we have studied the photodissociation dynamics of nitrobenzene, o-, m- and p-nitrotoluene at 266 nm. A new product channel was found for the nitrobenzene and m-, p-nitrotoluene.

The rotational temperatures and rotational energies of the OH products from the photodissociation of nitrobenzene and o-, m-, p- nitrotoluene were obtained from the experimental results, respectively. As is shown in Table 1, the rotational temperatures and rotational energies of m- and p-nitrotoluene are similar but higher than that of o-nitrotoluene, which may provide the evidence that m- and p-nitrotoluene in producing OH fragment have the similar dissociation mechanism, which have different dissociation mechanism from o-nitrotoluene. Moreover, the results of m- and p-nitrotoluene are also very different from that of nitrobenzene in the rotational temperatures and rotational energies, which indicates that methyl substitution effect leads to different internal state distributions of the OH photofragment and in further may result in different dissociation mechanism. Slight discrepancies of the rotational temperatures and rotational energies between m- and p-nitrotoluene probably originate from the different positions of the methyl substitution.

Table 1. Rotational temperatures (T_{rot}) and rotational energies (E_{rot}) of the nascent OH ($X^2\Pi_{3/2}$) from 266 nm photodissociation of nitrobenzene and three nitrotoluene isomers

	T_{rot} (K)	E_{rot} (kcal/mol)
nitrobenzene	850±150	1.34±0.32
o-nitrotoluene	800±120	1.43±0.28
m-nitrotoluene	1170±150	1.92±0.32
p-nitrotoluene	1010±150	1.78±0.32

It is well known that very fast nonradiative relaxation processes and nonemissive characters are widely existed in nitrobenzene and nitrobenzene derivatives through the IC and ISC after they are photoexcited. Recently, Takezaki *et al.* have investigated the photophysical dynamics of nitrobenzene and monosubstituted nitrobenzene with the picosecond time-resolved transient grating method [145-147]. They observed that the lifetime of the lowest excited singlet state S_1 is ≤10 ps and the lowest excited triplet state T_1 is about 480 ps [146-147]. The rapid internal conversion between T_2 and T_1 and the efficient intersystem crossing between S_1 and T_1 ($\Phi_{ISC} \geq 0.8$), T_1 and S_0 were reported both experimentally [146-147] and theoretically [148]. For better understanding this dissociation mechanism, we optimized the geometries, intramolecular hydrogen transfer tautomers, transition states, and products in the S_0 and T_1 states without any symmetrical restriction at the B3LYP/6-311+G(d,p) level. Based on the previous studies and our calculated results, we proposed one reasonable dissociation mechanism for the OH product pathway that after UV excitation, the parent nitrobenzene and nitrotoluene molecules first relax to the T_1 or S_0 states through the efficient intersystem crossing and internal conversion processes, followed by the intramolecular hydrogen transfer from the benzene ring to the nitro group, and subsequently break the N-OH bond to release OH. Considering the very efficient $S_1 \rightarrow T_1$ intersystem crossing ($\Phi_{ISC} \geq 0.8$) and the longer lifetime (~480 ps) [146-148] of the T_1 state, we predicted that the dissociation pathway was mainly via the T_1 state.

For o-nitrotoluene, the OH has been found in the previous experiment [78-83,103,149-150]. The dissociation mechanism was traditionally attributed to the intramolecular hydrogen transfer from the methyl group ($-CH_3$) to the nitro group ($-NO_2$) due to the so-called "ortho effect" [80-83]. Butler and Thomas[149] have concluded that the hydrogen atom in the OH product comes exclusively from the methyl group in the dissociation of o-nitrotouene. In present work, the observed LIF signals of the nascent OH fragment from the photodissociation of nitrobenzene, m- and p-nitrotouene are much weaker than that from o-nitrotoluene, and therefore this pathway is significantly very small for the photodissociation of nitrobenzene, m- and p-nitrotoluene, which may explain why the OH product has not previously been observed by the mass spectroscopy method. For nitrobenzene, m- and p-nitrotoluene, the hydrogen transfer from the benzene ring to the nitro group is inevitably existent. Similarly, this hydrogen transfer process is reasonable to occur in the photodissociation process of o-nitrotoluene. The hydrogen transfer from the methyl group to the nitro group for o-nitrotoluene has been calculated in the S_0 state by Xu and Lin [84]. For the significant ortho effect, hydrogen transfer between the methyl group and the nitro group is rather easier than that between the benzene ring and the nitro group, and therefore the hydrogen transfer from the methyl group to the nitro group would be dominant in the photodissociation of o-nitrotoluene. Consequently, the different hydrogen transfer pathway of o-nitrotoluene relative to m- and p-nitrotoluene results in the discrepancies of the rotational energies and rotational temperatures of o-nitrotoluene from m- and p-nitrotoluene. Of course, here we can't exclude that the OH products result from the two photon photodissociation process.

3.4. 2-Nitrophenol (2-NP) [107]

In contrast to photochemistry studies of nitrobenzene and nitrotoluene, less attention has paied to the nitrophenol. However, the nitrophenols strongly absorb the atmospherically relevant UV range 300-400 nm, corresponding to the S_0 to S_1 transition. Thus, the photochemical might be of importance for the atmosphere [93]. Another absorption band of 2-nitrophenol is located at 271.6 nm, which is assigned to a charge-transfer excitation from the benzene ring to the nitro group [88]. The molecular geometry of 2-nitrophenol has been determined by a joint investigation of gas-phase electron diffraction and ab initio molecular orbital calculations [87]. Chen et al. have studied the photo- or thermal-induced reactions of 2-NP with DFT-B3LYP/6-31G* level of theory [104]. Recently, Wang et al. obtained Resonance Raman spectra of 2-NP in cyclohexane solution [88]. Nagaya et al. [89] have used the matrix-isolation infrared spectroscopy of 2-NP to identify the conformation of the aci-nitro isomer of 2-NP, and examined the photoreaction mechanism with the aid of DFT and time-dependent DFT calculation. Formation of nitrous acid (HONO)/nitrite has been observed during the photolysis of nitrophenols in the liquid [90-92]. Very recently, Bejan et al. [93] have studied the photolysis of gas-phase 2-NP in a flow tube photoreactor upon irradiation. Formation of nitrous acid (HONO) from this photochemical process was detected for the first time and the corresponding mechanism was discussed [93]. In our experiment, we have studied the photodissociation dynamics of 2-NP at 266 and 355 nm.

The similar rotational state distributions of OH products were found in the photodissociation of 2-NP at both 266 and 355 nm, although the available energy at the 266 nm is 27 kcal/mol higher than that at the 355 nm. The rotational temperatures calculated from the rotational distribution of the nascent OH at 266 and 355 nm are about 900±40 K and 1070±80 K, respectively. The possible reason for the similar rotational distributions is that the dynamical constraint may exit in the 2-NP dissociation due to the relatively strong intramolecular hydrogen bonding. The higher value of the spin-orbit ratio (F_1/F_2) obtained at both wavelengths suggests a coupling between the dissociative states with a nearby triplet state. Such behaviour was also found in the photodissociation of molecules like o-nitrotoluene [150]. For 266-nm photolysis, the $\Pi_{3/2}$ state is populated little more predominately as compared to that at 355 nm. Moreover, at both wavelengths, the distributions of the Π^+ states are more preferential than that of the Π^- states, this indicates that the impulsive force acts along the dissociating bond. In the photodissociaton of acetylacetone at 266 and 291 nm, Yoon et al. have proposed that the OH rotation is not expected to be highly excited even when the impulsive force along the dissociating C-OH bond is considerable [151].

In the photodissociation of o-nitrotoluene, OH was produced and attributed to the so-called "ortho effect" (H-transfer from the CH_3-to the NO_2- group) [78, 85]. In the case of 2-NP, which contains OH group, the possible channel is by direct cleavage of the C-OH bond of 2-NP to produce the OH fragment. To determine whether this channel is energetic permission or not, the reaction energy of this OH production channel of 2-NP was calculated at B3LYP/6-311+G (d, p) level of theory with GAUSSIAN03 program package [141]. In this way, the reaction energy of this channel was calculated to be 113 kcal/mol, which is higher than the photon energy of 266 (107.5 kcal/mol) and 355 nm (80.5 kcal/mol). This indicates that the OH could not be produced from direct cleavage of C-OH bond of 2-NP at the photolysis wavelengths 266 and 355 nm. Thus, the proposed "ortho effect" may be applied to the photodissociation of 2-NP.

2-NP has a strong H bond between the H atom of hydroxyl group and the O atom of NO_2 group. Upon absorption of photon at 266 and 355 nm, the 2-NP was excited to the S_4 and S_1 states, respectively [88]. The H transfer may take place in either the excited electronic state or the ground electronic state. After H transfer and rearrangement, four possible aci-trio isomers of 2-NP would be formed. The 2-NP and four possible aci-nitro isomers of 2-NP were shown in Figure 6. Bejan et al. have observed the HONO product from 2-NP upon irradiation of 300-500 nm in the gas phase, which supports the formation of aci-nitro isomers. Nagaya et al.

have studied the photoisomerization of 2-NP at 248 nm and aci-nitro isomer of 2-NP was found based on the observed FTIR bands of product [89]. Both of them have assumed that the H transfer occurs on the excited energy surface. However, Chen et al. have also reported the ground state H transfer and the corresponding aci-nitro isomers were calculated using ab initio method [104]. In the case of o-nitrotoluene, the experimental and theoretical studies have confirmed that the H transfer and the following OH production occurred on the ground energy surface [152-154]. According to the present experimental results and the above discussions, the mechanism for OH production from the photodissociation of 2-NP at 266 and 355 nm should be a multi-step reaction like that of o-nitrotoluene. The first step is H transfer after excitation at 266 and 355 nm and the second step is to produce OH fragment from the aci-nitro isomers of 2-NP.

3.5. Benzoic Acid Monomer (BAM) [125,126]

BAM is the simplest member of the aromatic carboxylic acid family. A phenyl substitution may change the dissociation mechanism of aliphatic carboxylic acids, due to the conjugation interaction between the phenyl ring and the carboxylic group in aromatic carboxylic acids. In a recent work, Fang and co-workers [155] have mapped the potential energy surfaces of some low-lying excited states of BAM using CASSCF (complete active space self-consistent field) methods. They investigated the dissociation dynamics occurring on four different potential energy surfaces. They proposed that the α C-O bond cleavage started from the T_2 state and lead to the fragments of C_6H_5CO and OH, which was predicted to be the most possible channel upon photoexcitation of the BAM at 270 nm or shorter wavelengths[155].

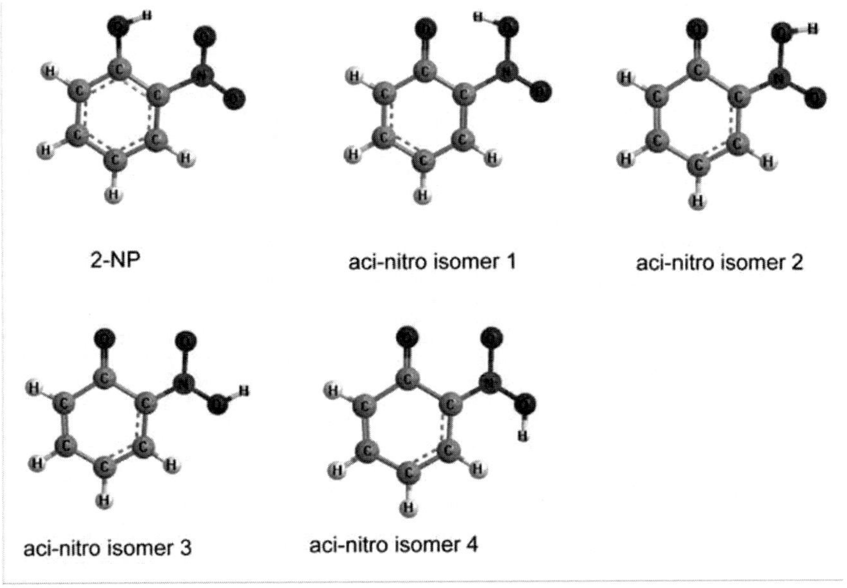

Figure 6. Schematic structures of the 2-NP and four possible aci-nitro isomers, adapted from the Ref. 89.

Generally, benzoic acid exists in the form of dimer at low temperatures [156], which is well documented by Baum and McClure [157]. When benzoic acid is seeded in a beam by using a heated nozzle and bubbling with the carrier gas, a large concentration of monomer beam can be produced. In this experiment, we have heated the sample and nozzle to 370 K for producing a beam with high concentration benzoic acid monomer. For distinguishing the monomer from dimer of benzoic acid, we have measured the fluorescence excitation spectrum from 279 to 281nm of the gas mixture used in our experimental condition. The spectrum is shown in Figure 7. We do not find the fluorescence peak at 279.9 nm (arrow position) which was assigned as the electronic transition origin of the benzoic acid dimer [158]. The result indicated clearly there was no benzoic acid dimer in our experimental condition. The observed LIF excitation spectra in Figure 7 was carefully analyzed and attributed to rotational spectra of the OH $A^2\Sigma^+$ (υ′ =1) ←$X^2\Pi$ (υ″ =0) R_{21} branch (Hund's case (a)), the detailed assignment is listed on the top of the spectra. The OH fragment was produced from the photodissociation of benzoic acid monomer.

The formation of OH radicals was observed on excitation of BAM at the range of 280-294 nm. The rotational state distributions of OH at 280 and 294 nm were also given in Fig 3. The rotation state distributions were peaked at J'' = 3.5 and extended to J'' = 13.5 at both wavelengths. The similar trend was also found at other wavelengths. The results of the OH rotational state distribution from the photolysis at different wavelengths indicate the internal energy of OH does not change obviously with the different excited energy, which imply the dissociation state should be one repulsive state or going on one potential energy surface with a barrier, the residual energy would be distributed into the translation energy.

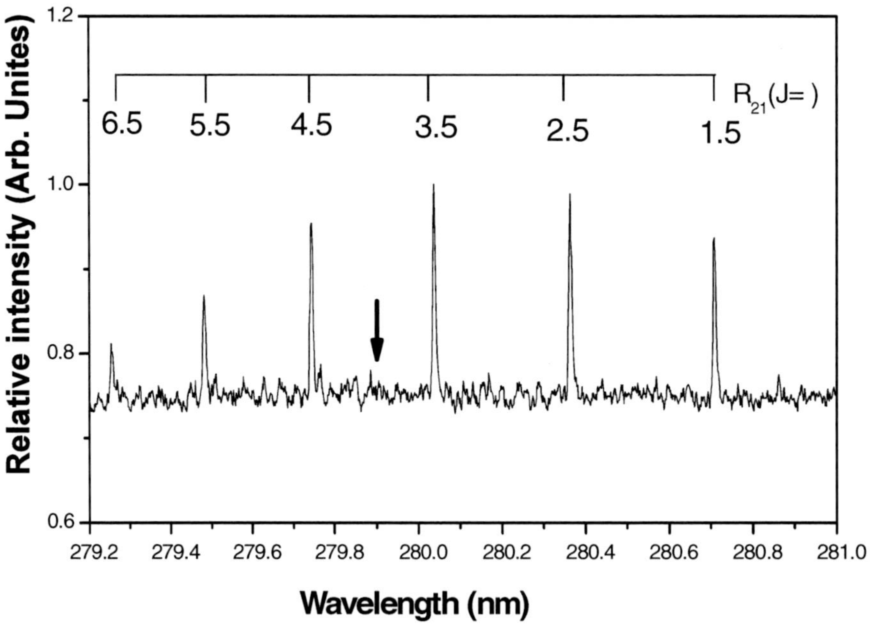

Figure 7. A portion of LIF excitation spectrum in the range 279.2 nm to 281 nm. The spectra assignment (top) is the LIF of OH ($A^2\Sigma^+$ (υ′ =1) ←$X^2\Pi$ (υ″ =0)) produced from the benzoic acid monomer photodissociation. The arrow position indicates the origin of electronic transition of benzoic acid dimer.

Table 2. The internal energy of BAM calculated with the B3LYP method

	Internal energy(kcal/mol) at different temperature			
Basis set	370K	350K	320K	298K
6-31G(d, p)	6.65	6.08	5.06	4.46
6-311+G(d, p)	6.68	6.02	5.10	4.48
6-311++G(df, pd)	6.67	5.99	5.09	4.47
6-311G (2df, 2dp)	6.63	5.98	5.06	4.45

Two possible pathways are responsible for the OH formation. The first pathway is direct dissociation of the excited parent molecule C_6H_5COOH to form C_6H_5CO and OH; the second pathway involves C-C single bond cleavage resulting in C_6H_5 + COOH and a secondary decomposition of COOH through C-O bond dissociation to form OH and CO. On the basis of thermochemical data, the C-C bond and HO-CO bond cleavage were estimated to be endothermic by 105.2 [159] and 36.0 kcal/mol [160], respectively. Thus the needed minimal energy for the second pathway to produce OH fragment was 141.2 kcal/mol. The energy range of photolysis laser is from 96.92 to 102.1 kcal/mol. In present experiment, the sample and the stainless-steel pipe were heated and maintained at 370 K, so the internal energy of BAM should be considered. The internal energies of BAM at different temperatures were calculated by ab initio calculations utilizing the GAUSSIAN 03 program package [141]. The calculations were performed with the B3LYP theory, and using 6-31G (d, p), 6-311+G (d, p), 6-311+G (df, p) and 6-311G (2df, 2dp) basis sets respectively. The calculated results are listed in Table 1. As shown in Table 2, the internal energies of BAM at same temperature calculated by different basis set are approximately same values, which indicated that the impact of basis sets can be neglected. The internal energy of BAM was roughly estimated to be from 5.6 to 6.6 kcal/mol in our experiment condition, considering the deviation of the sample temperature in the photolysis region and the S_2 state electronic origin of BAM. So, the maximal total energy for photodissociation of BAM is 108.7 kcal/mol, which is not enough for the second path to form OH product. However, α C-O bond cleavage was predicted to be endothermic by 98.7 kcal/mol [155] which was less than the total photodissociation energy. Therefore, the observed OH product in photodissociation of BAM in our experiment is due to direct α C-O bond cleavage.

Establishment of the energy threshold for the production of OH from BAM is an essential ingredient to understand the dynamics of its photochemical process. We get the threshold for OH production from BAM by comparing the LIF intensity of Q_1 (3.5) rotational line at each photolysis wavelength. The LIF intensity of the Q_1 (3.5) rotational line at each photolysis wavelength are normalized with respect to pressure, intensities of photolysis and probe laser. The LIF intensity of the Q_1 (3.5) rotational line with the function of photolysis wavelengths are plotted in Fig 6. The LIF intensity of the Q_1 (3.5) rotational line decreased rapidly from 280 to 286 nm and slowly from 286 to 295 nm. The LIF signal of the OH Q_1 (3.5) rotational line can be observed at 294 nm but no LIF signal of OH at 295 nm. It is reasonable to question that the lack of OH LIF signal at 295 nm is due to the reduction in absorption coefficients of BAM. We examined the UV absorption spectra of BAM [161] carefully but the absorption efficient is almost the same at both 294 and 295 nm photolysis wavelengths. Actually, the LIF signal of OH vanished when the photolysis wavelength turned to 295 nm in our experiment. As we did not probe the LIF signal of OH at the photolysis wavelength between 294-295 nm, thus the 294 and 295 nm (97.25 and 96.92 kcal/mol) are attributed to the upper and lower limit of the photodissociation threshold for OH producing from BAM. By adding the internal energy 5.6-6.6 kcal/mol, considering the deviation of the sample temperature in the photolysis region and the S_2 state electronic origin of BAM, the lower and upper limit of the photodissociation threshold for producing OH radical from BAM are estimated to be 102.5 and 103.9 kcal/mol. Fang and coworkers [155] have predicted that α C-O bond of BAM to produce OH fragment upon photoexcitation at 270 nm (105.9 kcal/mol) or shorter wavelengths. Our estimated energy threshold is about 3 kcal/mol less than the calculated result, which indicates a good agreement between the calculated result and our experimental results.

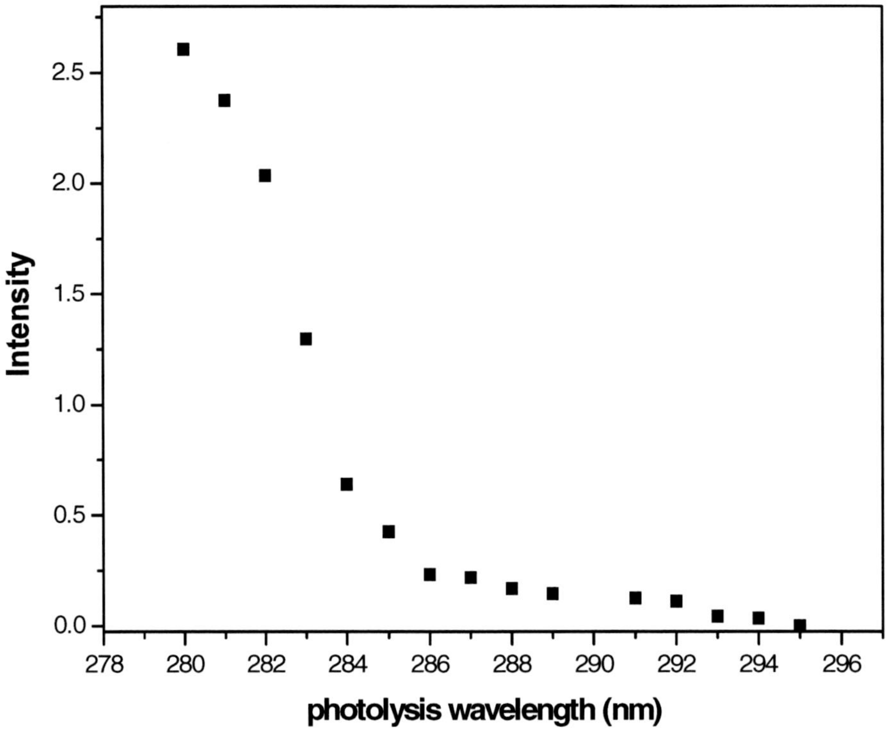

Figure 8. The LIF intensities of OH Q_1(3.5) rotational line from the photodissociation of BAM at different photolysis wavelength.

One possible reason for no LIF signal of OH observed at 295 nm is that a barrier for OH exit channel and the total photodissociation energy was not higher than the barrier for producing OH fragment. The other possible reason for no LIF signal of OH observed at 295 nm is that the BAM was excited to S_1 state not S_2 state. The electronic origin of S_2 state is located at 35 924 cm^{-1}(102.7 kcal/mol) [163]. Considering the lower limit of energy threshold, there exists the possibility that the total photodissociation energy at 295 nm was little below S_2 state. According to the calculated result [155] the oscillator strength of the $S_0 \rightarrow S_2$ transition is 75 times larger than that for the $S_0 \rightarrow S_1$ absorption, implying that a few BAM was populated to S_1 state at 295 nm. Thus, the dissociation from S_1 state or other lower states decayed from S_1 state to produce OH would be negligible. This may be responsible for lacking of LIF signal of OH at 295 nm. Which one is more reasonable between these two possible reasons is based on whether the height of barrier for producing OH fragment is higher than the electronic origin of S_2 state or not.

In present experiment, by selecting the photolysis wavelength, we can not only get the information about the threshold but also the dissociative state for producing OH radical from photodissociation of BAM. In Ref [125], we have carried out the photodissociaton at 306-315 nm but no LIF signal of OH was observed which indicated that the OH fragment could not produced on the S_0 state. The calculated results [155] have shown that the S_1 and T_2 states were related to the production of OH. Relative to the S_0 zero level, the energy barriers were 107.1 and 102.3 kcal/mol for producing OH from the S_1 and T_2 potential energy surfaces, respectively[155]. Considering the internal energy of BAM, if the photolysis wavelength is shorter than 284.5 nm (100.5 kcal/mol), the total energy of photolysis is above the barriers on S_1 and T_2 states. Thus BAM has the ability to pass the barriers on S_1 and T_2 states to produce OH radical. However, if the wavelength of the photolysis laser is longer than the 284.5 nm, the total energy of photolysis is below the barrier on S_1 state but higher than the barrier on T_2 state, the OH is produced only from the T_2 state. By comparing the state distribution of OH produced from different photodissociation energy, we can get the information about the dissociative state. The state distributions of OH at each photolysis wavelength were analyzed carefully and we found that they are essentially the similar distribution, even compared to the state distribution of OH at 266 nm. Hence, the T_2 state is proposed to be the dissociative state for photodissociation of BAM at 280-294 nm.

Now we can get a picture of the photodissociation mechanism of BAM. The BAM was populated to S_2 state when excited at 280-294 nm and then dissociated from the T_2 state. The T_2 state can be formed through two different ways. One is a direct way, through intersystem crossing from S_2 state to T_2 state. The other is an indirect way, which involves a three-surface intersection $S_1/T_1/T_2$ calculated by Fang and coworkers [155]. The three-surface intersection $S_1/T_1/T_2$ lied between the T_1 and S_1 states, and successfully elucidated the photophysical process of BAM. So the indirect way is that the prepared S_2 state of BAM can through internal conversion to S_1 state and then crossing to T_2 state through the three-surface intersection $S_1/T_1/T_2$.

4. Summary and Conclusion

In this review, the dynamics of OH formation have been presented in photoexcited Methyl nitrite, n-butyl nitrite, nitromethane, nitroethane, nitrobenzene, o-, m-, p-nitrotoluene, nitrophenol and benzoic acid. The internal distribution, spin-orbit ratios and the ratio of Λ−doublet states were obtained in these photodissociation processes. For photodissociation of alkyl nitrites and nitro compounds studied in our experiment, the H transfer was proposed to interpret the OH production.

At 266 nm photolysis of CH_3ONO, the rotational state distribution of OH was Boltzmann behavior but was non-Boltzmann behavior at 355 nm. In the photodissociation of $CH_3(CH_2)_3ONO$ at 266 nm, the OH fragment was found to be vibrationally excited. The rotational state distribution of OH v"=0 state was non-Boltzmann distribution, but the rotational state distribution of v"=1 state was Boltmann behavior. The DFT calculations corresponding to the proposed mechanism were carried out for the CH_3ONO molecule, which was consistent with the experimental results. The findings of this study, especially photolysis pathway and mechanisms, can be utilized to predict the behavior of other alkyl nitrites molecules ($C_nH_{2n+1}ONO$, $n \geq 1$).

For the photodissociation dynamics of nitromethane and nitroethane, different rotational state distribution have been found. The OH produced from nitromethan and niroethane were through a four-membered and five-membered ring transion states, respectively.

For nitrobenzene and nitrotoluene molecules, one possible dissociation mechanism for the OH product pathway was proposed that after UV excitation, the parent nitrobenzene and nitrotoluene molecules first relax to the T_1 or S_0 states through the efficient intersystem crossing and internal conversion processes, followed by the intramolecular hydrogen transfer from the benzene ring to the nitro group, and subsequently break the N-OH bond to release OH. For o-nitrotoluene, two possible hydrogen transfer pathways were confirmed: one is from the methyl group to the nitro group, and the other is from the benzene ring to the nitro group. Due to the ortho effect, the former one (between the methyl group and the nitro group) would be dominant in the dissociation process. For 2-nitrophenol, we have proposed that the OH fragments were produced through multi-step reactions from photodissociation at 266 and 355 nm.

For benzoic acid monomer, BAM is excited at 280-294 nm, the initially prepared S_2 state undergoes direct and indirect transitions to T_2 state and then the dissociation takes place along the T_2 state potential energy surface. Moreover, the energy threshold for producing OH from BAM is estimated to be 102.5-103.9 kcal/mol.

ACKNOWLEDGEMENT

This work was supported by NKBRSF (2007CB815202), NSFC (20333050, 20573110).

REFERENCES

[1] R. K. Sparks, K. Shobatake, L. R. Carlson, Y. T. Lee, J. Chem. Phys, *78*, 3838 (1981).
[2] H. S. Johnston, *Annu. Rev. Phys Chem. 43*, 1 (1992).
[3] S. Solberg, N. Schmidbaure, A. Semb et al., *J. Am. Chem. Soc. 23*, 301(1996).
[4] L. A. Barrie, J. W. Bottenheim, R. C. Schnell, P. J. Crutzen, R. A. Rasmussen, *Nature. 334*,138 (1988).
[5] N. J. Turro, *Modern Molecular Photochemistry.* University Science Books: Sausalito, CA, USA (1991).
[6] R. Schinke, *Photodissociation dynamics.* Cambridge University Press (1993).
[7] K. L. Han, G. Z. He, J. *Photochem. Photobiol. C: Photochem. Rev. 8*, 55 (2007).
[8] S. J. Baek, C. R. Park, H. L. Kim, J. *Photochem. Photobiol. A: Chem, 104*, 13 (1997).
[9] S. J. Singer, K. F. Freed, Y. B. *Band, Adv .Chem. Phys, 61*, 1 (1985).
[10] M. N. R.Ashfold, J. E. Baggott, *Molecular Photodissociation dynamics.* Royal Society of Chemistry, London (1987).
[11] H. Finke, H. Spiecker, P. Andresen, *J. Chem. Phys. 110,* 4777 (1999).
[12] P. Farmanara, V. Stert, W. Radloff, *Chem. Phys. Lett. 303,* 521 (1999).

[13] H. Ihm, K. Scheer, H. Celio, *J. M. White, Langmuir 17,* 786 (2001).
[14] F. Lahmani, C. Lardeux, D. Solgadi, *Chem. Phys. Lett. 102,* 523 (1983).
[15] O. B. D'azy, F. Lahmani, C. Lardeux, D. Solgadi, *Chem. Phys. 94,* 247 (1985).
[16] F. Lahmani, C. Lardeux, D. Solgadi, *Chem. Phys. Lett. 129,* 24 (1986).
[17] M. Hippler, F. A. H. Al-Janabi, J. Pfab, *Chem. Phys. Lett. 192,* 173 (1992).
[18] M. Hippler, M. R. S. McCoustra, J. Pfab, *Chem. Phys. Lett. 198,* 168 (1992).
[19] M. R. S. McCoustra, M. Hippler, J. Pfab, *Chem. Phys. Lett. 200,* 451 (1992).
[20]B. A. Keller, P. FELDER, J. R. Huber, *Chem. Phys. Lett. 124,* 135 (1986).
[21] U. Brühlmann, M. Dubs, J. R. Huber, *J. Chem. Phys. 86,* 1249 (1987).
[22] H. U. Suter, U. Brühlmann, and J. R. Huber, *Chem. Phys. Lett. 171,* 63 (1990).
[23] E. Kades, M. Rösslein, U. Brühlmann, J. R. Huber, *J. Phys. Chem. 97,* 989 (1993).
[24] M. Nonella and J. R. Huber, *Chem. Phys. Lett. 131,* 376 (1986).
[25] S. Hennig, V. Engel, R. Schinke, M. Nonella, J. R. Huber, *J. Chem. Phys. 87,* 3522 (1987).
[26] R. Schinke, S. Henning, A. Untch, M. Nonella, J. R. Huber, *J. Chem. Phys. 91,* 2016 (1989).
[27] M. Nonella, J. R. Huber, A. Untch, R. Schinke, *J. Chem. Phys. 91,* 194 (1989).
[28] F. Lahmani, C. Lardeux, M. Lavollée, D. Solgadi, *J. Chem. Phys. 73,* 1187 (1980).
[29] G. Inoue, M. Kawasaki, H. Sato, T. Kikuchi, S. Kobayashi, T. Arikawa, *J. Chem. Phys. 87,* 5722 (1987).
[30] M. Hippler, M. R. S. McCoustra, J. Pfab, *Chem. Phys. Lett. 198,* 168 (1992).
[31] X. Wang, Y. Wang, C. Lv, M. Kubo, A. Miyamoto, J. Photochem. *Photobio. A: Chem. 187,* 119 (2007).
[32] J. W. Winniczek, R. L. Dubs, J. R. Appling, V. McKoy, M. G. White, *J. Chem. Phys. 90,* 949 (1989).
[33] H.-M. Yin, J.-L. Sun, Y.-M. Li, K.-L. Han, G.-Z. He, S.-L. Cong, *J. Chem. Phys. 118,* 8248 (2003).
[34]F. Tuck, *J. Chem. Soc. Faraday Trans.* II *73,* 689 (1977).
[35] T. Ebata, H. Yanagishita, K. Obi, I. Tanaka, *Chem. Phys. 69,* 27 (1982).
[36] B. A. Keller, P. Felder, J. R. Huber, *J. Phys. Chem. 91,* 1114 (1987).
[37] R. Lavi, D. Schwartz-Lavi, I. Bar, S. Rosenwaks, *J. Phys. Chem. 91,* 5398 (1987).
[38] D. Schwartz-Lavi, S. Rosenwaks, *J. Chem. Phys. 88,* 6922 (1988).
[39] J. M. Mestdagh, M. Berdah, I. Dimicoli, M. Mons, P. Meynadier, P. d'Oliveira, F. Piuzzi, J. P. Visticot, C. Jouvet, C. Lardeux-Dedonder, S. Martrenchard-Barra, B. Soep, D. Solgadi, *J. Chem. Phys. 103,* 1013 (1995).
[40] K. J. Castle, J. Abbott, X. Peng, W. Kong, *Chem. Phys. Lett. 318,* 565 (2000).
[41] C. S. Effenhauser, P. Felder, J. R. Huber, *J. Phys. Chem. 94,* 296 (1990).
[42] E. Kades, M. Rösslein, J. R. Huber, *J. Phys. Chem. 98,* 13556 (1994).
[43] D. Schwartz-Lavi, I. Bar, S. Rosenwaks, *Chem. Phys. Lett. 128,* 123 (1986).
[44] K. J. Castle, W. Kong, *J. Chem. Phys. 112,* 10156 (2000).
[45] H.-M.Yin, J.-L.Sun, Y.-M. Li, K.-L.Han, G.-Z.He, *J. Chem. Phys. 118*, 8248 (2003).
[46] X.F. Yue, J. L. Sun, Z. F. Liu, Q. Wei, K. L. Han, *Chem. Phys. Lett. 426*, 57 (2006).
[47] X.F. Yue, J. L. Sun, H. M. Yin, Q. Wei, K. L. Han, submitted to *J. Chem. Phys.*
[48] X. F. Yue, J. L. Sun, C. H. Zhou, S. B. Chen, H. M.Yin, K.L. Han, *Chem. Phys. Lett. 452*, 14 (2008).
[49] A. M. Wodtke, E. J. Hinsta, Y. T. Lee, *J. Phys. Chem. 90,* 3549 (1986).

[50] W. Tsang, D. Robaugh, W. Mallard, *J. Phys. Chem. 90,* 5968 (1986).
[51] N. C. Blais, *J. Chem. Phys. 79,* 708 (1983).
[52] B. N. Kondridov,, G. D. Kozad, V. M. Raikova, A. V. Starshinov, *Sov. Phys. Dokl. 233*, 402 (1997)
[53] H. S. Kwok, G. Z. He, R. K. Sparks, Y. T. Lee, *Int. J. Chem. Kinet. 13,* 1125 (1981).
[54] A. A. Turnipseed, G. L. Vaghjiani, J. E. Thompson, A. R. Ravishankara, *J. Chem. Phys. 96,* 5887 (1992).
[55] K. Nojima, K. Fukaya, S. Fukui, S. Kanno, *Chemosphere. 2*, 77 (1975).
[56] G. Rippen, E. Zietz, R. Frank, T. Knacker, W. klöpffer, *Environ. technol. lett. 8*, 475 (1987).
[57] A. Kawai, S. Goto, Y. Matsumoto, H. Matsushita, *Jap. J. Ind. Health, 29*, 34 (1987).
[58] M. A. J. Harrison, S. Barra, D. Borghesi, D. Vione, C. Arsene, R.I. Olariu, *Atmos. Environ. 39*, 231 (2005).
[59] N. S. Bayliss, E. G. McRae, *J. Phys. Chem. 58*, 1006 (1954).
[60] S. Nagakura, *Mol. Phys. 3*, 152 (1960).
[61] N. C. Blais, *J. Chem. Phys. 79,* 1723 (1983).
[62] L. J. Butle, D. Krajnovich, Y. T. Lee, G. Ondrey, R. Bersohn, *J. Chem. Phys. 79*, 1708 (1983).
[63] K. Q. Lao, E. Jensen, P. W. Kash, L. J. Butler, *J. Chem. Phys. 93*, 3958 (1990).
[64] D. B. Moss, K. A. Trentelman, P. L. Houston, *J. Chem. Phys. 96*, 237 (1992).
[65] E. A. Wade, K. E. Reak, S. L. Li, S. M. Clegg, P. Zou, D. L. Osborn, *J. Phys. Chem. A. 110*, 4405 (2006).
[66] K. J. Spears, S. P. Brugge, *Chem. Phys. Lett. 54*, 373 (1978).
[67] H. S. Kwok, G. Z. He, R. K. Sparks, Y. T. Lee, Int. *J. Chem. Kinet. 13*, 1125 (1981).
[68] P. E. Schoen, M. J. Marrone, J. M. Schnur, L. S. Goldberg, *Chem. Phys. Lett. 90*, 272 (1982).
[69] S. Zabarnick, J. W. Fleming, A. P. Baronavski, *J. Chem. Phys. 85*, 3395 (1986).
[70] G. D. Greenblatt, H. Zuckerman, Y. Haas, *Chem. Phys. Lett. 134*, 593 (1987).
[71] M. S. Park, K. H. Jung, H. P. Upadhyaya, H. R. Volpp, *Chem. Phys. 270*, 133 (2001)
[72] H. L. Gregory, *Ph.D. Dissertation,* University of Wisconsion- Madison, (1992).
[73] D. B. Galloway, A. Bartz, L. G. Huey, F. F. Crim, *J. Chem. Phys. 98,* 2107 (1993).
[74] Y.-M. Li, J.-L. Sun, H.-M. Yin, K.-L. Han, G.-Z. He, *J. Chem. Phys. 118,* 6244 (2003).
[75] Y. He, A. Gahlmann, J. S. Feenstra, S. T. Park, A. H. Zewail, *Chem. Asian. J. 1-2,* 56 (2006).
[76] D. B. Galloway, T. Glenewinkel-Meyer, J. A. Bartz, L. G. Huey, F. F. *Crim. J. Chem.* Phys. *100,* 1946 (1994).
[77] C. Kosmidis, K. W. D. Ledingham, A. Clark, A. Marshall, R. Jennings, J. Sander, R. P. Singhal, *Int. J. Mass Spectrom. and Ion Processes 135,* 229 (1994).
[78] C. Kosmidis, K. W. D. Ledingham, H. S. Kilic, T. McCanny, R. P. Singhal, A. J. Langley, W. Shaikh, *J. Phys. Chem.* A *101,* 2264 (1997).
[79] A. Marshall, A. Clark, K. W. D. Ledingham, J. Sander, and R. P. Singhal, Int. J. Mass Spectrom. *Ion Processes 125,* R21 (1993).
[80] K. J. Castle, J. E. Abbott, X. Peng, W. Kong, *J. Phys. Chem.* A *104,* 10419 (2000).
[81] J. E. Abbott, X. Peng, W. Kong, *J. Chem. Phys. 117,* 8670 (2002).
[82] J.-D. Shao, T. Baer, *Int. J. Mass Spectrom. and Ion Processes 86,* 357 (1988).

[83] C. Kosmidis, A. Marshall, A. Clark, R. M. Deas, K. W. D. Ledingham, R. P. Singhal, Rapid Commun. *Mass Spectrom. 8,* 607 (1994).
[84] S. C. Xu, M. C. Lin, *J. Phys. Chem. B 109,* 8367 (2005).
[85] S.A. McLuckey, G.L. Glish, *Org. Mass Spectrom. 22*, 224 (1987).
[86] K.W.D. Ledingham, H.S. Kilie, C. Kosmidis, R.M. Deas, A.Marshall, T. McCanny, R.P. Singhal, A.J. Langley, W. Shaikh, *Rapid Commun. Mass Spectrom. 9*, 1522 (1995).
[87] K.B. Borisenko, C.W. Bock, I. Hargittai, *J. Phys. A . 98*, 1442 (1994).
[88] Y. Q. Wang, H. G. Wang, S. Q. Zhang, K. M. Pei, X. M. Zheng, D. L. Phillips, *J. Chem. Phys. 118*, 8248 (2003).
[89] M. Nagaya, S. Kudoh, M. Nakata, *Chem. Phys. Lett. 427*, 67 (2006).
[90] M.I.O. Ishag, P.G.N. Moseley, *Tetrahedron. 33*, 3141 (1977).
[91] A. Alif, J. -F. Pilichowski, P. Boule, *J. Photochem. Photobiol.* A. *59*, 209 (1991).
[92] C. Bing, Y. Chun, G.N. Khang, *J. Environ. Sci.* (China). *17*, 598 (2005).
[93] I. Bejan, Y.A.E. Aal, I. Barnes, T. Benter, B. Bohn, P. Wiesen, J. Kleffmann, *Phys. Chem. Chem. Phys. 8*, 2028 (2006).
[94] M. J. S. Dewar, J. P. Ritchie J. Alster, *J. Org. Chem. 50,* 1031 (1985).
[95] M. L. McKee, *J. Phys. Chem. 93*, 7365 (1989).
[96] M. T. Nguyen, H. T. Le, B. Hajgato, T. Veszpremi, M. C. Lin, *J. Phys. Chem.* A. *107,* 4286 (2003).
[97] W. F. He, T. J. He, D. M. Chen, F. C. Liu, *J. Phys. Chem. A. 106,* 7294 (2002).
[98] C. Mijoule, S. Odiot, S. Fliszar, J. M. Schnur, *J. Mol. Struct.: THEOCHEM 149*, 311 (1987)
[99] S. Roszak, J. J. Kaufman, *J. Chem. Phys. 94*, 6030 (1991).
100] J. F. Arenas, J. C. Otero, D. Peláez, J. Soto, *J. Chem. Phys. 119*, 7814 (2003).
[101] J. F. Arenas, J. C. Otero, D. Peláez , J. Soto, *J. Chem. Phys. 122*, 084324 2005.
[102] S. C. Chen, S. C. Xu, E. Diau, M. C. Lin, *J. Phys. Chem.* A *.110,* 10130 (2006).
[103] M.-F. Lin, Y. T. Lee, C.-K. Ni, S. Xu, M. C. Lin, *J. Chem. Phys. 126,* 064310 (2007).
[104] P. C. Chen, Y. C. Chieh, *Chem. Phys. Lett. 372*, 147 (2003).
[105] X. F. Yue, J. L. Sun, Q. Wei, H. M Yin, K. L. Han, *Chinese Journal of Chemical Physics. 20*, 401 (2007).
[106] X. F. Yue, J. L. Sun, Q. Wei, H. M. Yin, K. L. Han, to be published
[107] Q. Wei, H. M. Yin, J. L. Sun, X. F. Yue, K. L. Han submitte to *Chem. Phys. Lett.*
[108] P. D. Naik, H. P. Upadhyaya, A. Kumar, A. V. Spare, J. P. Mittal, *J. Photochem. Photobiol.*C. *3,* 165 (2003).
[109] C. K. Ni, Y. T. Lee, I*nt. Rev. Phys. Chem. 23,* 187 (2004).
[110] H. Baba, M. J. Kitamura, *Mol. Spectrosc. 41,* 302 (1972).
[111] S. S. Hunnicutt, L. D. Waits, J. A. Gues*t, J. Phys. Chem. 93*, 5188 (1989).
[112] S. S. Hunnicutt, L. D. Waits, J. A. Guest, *J. Phys. Chem. 95*, 562 (1991).
[113] H. T. Kong, S. K. Shin, S. K. Kim, H. L. Kim, C. R. Park, *J. Phys. Chem.* A. *105*, 6775 (2001).
[114] P. D. Naik, H. P. Upadhyaya, A. Kumar, A. V. Spare, J. P. Mittal, *Chem. Phys. Lett. 116*, 340 (2001).
[115] M. F. Arendt, P. W. Browning, L. J. Butler, *J. Chem. Phys. 103,* 5887 (1995).
[116] D. C. Kitchen, N. R.Forde, L. J. Butler, *J. Phys. Chem*. A. *101,* 6603 (1997).
[117] J. C. Owrutsky, A. P. Baronavski, *J. Chem. Phys. 111*, 7329 (1999).

[118] H. M. Su, F. N. Kong, W. H. Fang, R. Z. Liu, J. Chem. Phys. *113,* 1891 (2000).
[119] A. Kumar, H. P. Upadhyaya, P. D. Naik, *J. Phys. Chem.* A. *108,* 6275 (2004).
[120] S. Dhanya, A. Kumar, H. P. Upadhyaya, P. D. Naik, R. D. Saini, *J. Chem. Phys. 118,* 10093 (2003).
[121] K. K. Pushpa, H. P. Upadhyaya, A. Kumar, P. D. Naik, P. N. Bajaj, *J. Chem. Phys. 120*, 6964 (2004).
[122] H. P. Upadhyaya, A. Kumar, P. D. Naik, A. V. Sapre, J. P. Mittal, *J. Chem. Phys. 117,* 10097. (2002).
[123] A. Kumar, H. P. Upadhyaya, P. D. Naik, D. K. Maity, J. P.Mittal, *J. Phys. Chem.* A. *106,* 11847 (2002).
[124] A. Kumar, P. D. Naik, *Chem. Phys. Lett. 422,* 152 (2006).
[125] Q. Wei, J. L. Sun, X. F. Yue, H. M. Yin, K. L. Han. *Chem. Phys. Lett. 448,* 11 (2007).
[126] Q. Wei, J. L. Sun, X. F. Yue, S. B. Cheng, C. H. Zhou H. M. Yin, K. L. Han, *J. Phys. Chem. A, accepted*
[127] J. Luque, D. R. Crosley, *J. Chem. Phys. 109,* 439 (1998).
[128] I. Kovacs, *Rotational Structure in the Spectra of Diatomic Molecule,* ADAM HILGER LTD, London, p. 130 (1969).
[129] G. Herzberg, Molecular Spectra and Molecular Structrue. *I. Spectra of Diatomic Molecules,* Van Nostrand, New York, p. 201 (1950).
[130] P. Andresen, G. S. Ondrey, B. Titze, E. W. Rothe, *J. Chem. Phys. 80,* 2548 (1984).
[131] M. C. Yoon, Y. S. Choi, S. K. Kim, *J. Chem. Phys. 110,* 11850 (1999).
[132]F. R. Cordell, J. E. Boggs, A. Skanche, *J. Mol. Struct. 64,* 57 (1980).
[133] H. U. Suter, M. Nonella, *J. Phys. Chem.* A. *101,* 5580 (1997).
[134] P. Tarte, *J. Chem. Phys. 20,* 1570 (1952).
[135] P. N. Ghosh, A. Bauder, H. H. Günthard, *Chem. Phys. 53,* 39 (1980).
[136] P. Felder, T. K. Ha, A. M. Dwivedi, H. H. Günthard, Spectrochim. *Acta 37A,* 337 (1981).
[137] L. H. Piette, J. D. Ray, R. A. Ogg, *J. Chem. Phys. 26,* 1341 (1957).
[138] P. Felder, H. H. Günthard, *Chem. Phys. 71,* 9 (1982).
[139] D. A. Southern, D. H. Levy, G. M. Florio, A. Longarte, T. S. Zwier, *J. Phys. Chem. A 107,* 4032 (2003).
[140] A. L. Sobolewski, W. Domcke, *J. Phys. Chem.* A. *108,* 10917 (2004).
[141] M. J. Frisch, et al. Gaussian 03, Revision C.02, Gaussian, Inc., Wallingford CT, 2004.
[142] G. Radhakrishnan, T. Parr, C. Wittig *Chem. Phys. Lett. 111,* 25 (1984).
[143] R. P. McLaughlin, W. A. Donald, D. Jitjai, Y. Zhang, *Spectrochim. Acta 67A,* 178 (2007).
[144] S. Nagakura, M. Kojima, Y. Maruyama, *J. Mol. Spectrosc. 13,* 174 (1964).
[145] M. Takezaki, N. Hirota, M. Terazima, Prog. Nat. Sci. *6,* S-453 (1996).
[146] M. Takezaki, N. Hirota, M. Terazima, *J. Phys. Chem.* A. *101*, 3443 (1997).
[147] M. Takezaki, N. Hirota, M. Terazima, *J. Chem. Phys. 108*, 4685 (1998).
[148] M. Takezaki, N. Hirota, M. Terazima, H. Sato, T. Nakajima, and S. Kato, *J. Phys. Chem.* A. *101*, 5190 (1997).
[149] A. R. Butcher, C. B. Thomas, *Org. Mass Spectrom. 14,* 448 (1979).
[150] Sumana SenGupta, Hari P. Upadhyaya, Awadhesh Kumar, Suresh Dhanya, Prakash D. Naik, *Parmanand Bajaj Chem. Phys. Lett. 452,* 452 (2008).
[151] M. C. Yoon, Y. S. Choi, S. K. Kim, *J. Chem. Phys. 110,* 11850 (1999).

[152] M. F. Lin, Y. T. Lee, C.-K. Ni, S. Xu, M. C. Lin, *J. Chem. Phys. 126,* 64310 (2007).
[153] S. SenGupa, H. P. Upadhyaya, A. Kumar, S. Dhanya, P.D. Naik, *Chem. Phys. Lett. 426,* 57 (2006).
[154] S. C. Chen, S.C. Xu, E. Diau, M.C. Lin, *J. Phys. Chem.* A. *110,* 10130 (2006).
[155] J. Li, F. Zhang, W. H. Fang, *J. Phys. Chem.* A. *109,* 7718 (2005).
[156]G. Allen, J.G. Watkinson, Webb, K. H. Spectrochim. *Acta. 22,* 307 (1966).
[157] J. C. Baum, D. S. McClure, *J. Am. Chem. Soc. 101,* 2335 (1979).
[158] Y. Tomioka, H. Abe, N. Mikami, M. Ito, *J. Phys. Chem. 88,* 2263 (1984).
[159] D. Feller, D. A. Dixon, J. S. Fransico, *J. Phys. Chem.* A. *107,* 1604 (2003).
[160] S. W. Benson, *Thermochemical Kinetics;* John Wiley and Sons: New York, 1968.
[161] J. C. Baum, D. S. McClure, *J. Am. Chem. Soc. 101,* 2336 *(1979).*
[162] G. Meijer, M. S. Vries, H. E. Hunziker, Wendt, H. E. *J. Phys. Chem. 94,* 4394 (1990).

In: Environmental Chemistry Research Progress
Editors: P. Robinson and R. Gallo
ISBN 978-1-60741-055-3

Chapter 3

INTELLIGENT AIR QUALITY MONITORING USING CONNECTIONIST MODELS

Ali Elkamel[1,*] and Sabah A. Abdul-Wahab[2,†]

[1]Department of Chemical Engineering, University of Waterloo, 200 University Avenue West, Waterloo, ON, N2L 3G1, Canada

[2]College of Engineering, Sultan Qaboos University, P.O. Box: 33, Al-Khod, Postal Code:123, Muscat, Sultanate of Oman

ABSTRACT

The use of Artificial Neural Networks (ANNs) for emission estimation and forecasting has been recently proven to be promising. For instance, a neural network approach can account for the synergistic effects that arise from the complex interactions of several variables that affect atmospheric dispersion of gases. Unlike other modeling approaches, artificial neural networks are able to provide reliable estimates of emission rates based solely on limited information that is not known with as much certainty. The deterministic models, for instance, would require information regarding atmospheric stability, dispersion coefficients in the lateral and vertical directions, reaction mechanisms, and kinetic data. These information are usually either not well understood or only known with a certain degree of uncertainty. The objective of the present chapter is to offer a background on the use of artificial neural networks in predicting pollutants concentrations. The chapter will start with an introduction to ANNs and will discuss different algorithms used for training them. The chapter will then present several examples that will illustrate the predictive performance of the prepared network models when compared against linear and non-linear regression models, commercial simulators, and measured data.

* E-mail: aelkamel@uwaterloo.ca
† E-mail: sabah1@squ.edu.om

1. INTRODUCTION

Air pollution is woven throughout the fabric of modern life. It is a by-product of human activities including transportation, production of goods and services, and the generation of energy. The major cause of air pollution is combustion, which is essential to humans. When perfect combustion takes place, the hydrogen and carbon in the fuel combine with oxygen from the air to produce heat, light, carbon dioxide, and water vapor. However, due to the presence of impurities in the fuel, poor fuel to air ratio, or inappropriate combustion temperature, the formation of a number of products such as carbon monoxide, sulfur oxides, nitrogen oxides, fly ash, and unburned hydrocarbons occurs (Seinfeld and Pandis, 1998).

Air pollution is an important public concern and scientists are continuously concerned with devising systems that can monitor pollutants in the atmosphere. Traditionally, the installation of instrumentation systems have been the major source of measuring emissions to the atmosphere. Mathematical models have also been used for such purposes. An idealistic approach would be to construct a deterministic model which depends on chemical processes and atmospheric movements. Although kinetic mechanisms that describe these processes are being formulated to estimate the amount and type of emissions, their complexities coupled with the uncertainties in assigning values to parameters restrain their use (Zanetti,1990). Especially, it is still not feasible to represent explicitly the chemistry of hundreds of organic species present. Instead, species are often lumped in terms of type of molecules (molecular-lumping approach, Leone and Seinfeld, 1986) or in terms of type of carbon bonds (structural-lumping approach, Whitten et al, 1980 and Gery et al., 1989). Even after structural lumping, the number of reactions exceed several hundreds (Gery et al.(1989)). While such kinetic mechanisms can be studied in smog chambers, their use in actual circumstances is not yet mature. These reaction kinetic information need to be coupled with meteorological data which can use both Lagrangian and Eulerian modeling (Zannetti,1990, Fatehifar et. al., 2007). In the Lagrangian modeling approach, the model considers a column of air which extends from the surface through a certain height above the surface. The column moves, and as a result, is subjected to fresh emissions, dilution and chemical reactions. In the Eulerian approach, emissions enter the control volume and are transformed through a series of reactions. Transport in and out of the control volume is influenced by the surrounding meteorology. Both approaches are used in the Emprical Kinetic Modeling Approach (EKMA) which is considered to be an effective tool for designing emission control strategies (Kinosian,1982).

Models (often known as black-boxes) which depend on the statistical analysis of air quality data represent an alternative approach for predicting emission levels in terms of meteorological variables and other properties of the emission sources (Kuntasal and Chang, 1987; Abdul-Wahab et al.,1996). For instance, Abdul-Wahab et al.(1996) developed a statistical-empirical model from a measured ambient air quality data which was based on obtaining the functional relationship between ozone level and the various independent variables by using a stepwise multiple regression procedure. Their method consisted of combination of forward selection and backward elimination multiple regression procedures. This procedure automatically selects parameters that are of most importance and eliminate those that are of least importance. Out of 13 independent primary pollutants and the meteorological conditions, they were able to eliminate 8 variables and correlated ozone level to concentrations of NO_x , non-methane hydrocarbons (NMHC), sulphur dioxide, wind speed and direction, and solar energy. Multiple linear regression, also known as ordinary least squares, is a commonly used method to obtain a linear input-output model for a given data set. Often one or more of the inputs can be transformed first into a nonlinear form. For instance, the predictive correlation of Abdul-Wahab et al.(1996) based on the highest R^2 of several alternative transformations, consisted of correlating logarithm of ozone concentration to linear combinations of three concentrations, wind speed and direction and logarithm of solar energy. Although there are now several nonlinear multivariate statistical methods that are able to approximate any nonlinear relationship, the assumption of functional dependency is a serious drawback of such procedures (Utojo and Bakshi, 1995). An alternative approach would be the use of neural networks for extracting information -without any assumption- from measured data in the form of predictive input-output models, or as a way of efficiently representing the input space.

Artificial Neural Networks (ANNs), also classified as black box models, can handle multivariable problems. ANNs are based on a principle stating that a system of highly interconnected simple processing elements can learn complex interrelationships between independent and dependent variables (Boughman and Liu, 1995). Prediction has been one of the most popular applications in engineering for problems that have previously been solved using multivariate statistical methods, such as multiple linear regression.

Applications of ANNs to enviromental problems are recently becoming available (Pasini and Potesta (1996, Fausett,1998, Mayora-Ibarra et al.,1998, Boznar et al., 1993, Schumuller, 1989, Rege and Tock, 1996)). The objective of this chapter is to illustrate the use of ANNs for predicting and forecasting emission levels. Neural network models have been applied recently to model source emissions. Although the neural network methods are inherently mathematical, analogies can be made to biological learning processes. The traditional regression methods usually provide a model that is calculated by using a set of equations. In the neural network method, the model constants are not calculated, but are determined through an iterative process. The next section will provide a brief tutorial on ANN models, how to efficiently construct and test them. The following sections will provide several applications.

2. Artificial Neural Networks: A Tutorial

The development of artificial neural networks was inspired by the structure of the human brain. The greatest advantage of a neural network is its ability to model complex nonlinear relationships without any assumptions about the nature of relationships. There are many models of neural networks of which two are extensively used: feed-forward and feedback. For the feed-forward architecture (used in this paper), the neurons are arranged in layers. Neurons are arranged in parallel layers with each neuron forming weighted connections to all layers. In between lies at least one layer termed the hidden layer. Hidden layers act as feature detectors and according to universal approximation theory (Tambe et al., 1996), a network with a single hidden layer with sufficiently large number of neurons can map any input to any output with an arbitrary degree of accuracy. Additionally, there is a "bias" neuron that is connected to all neurons in the hidden layer and the output layer. The bias neuron accepts no inputs and supplies an invariant output (Elkamel, 1998). Bias neurons are usually added to the hidden and output layers to preserve the universal approximation of the network (Sarle, 1997). Figure 1 shows a typical three-layered ANN with 'I' input neurons, 'H' hidden neurons and 'O' output neurons. The number of input neurons and the number of output neurons are determined by the nature of the problem under study. The number of neurons in the hidden layer is often determined by the required degree of accuracy and is therefore a parameter in formulating an ANN model.

The input neurons transmit their inputs to the neurons in the hidden layer, thus acting as distribution channels. The neurons in the hidden and output layers calculate their inputs by performing a weighted summation of all the outputs they receive from the preceding layer. Their outputs, on the other hand, are calculated by transforming their inputs using a non-linear transfer function or learning laws which are in the form of equations that modify all or some of the weights in response to the input signals and the values provided by the transfer function. There are many used types of transfer functions: linear, saturating linear, positive linear, log-sigmoid, hyperbolic tangent sigmoid, and competitive. In addition, the bias associated with the neuron can be looked at as a weight with a constant input of 1. So, learning laws specify how to compute appropriate weight changes, and so how the network adapts itself and learns.

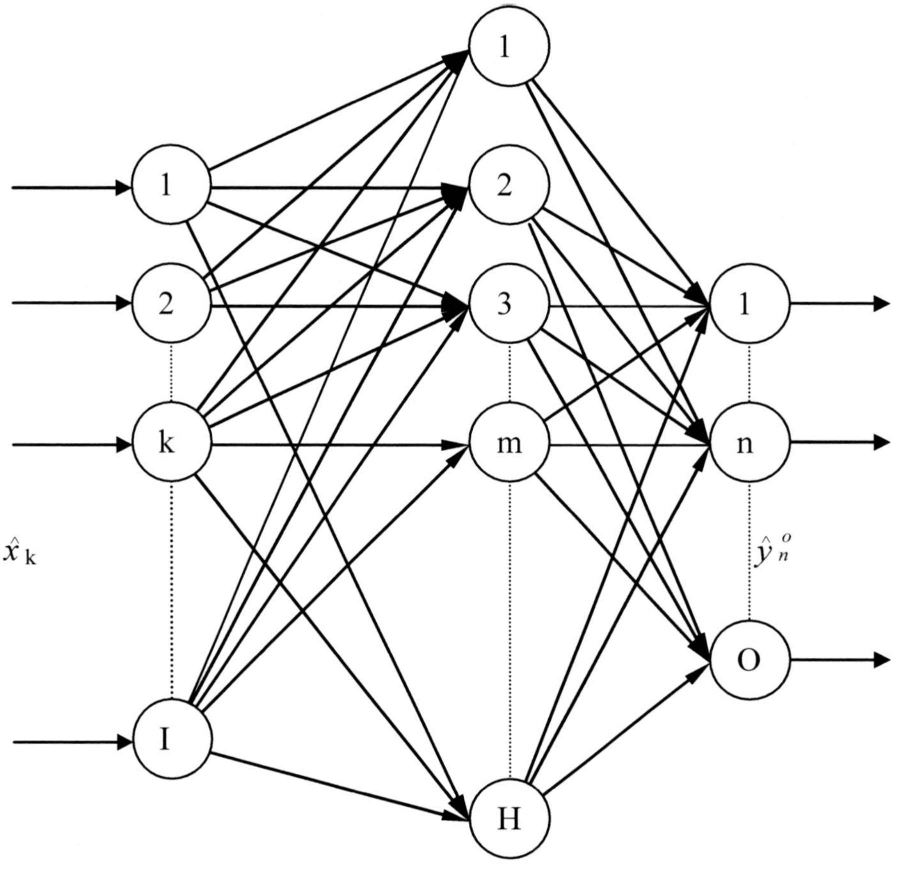

Figure 1. Architecture of a one hidden layer feedforward neural network.

The most widely used transfer function is the 'S'-shaped logistic sigmoid transfer function. This transfer function is continuously differentiable, monotonic, symmetric, bounded between 0 and 1, and its derivative can be expressed in terms of the function itself. The function and its derivative are

$$f(x) = \frac{1}{1+e^{-x}} \tag{1}$$

$$\frac{d}{dx}(f(x)) = f(x)(1-f(x)) \tag{2}$$

This necessitates the normalization of input-output data so that it is in the same range of the transfer function used. The normalization of inputs avoids also overflows due to very large or very small weights. Data is usually normalized using

$$\hat{x} = \frac{x - x_{\min}}{x_{\max} - x_{\min}} \tag{3}$$

where $\hat{x}$ is the normalized value; x_{min} and x_{max} are respectively the minimum and maximum values of z which may be an input or an output variable.

Neural networks learn by modifying the weights on their synapses. This training process continues until the network is able to produce accurate expected output when given particular inputs, and this is achieved through the accurate selection of the weights. So, the goal of the training is for the network to learn the complex nonlinear mapping between input and output data, so that the network is able to interpolate or extrapolate data not included in the training set, and this is often called network's generalization. Training is stopped when the performance stops improving. The network is then tested with the data set that was not used during the training phase. The testing technique used to evaluate the multi-layer neural network is called cross-validation.

Neural networks either learn by supervised or unsupervised training. In supervised training, the network is fed with both input and desired output data. The network learns how the inputs may be related to the output. During the training procedure, the network compares its own output with the target answers, corrects any difference, and tries again, adjusting its weights until the output error reaches an acceptable level. More precisely, the weights are iterated in a manner that the performance index of the network expressed in vector form in terms of the error for the input x_i is:

$$E(w,b) = \sum_{k=1}^{N} \hat{E} = \sum_{k=1}^{N} (T_k - a_k^r)^T \ (T_k - a_k^r) \tag{4}$$

where T_k is the target output of the input x_k. N is the total number of samples. a_k^r is the output of an r-layer feed-forward network. w and b are respectively the weights and the bias of the network. The error in Equation (4), most commonly used, is defined in the least square sense. However, other types of performance indices can be defined, for example, in the least p^{th} sense, or minimax. For a single input/output pair, the squared error in (1) is $\hat{E}(w,b) =$

$$(T_k - a_k^r)^T (T_k - a_k^r) \tag{5}$$

For the general case of an r layer multilayer network, where each layer has n_i ($i = 1, 2\ldots r$) neurons, the mathematical governing equations are given below:

$$o^i = w^i a^{i-1} + b^i \ , \ i = 1, 2\ldots r \tag{6}$$

$$a^i = f^i(o^i) \ , \ i = 1, 2\ldots r \tag{7}$$

where a^0 is the n_0-dimensional input vector, a^i and f^i are the n_i-dimensional output and the activation function vectors in the ith layer. w^i and b^i are

$$w^i = \begin{bmatrix} w_{11}^i & \cdots & w_{1n_{i-1}}^i \\ \vdots & \ddots & \vdots \\ w_{n_i 1}^i & \cdots & w_{n_i n_{i-1}}^i \end{bmatrix} ; \ b^i = \begin{bmatrix} b_1^i \\ \vdots \\ b_{n_i}^i \end{bmatrix}$$

The back-propagation algorithm is based on a gradient search technique and computes the weight adjustments using the generalized data learning rule. The gradient descent algorithm enhanced by the positive-valued momentum adaptation α is:

$$\Delta w^i = -\eta \frac{\partial \hat{E}}{\partial w^i} + \alpha (\Delta w^{i-1}) \tag{8}$$

$$\Delta b^i = -\eta \frac{\partial \hat{E}}{\partial b^i} + \alpha (\Delta b^{i-1}) \tag{9}$$

η is the positive-valued learning rate, and Δ is the forward difference operator. The sensitivity through the neural network is computed as:

$$\frac{\partial \hat{E}}{\partial w^i} = \sum_{j=1}^{n_i} \frac{\partial \hat{E}}{\partial o_j^i} \frac{\partial o_j^i}{\partial w^i} = \delta^i a^{(i-1)T} \tag{10}$$

$$\frac{\partial \hat{E}}{\partial b^i} = \delta^{iT} \tag{11}$$

δ^i represents the local gradients at the ith layer, and is defined as the gradient of the squared errors $\hat{E}$ with respect to the output o^i

$$\delta^i = \nabla_{o^i} \hat{E} = [\delta_1^i \; \delta_2^i \; \cdots \; \delta_{n_i}^i]\;;\; i = 1, 2 \ldots r \tag{12}$$

The recurrence relation of the gradient δ^i is given by:

$$\delta^i = F^i(o^i)\; w^{(i+1)T} \; \delta^{i+1} \;;\; i = 1, 2 \ldots r-1 \tag{13}$$

$$\delta^r = -\; F^r(o^r)\,(T_k - a_k^r) \tag{14}$$

where $F^i(o^i)$ is a diagonal matrix

$$F^i(o^i) = \begin{bmatrix} \frac{df^i(o_1^i)}{do_1^i} & 0 & \cdots & 0 \\ 0 & \frac{df^i(o_2^i)}{do_2^i} & \cdots & 0 \\ \vdots & 0 & \ddots & \\ 0 & 0 & \cdots & \frac{df^i(o_{n_i}^i)}{do_{n_i}^i} \end{bmatrix}$$

During training, an epoch represents one forward and backward propagation through all training samples. Many epochs are required to minimize the error for a complete training.

As mentioned earlier, the development of an artificial neural network model usually consists of two steps. The first step is a training stage where the network is subjected to a training set of input-output patterns. The second step is a testing stage where the performance of the network is tested on patterns that have not been "seen" by the network during the training stage. More details about these two stages in neural network development will be discussed later when we illustrate the prediction of emission levels for different case studies.

3. Case Studies

3.1. Predicting Ozone Levels

It is a well known phenomenon that the formation of ozone results from the complex chemical interaction of oxides of nitrogen (NO_x) and organic species in the presence of solar radiation (Gery et al.,1989). These primary pollutants are emitted from both urban and industrial sources. Ozone can cause health hazards and it is also a highly reactive chemical, capable of attacking surfaces, fabrics and rubber materials. It is also toxic to certain crops. There is therefore considerable interest in determining the concentration of ozone in the lower atmosphere because of its effect on oxidizing photochemical smog. It has been reported that ozone is the most important index substance of photochemical smog and it has been recognized as one of the key pollutant degrading the air quality. Ozone is a secondary pollutant and its concentration is known to depend on both meteorological factors and the concentration of other chemicals.

In this first case study, we will illustrate the preparation of an artificial neural network model that is able to predict ozone concentrations as a function of meteorological conditions and precursor concentrations. This case is based on the work of Elkamel et. al.., 2001. The network is trained using data collected during a period of 60 days nearby an industrial area.

3.1.1. Data Collection

The most important step in developing an artificial neural network is to assemble data that can be used for training the network. In addition, the inputs and outputs of the network need to be decided upon. The focus of the data collection step in this case study is therefore to generate data for ozone as a function of these factors. The network is based on the data set of Elkamel et. al. (2001) which was obtained using a mobile laboratory located nearby an industrial site and which can measure 14 different parameters as listed in Table 1 (see Abdul-Wahab et al. (1996) for more details about the location of the industrial area and the layout of its different plants). Details of the mobile laboratory's chemical sensors and meteorological instruments are summarized in Tables 1 and 2, respectively. The sampling inlets were located on top of the laboratory 10 m above the ground. The measurements of chemical concentrations and meteorological conditions were recorded every 5 minutes. Methane and non-methane hydrocarbons were measured by gas chromatography using flame ionization detector (Model MAS-1030A, Mine Safety Appliances Co.) which had a lower detection limit of 0.05 ppm. Carbon monoxide and carbon dioxide concentrations were measured by an instrument based on non-dispersive infra-red absorption (Model 48 and Model 41/41H of Thermo Environmental Instruments respectively). The detection limits for carbon monoxide and carbon dioxide were 0.1 ppmv and 5 ppb, respectively. The NO_x were measured with a detection limit of 0.5 ppb (Thermo Environmental Instrument, Model 42). Pulsed fluorescent by Thermo Environmental (Model 43 A) which had a detection limit of 1 ppb were used to measure SO_2 concentrations. The ozone concentrations were measured by a non-dispersive UV photometer (MonitorsLabs, Model ML 9812) with a detection limit of 1.0 ppbv. Suspended dust was measured gravimetrically (TEOM Series 1400a). This was a realtime device for measuring the particulate concentrartion for sizes smaller than 10 μm in diameter.

It was a filter based mass monitor which composed of a TEOM sensor and a TEOM control unit and had a detection limit of 5 μg/m^3.

Table 1. Details of the chemical sensors positioned in the mobile laboratory

Species	Method	Measurement Range	Accuracy
Hydrocarbons	GC-Flame ionisation	0-20 ppm	± 0.2 ppm
Carbon monoxide	Non-dispersive Infrared	0-20 ppm	± 0.1 ppm
Carbon dioxide	Non-dispersive Infrared	0-1000 ppm	± 1%
NO and NO_x	Chemiluminence	0-1000 ppb	± 0.5 ppb
Sulfur dioxide	Fluorescence	0-2000 ppb	± 1 ppb
Ozone	Non-dispersive UV photometer	0-1000 ppb	± 1 ppb
Suspended dust (PM-10)	Gravimetric	0-5000 μg/m^3	± 5 μg/m^3

Table 2. Details of the meteorological parameters recorded by the mobile laboratory

Parameter	Technique	Range	Accuracy
Wind speed	Speedometer with a reed switch	0.4-76.7 m/s	± 0.2 m/s
Wind direction	Compass with potentiometer	0-360 degrees	± 5 degrees
Solar energy	High sensitive thermistor bridge	0-2000 W/m^2	± 20 W/m^2
Temperature	Ohmic half-bridge	273-333 K	± 0.1 % range
Relative humidity	Capacitive polymer	0-100 %	± 3% RH
Air pressure	Silicon chip as a Sensing element	918-1092 mbar	±0.2 mbar

Table 3 shows the variables measured by the mobile laboratory and their minimum, maximum and mean values. The data was collected on various days of different meteorological conditions and also both during the day and night hours in an effort to encompass a wide variation in data. The number of complete data points with values of all 14 varibles recorded was 2376.

3.1.2. Neural Network Modeling and Testing

The data collected for the variation of ozone as a function of meteorological conditions and concentrations of other chemicals is used to train an artificial neural network. In this training process, the data is divided into two sets: one set (containing about 85% of the data) to be used for training the network and one set for testing the network. A random number generator is usually employed to randomly select the two sets. Since the ozone concentration can be a function of 13 other parameters, the architecture of the ANN should consist of an input layer of 13 neurons and an output layer of one neuron only corresponding to ozone concentration. All variables should be normalized using their minimum and maximum values according to equation (3) and in this case study using the values in Table 3.

Table 3. Variables measured in case study 1 and their ranges

Variable	Average value	Minimum value	Maximum value
Non-methane Hydrocarbon (ppm)	0.05	0.00	1.61
Carbon dioxide (ppm)	399.8	302.0	545.5
Methane (ppm)	2.23	1.29	11.12
Carbon monoxide (ppm)	0.35	0.12	8.39
Sulfur dioxide	21.26	2.00	989.00
NO (ppb)	8.6	0.0	199.5
NO_2 (ppb)	15.81	0.0	74.5
Ozone (ppb)	23.09	0.0	77.5
Temperature (K)	295.7	287.8	304.8
Relative humidity (%)	56.8	13.1	84.7
Suspended dust ($\mu g/m^3$)	67.6	0.0	1165.0
Solar energy (kW/m^2)	0.22	0.00	1.08
Wind direction (N degree)	164.9	4.86	355.3
Wind speed (m/s)	3.55	0.44	10.8

The decision about the architecture of the network is often made on a trial and error basis. First, a network with one hidden layer only and with a fixed number of neurons is chosen. The predictions of the network are then compared to actual measurements. The number of neurons is increased slowly and the predictions of the network are checked continuously. If the predictions are still poor, then one opts to use one additional hidden layer. For each architecture used, the network is trained to learn the input-output pattern fed to it. Training means the determination of the connection weights among the various neurons. Various algorithms can be used to train a neural network as discussed in section 2. In this case study, the backpropagation algorithm is used. When the neural network is trained with this algorithm, the errors between the estimated outputs from the network and the actual outputs are calculated and propagated backward through the net. These errors are used to update the connection weights among the neurons. This operation is repeated until the network outputs are within a pre-specified tolerance of the actual outputs (refer to section 2 for more details).

It was found that a network with one hidden layer of 25 neurons gives sufficiently accurate predictions. The use of more neurons or more hidden layers did not result in improved accuracies. Figure 2 shows a cross plot of predicted and measured ozone concentrations for the training set and Figure 3 shows a plot of the predicted and actual ozone concentrations versus time. The average absolute error for this data set was 11.1% and the maximum error and minimum error were 68.5% and 0.0086%, respectively. For the testing data set, the average absolute error, the maximum error, and the minimum error are 12.5%, 57.8%, and 0.02%, respectively.

In order to test the prepared network, the second data set is used and the predictions of the trained network are compared with the measurements of this set. This validation step is important to check the generalization characteristics of the prepared neural network. In this case, the average absolute error, the maximum error, and the minimum error were 12.5%, 57.8%, and 0.02%, respectively. Figure 4 shows a cross plot of the predicted and measured ozone concentrations for this data set. The data falls close to the 45° line indicating that the predictions are also good for this testing set.

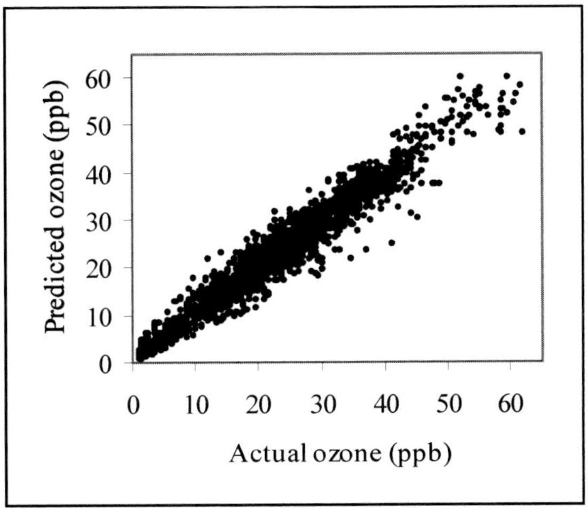

Figure 2. Cross plot of predicted and actual ozone levels for the training data set.

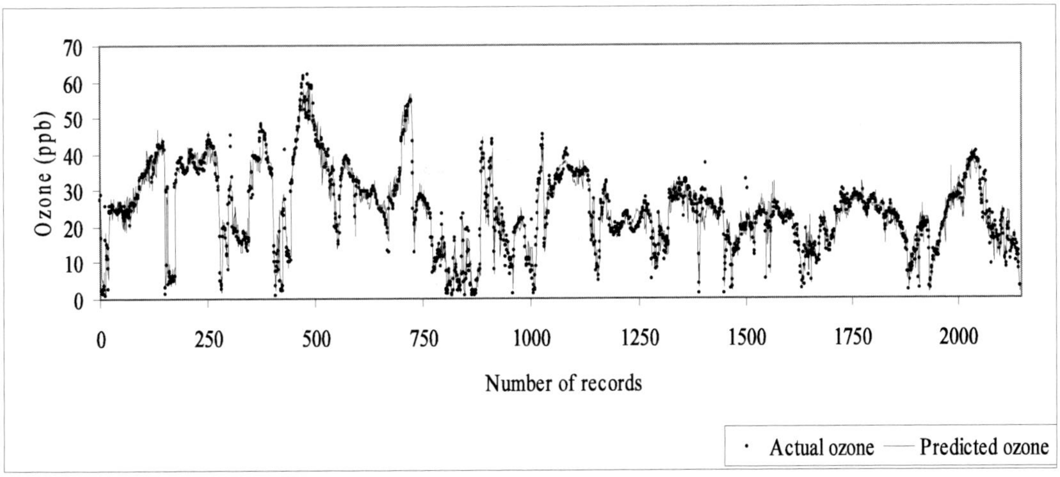

Figure 3. Comparison of predicted and measured ozone levels for the training data set.

3.1.3 Comparison with Regression Models

The predictions of the above NN model are compared here against a linear (equation 15) and nonlinear (equation 16) regression models of the data. Equation 15 is the best linear fit to the data. It has an average error of 37.4%, a maximum error of 986%, and a minimum error of 0.004%. The best available nonlinear model (Elkamel et. al., 2001) has an average percent error, a maximum error, and a minimum error of 12.84, 511, and 0.02%, respectively. The cross plots for these two models are given in Figures 5 and 6. In the nonlinear model (equation 15), the notation log stands for natural logarithm.

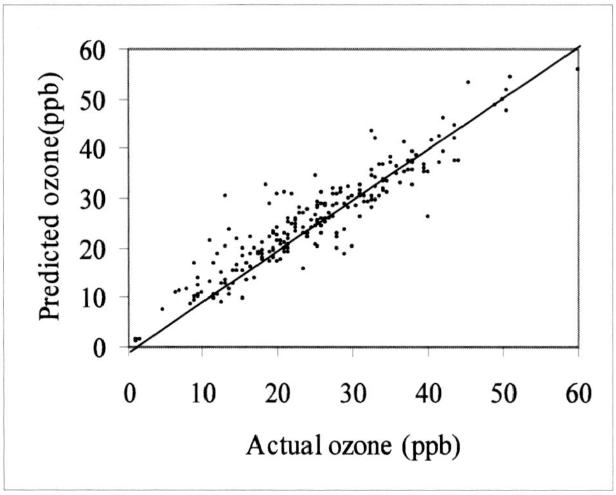

Figure 4. Cross plot of predicted and actual ozone levels for the testing data set.

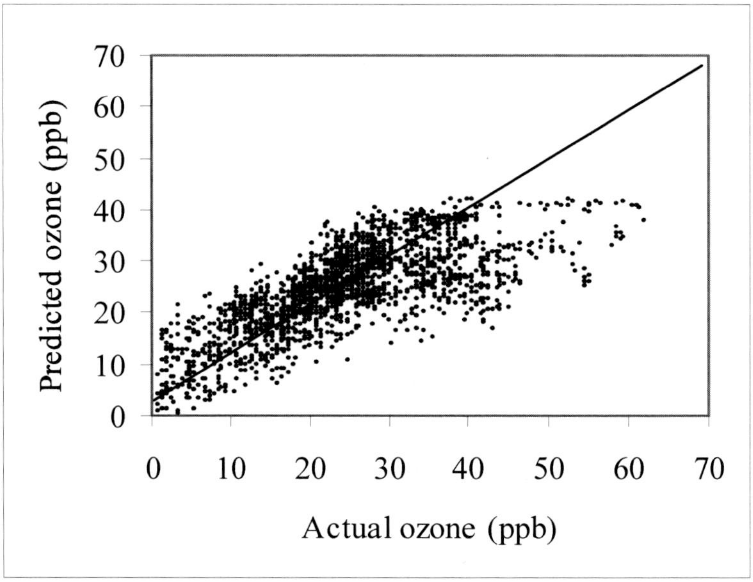

Figure 5. Cross plot of predicted and actual ozone levels for the linear regression model.

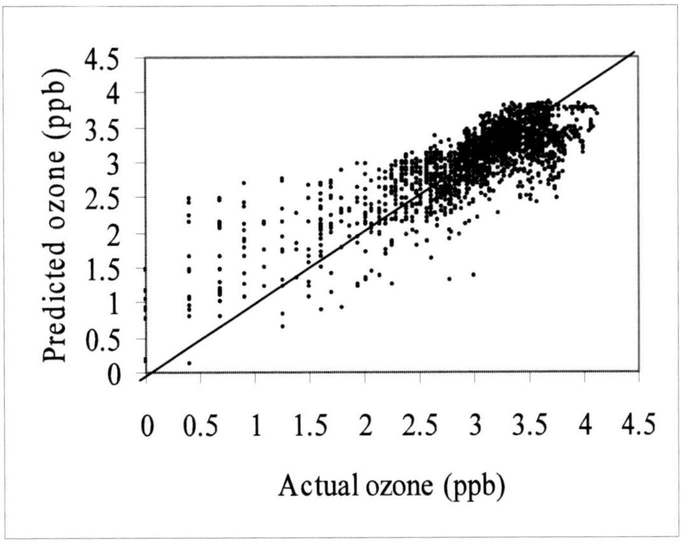

Figure 6. Cross plot of logarithm of predicted and actual ozone levels for the nonlinear regression model for the training data set.

$$[O_3] = 59.5 + 0.329[CH_4] - 9.92[NMHC] + 0.570[CO] + 0.0208[CO_2] - 0.0960[NO] - 0.385[NO_2] + 0.00991[SO_3] - 0.068WS - 0.0272WD - 1.10T - 0.218RH + 13.6S + 0.00298D \quad (15)$$

$$\log[O_3] = 4.74 - 0.0391[CH_4] - 0.510[NMHC] + 0.0023[CO] + 0.000268[CO_2] - 0.0105[NO] - 0.0268[NO_2] + 0.00113[SO_3] - 0.00256WS - 0.000985WD - 0.0283T - 0.0109RH + 0.503S + 0.000192D \quad (16)$$

It is clear that the artificial neural network model gives better predictions than the two models (See Figures 2 and 4 and contrast them with Figures 5 and 6). Another advantage of the neural network model is that it is able to handle noisy and imprecise data, it can be retrained frequently whenever new data is available, and can also predict the concentration of several pollutants at a time. Another advantage of neural network modeling is its ability to assess the importance of each of the input variables using the network weights (see Elkamel et. al., 2001).

3.2. Pollution from Mobile Sources

Here we consider a case study that deals with predicting pollution levels due to mobile sources. It is necessary to identify and periodically update, the respective contribution to concentrations of atmospheric pollutants in residential areas, from stationary sources (including industry), but also from motor vehicles. A periodic assessment should also be made of the influence on air quality trends due to reductions in emission from motor vehicles through improved traffic flow and progressive replacement of existing vehicles by new ones meeting improved emission control standards. This is essential in evaluating the need for improved transportation schemes and new types and levels of emission control.

Table 4. Variables measured in case study 2 and their ranges

Variable	Average value	Minimum value	Maximum value
Hydrocarbon (ppm)	2.44	1.72	10.3
Carbon monoxide (ppm)	1.93	0.110	16.2
NOx (ppb)	140.1	4.92	1157.0
Temperature (K)	19.6	5.74	34.5
Relative humidity (%)	48.12	14.60	93.80
Solar energy (kW/m^2)	0.26	0.03	1.13
Wind direction (N degree)	207.78	30.70	343.00
Wind speed (m/s)	2.10	0.47	5.54
Number of Vehicles	103	5	230

The analysis of pollution due to motor vehicles is a complex problem that depends on various factors including diverse road and climatic conditions by motorists with a wide variety of living and driving habits.

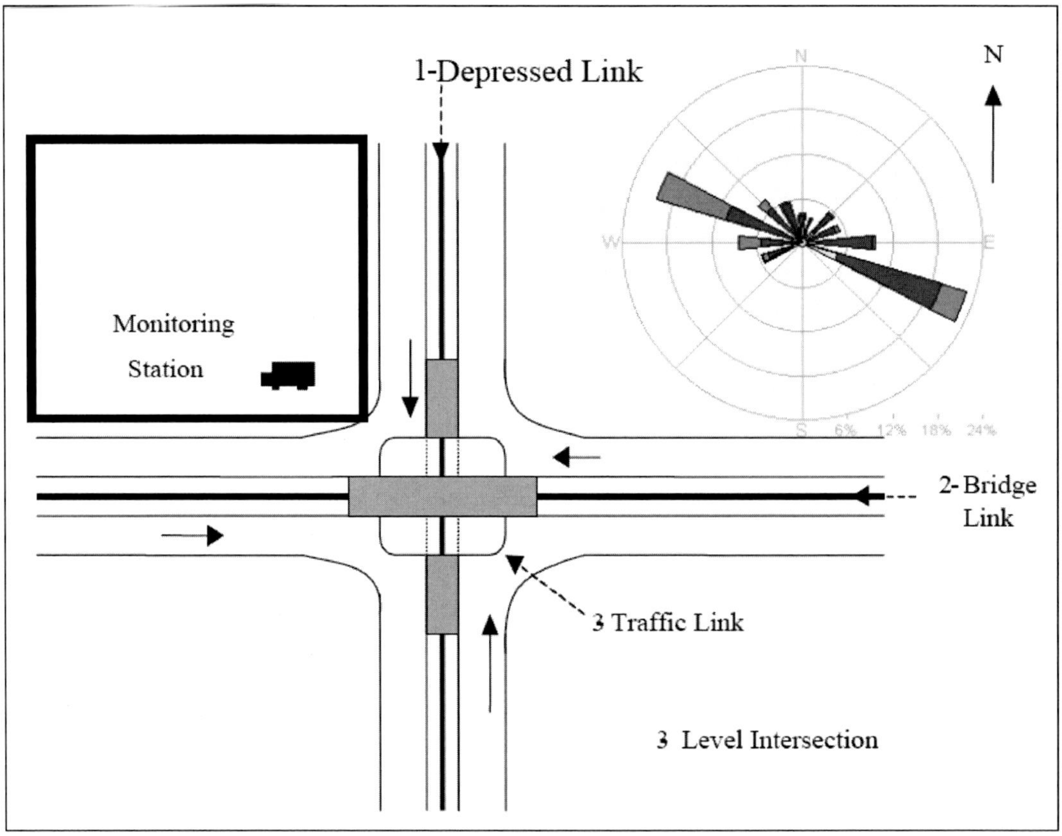

Figure 7. The location of monitoring station nearby traffic intersection.

In this case study, the vehicular carbon monoxide (CO), volatile organic compounds (COCs), and nitrogen oxides (NOx) emissions at a cross-road intersection are modeled using ANNs and the predictions are compared to the measured data and also to an existing simulator (Mobile5). This simulator utilizes emission rates, intersection road dimension, traffic variables, and meteorological conditions to estimate emissions

3.2.1. Data Collection

The data collected at a road intersection with heavy traffic by Mohesen (2000) is used here to prepare an artificial neural network model. The intersection is shown in Figure 7 which indicates that the main dominant wind direction was South East (SE) and sometimes North West (NW). A mobile laboratory was located North West from the road intersection and was used to collect the data. Therefore, whenever SE wind is dominant maximum concentration will be observed. The details of the mobile laboratory's chemical sensors and meteorological instruments used to assemble the data set are the same as in case study 1. The data considers three major pollutants: carbon monoxide (CO), hydrocarbons (HC), and nitrogen oxides (NOx) as a function of six variables (wind speed, wind direction, temperature, relative humidity, solar intensity, and number of vehicles). Data was collected every five minutes on various days of different meteorological conditions and also both during the day and night hours in an effort to encompass a wide variation in data. The number of complete data sets with values for all 9 variables recorded is 1370. Table 4 shows the variables measured and their minimum, maximum and mean values.

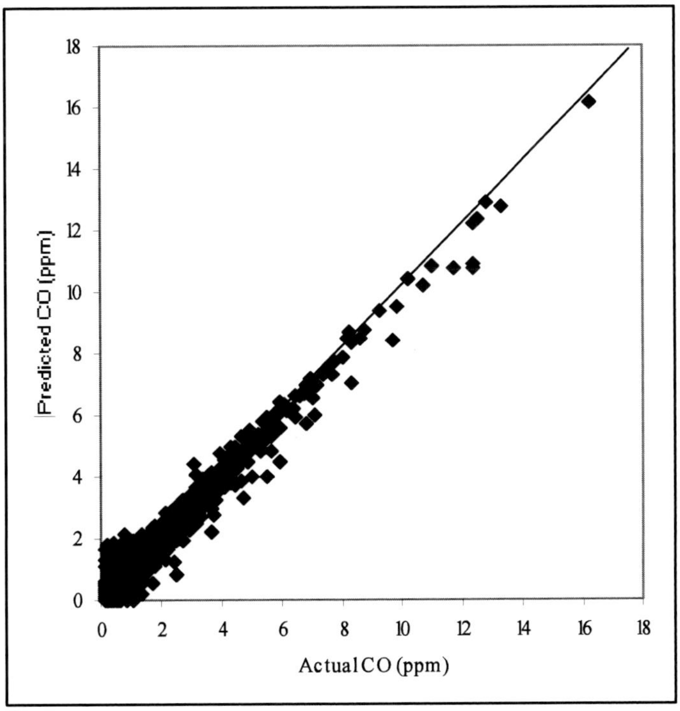

(a)

Figure 8. (Continued)

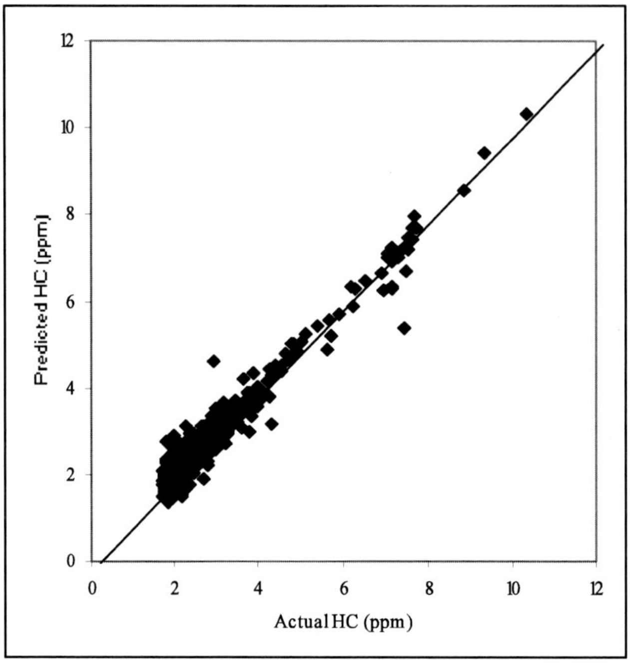

(b)

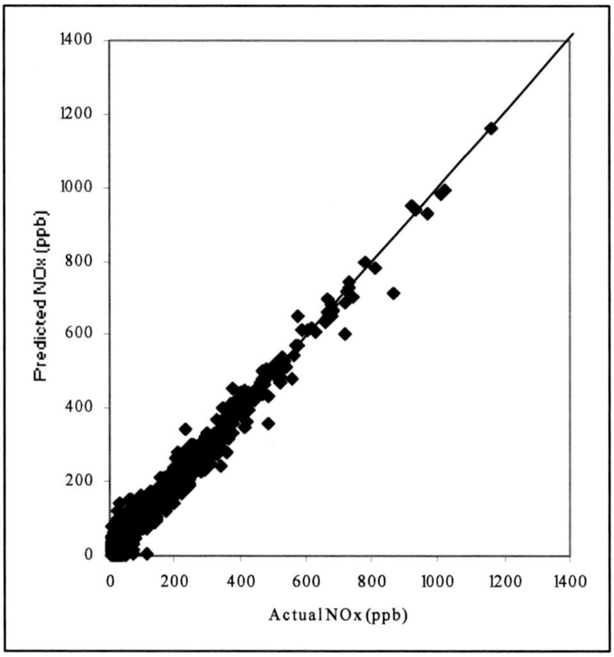

(c)

Figure 8. Cross plots of predicted and actual pollutants levels for the training data set.

3.2.2. Neural Network Modeling and Testing

As in case study 1, the data is divided into two sets: a training set and a testing set. The concentrations of CO, HC, and NOx are predicted as a function of the six variables wind speed, wind direction, temperature, relative humidity, solar intensity, and number of vehicles. The predictions of a one hidden layer neural network with twenty five neurons are satisfactory for both the training and testing data sets as shown in Figures 8 and 9, respectively. The CO coefficient of correlation (R^2), the average percent error, the maximum percent error, and the minimum percent error for the training data set are 0.967, 38.45, 101.7, and 0.015, respectively. For the testing data set, the values are 0.957, 35.5, 54.6, and 0.047. The values for HC and NOx for the training data sets are 0.949, 6.39, 56.4, 0.027, and 0.976, 45.38, 91.57, and 0.0081, respectively. For the testing data sets, the values of the above parameters for HC and NOx are 0.915, 4.62, 27.2, 0.013 and 0.960, 42.91, 45.77, and 0.0015, respectively.

3.3.3. Comparison with a Dispersion Model

The predictions of the NN models are compared with the commercial simulator MOBILE5. This simulator is an integrated set of routines from Trinity that is usually used in the analysis of air pollution impact of gasoline-fueled and diesel highway mobile sources. The program provides the user with a flexible analytical tool which can be applied to a wide variety of air quality planning functions. Fuel type and properties such as RVP (Reid Vapor Pressure) and volatility, vehicle average speed, vehicle maintenance period, and ambient temperature are input data needed to calculate the emission rates. The modeling behind this simulator is based on the assumption that in heavy traffic along a straight stretch of highway, the pollution problem may be modeled as a continuing infinite line source. When the wind direction is normal to the line of emission, the ground level concentration $C(x)$ at downwind distance x is developed assuming Gaussian or normal distribution (Wark and Warner, 1981)

$$C(x) = f(\, q,\, u,\, \textit{meteorological stability, number of vehicles}) \tag{17}$$

where q is the source strength per unit distance (g/m.s) and u is wind speed (m/s).

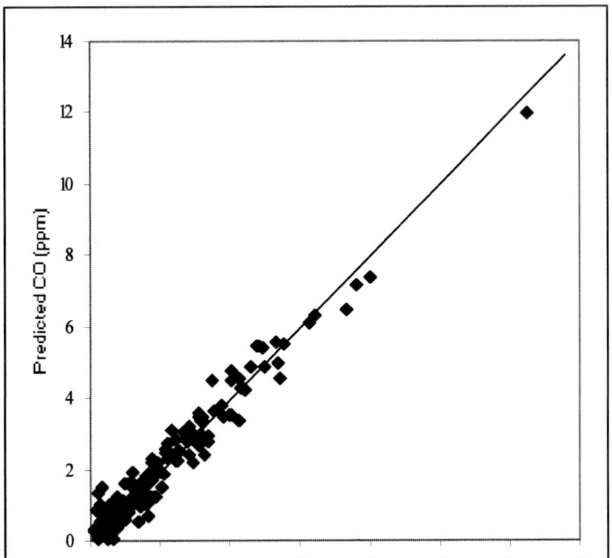

(a)

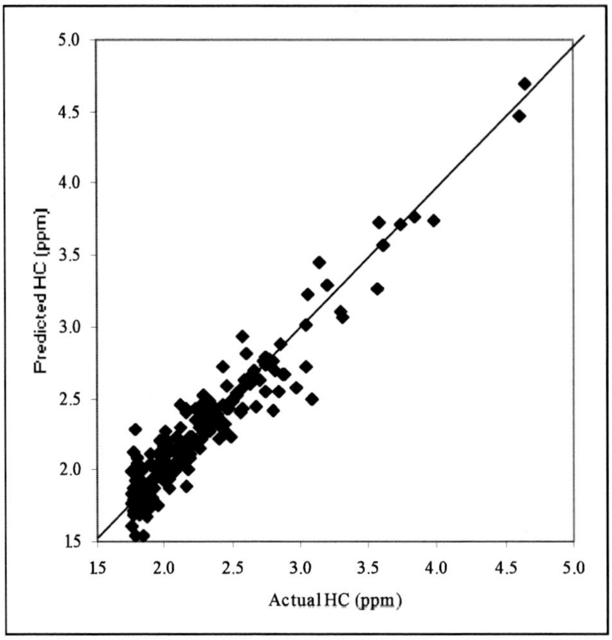

(b)

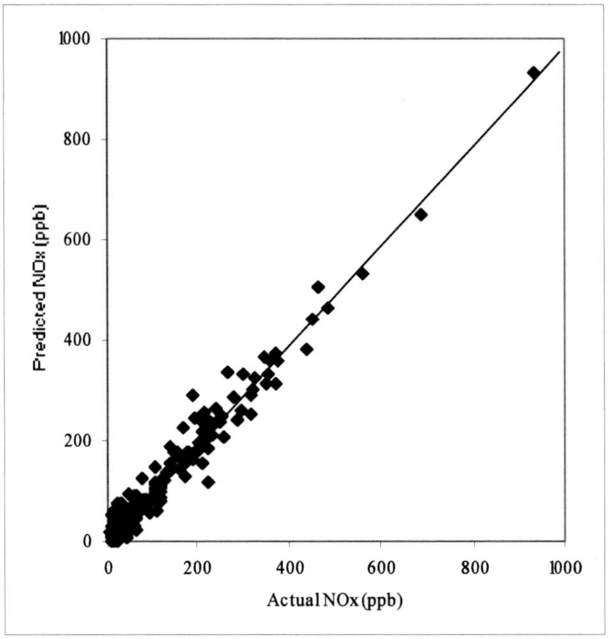

(c)

Figure 9. Cross plot of predicted and actual pollutant levels for the testing data set.

To determine the total Linear Source Strength (q_t) for a queuing link, the total number of lanes in the queue link and the percent of time that vehicles are estimated to be idling in the queue link must be considered. This is done by multiplying the Linear Source Strength for one lane (q_1) by the number of traffic lanes in the link and the percent of red time during the signal cycle. The total Linear Source Strength (q_t) for the queuing link is then calculated as follows:

$$q_t = VPH*EF * number\ of\ lanes * percent\ red\ time \tag{18}$$

where: C= hourly capacity (equation 19)
EF=Emissions factor (g/m) calculated from MOBILE5

It is assumed that the vehicles will be in the idling-mode of operation only during the Red Time phase of the signal cycle. The input parameters required to determine the queue length are: traffic volume of the link, signal cycle length, red time length, and clearance interval lost time.

The capacity of an intersection approach lane is determined by applying the effective green time to its saturation flow rate. Saturation flow rate represents the maximum number of vehicles that can pass through a given intersection approach lane. Effective green time is calculated by subtracting the amount of red time, start up delay and the time lost during the clearance interval from total signal cycle length. The clearance interval lost time represents the portion of the yellow phase that is not used by the motorists. Thus, the capacity of the intersection approach per lane is calculated as:

$$C = (Q_{sat}) * (t_{all} - t_{red} - t_{delay} - t_{loss}) / t_{all} \tag{19}$$

where:C = hourly capacity per lane [veh/hr/lane]
Q_{sat}= saturation flow rate [veh/lane/hr of green time]
t_{all} = cycle length [s]
t_{red} = length of red phase [s]
t_{delay}= start-up delay = [s]
t_{loss} =clearance interval lost time [s]

In the under-saturated condition (i.e., volume to capacity ratio, v/c, is less than 1), the number of vehicles queued (*Nu)* per lane at an intersection at the beginning of the green phase is estimated based on the following formula (Breeze Manual, 1994):

$$Nu = MAX\,[a * D + t_{red} / 2 * a,\ a * t_{red}] \tag{20}$$

where: a =vehicle arrival rate per lane [veh/lanes/s]
D =average vehicle approach delay [s/veh]
t_{red}=length of the red phase [s]

Since the queue generally reaches its maximum at the end of the red phase in under-saturated condition, these two components are added together in the first term to estimate the average queue at the beginning of the green phase. The average approach vehicle delay, D, in Equation (20) is estimated using the following formula for signalized intersection delay:

$$D = d * PF * Fc \tag{21}$$

where: d = average stopped delay per vehicle [s/veh]
PF = progression adjustment factor
Fc =stopped delay-to-approach delay conversion factor (= 1.3)

Table 5. Input data required for the program MOBILE5

Vehicle average speed
Vehicle miles traveled
Annual mileage accumulation rates by age and registration distribution by age that defines the composition and travel characteristic of the fleet.
Basic exhaust emission rates by model year.
Inspection and maintenance program
Effect of loading such as: air conditioning usage, extra loading, trailer towing and humidity correction for NOx emissions.
Anti-tampering program
Refueling emissions for gasoline vehicles.
Local area parameters, low altitude or high altitude area.
Ambient, minimum and maximum temperatures.
Specify pollutants : CO, NOx, and Hydrocarbons or select the three
Calculate idle emission factor
ASTM volatility class of fuel
Reid vapor pressure of gasolinc
Operating modes: cold start, stabilized or hot start
Vehicle type: non-catalyst or catalyst

The first term in Equation (21), *d*, the average stopped delay per vehicle for an assumed random arrival pattern for approaching vehicles, is estimated using the following formula (Breeze Manual, 1994):

$$d = (0.38)(t_{all})\frac{\left[1 - \frac{t_{green}}{t_{all}}\right]^2}{\left[1 - X\left(\frac{t_{green}}{t_{all}}\right)\right]} + 173X^2\left[(X-1) + \sqrt{(X-1)^2 + \frac{16X}{C}}\right] \quad (22)$$

where:

t_{green}= length of geen phase [s]
t_{all} = cycle length [s]
C = hourly capacity per lane [veh/hr/lane] (equation 3)
X= volume to capacity ratio V/C
V = hourly approach volume per lane [veh/hr/lane]

The first term of Equation (22) accounts for uniform delay. The second term of the equation accounts for additional delay due to random arrivals and/or occasional cycle failures.

The second term in Equation (21), the progression adjustment factor (*PF*), is included to account for the variation of stopped delay with traffic flow progression quality. Progression adjustment factors are determined using to following key variables:

PF = 1 for dense arriving at the beginning of the red phase
PF = 2 for dispersed arriving during the red phase.

PF = 3 for random arrivals
PF = 4 for dispersed arriving during the green phase.
PF = 5 for dense arriving at the beginning of the green phase.

The input variables needed for MOBILE5 are summarized in Table 5.

The emission rates calculated for the three pollutants HC, CO, and NOx at different days are shown in Figures 10 and are compared with the NN predictions. As can be seen in the Figure, the NN predictions fair well enough even though they are based only on six input variables as compared to the extensive input data needed for MOBILE 5 (Table 5).

3.3. Predicting Ground-Level Ozone through a Combined NN-principal Component Approach

Selecting a specific linear or a non-linear model to predict ozone concentration might not be enough to capture all the relationships. Principal component regression (PCR), for instance, cannot deal with non-linear relationships while the artificial neural network (ANN) models alone are not able to handle both linear and non-linear patterns equally well (Zhang, 1998; Zhang, 2003). Hence, by combining forecasts from PCR and ANN, forecasting accuracy is higher than individual forecasts.

In this case study, we will revisit the problem of ozone prediction discussed in Case Study 1 and we will employ an improved approach that combines the use of ANN with principle component regression (PCR). This case is based on the work of Al-Alawi et al., 2008. It is focused on modeling ground-level ozone as a function of meteorological variables and concentrations of primary pollutants. The idea of the study is based on using the method of principal components to select explanatory variables for multiple regression analysis, followed by applying an artificial neural network to the resulting residuals.

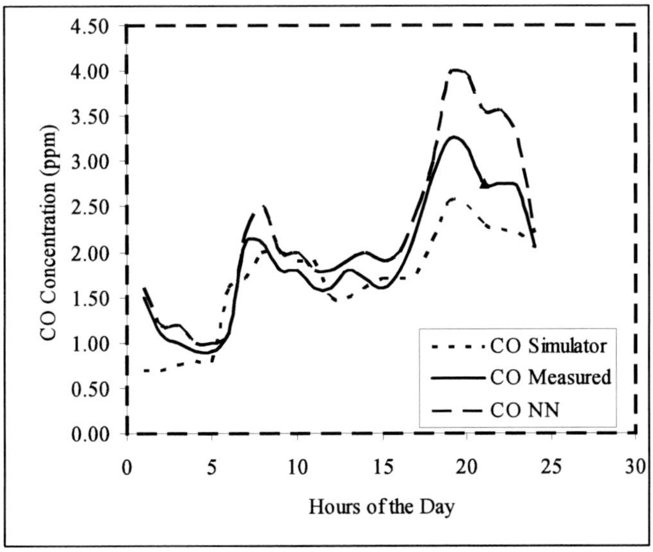

(a)

Figure 10 (Continued)

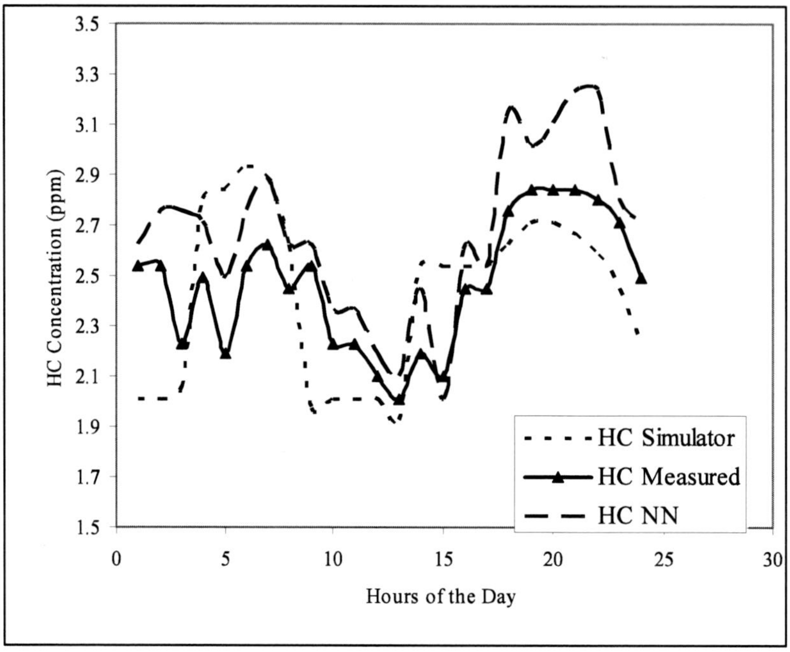

(b)

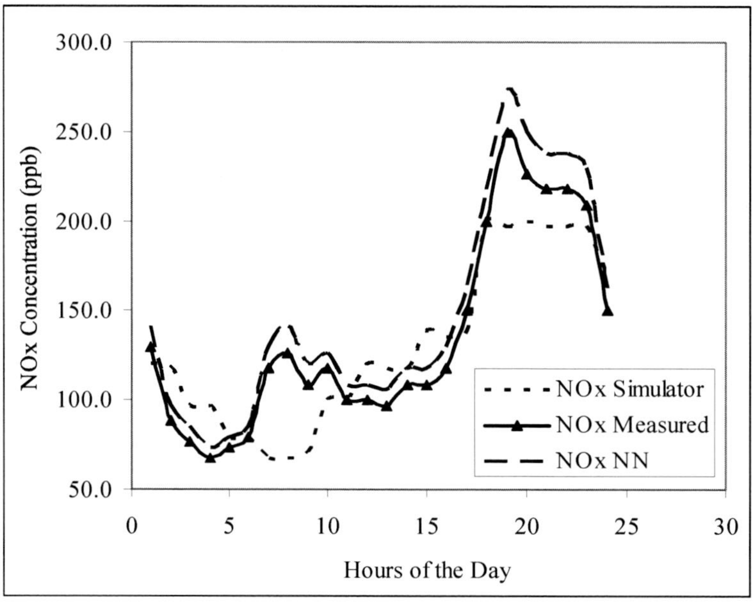

(c)

Figure 10. Comparison of the average daily predictions of the NN model and the MOBILE5 model with actual measurements of the three pollutants.

3.3.1. Data Collection

The data was collected using a mobile laboratory located in a residential area. The area was expected to be affected by the pollutants, which are discharged continuously from the traffic load. The mobile laboratory was fitted with chemical monitors and meteorological sensors. All the units, which were installed inside the laboratory, were operated automatically. Instantaneous measurements were recorded continuously every 5 minutes. Pollutants measured included methane (CH_4), non-methane hydrocarbons (NMHC), carbon monoxide (CO), carbon dioxide (CO_2), nitrogen oxide (NO), nitrogen dioxide (NO_2), sulphur dioxide (SO_2) and ozone (O_3). Meteorological parameters monitored simultaneously included wind speed and direction, air temperature, relative humidity and solar radiation. Accordance to Abdul-Wahab and Al-Alawi (2002), the selected residential area was occasionally subjected to ambient ozone concentrations exceeding the National Ambient Air Quality Standard (NAAQS) of 80 ppb and the results confirmed that high ozone events occur mainly in summer. Based on these findings, only the air quality data that were collected during the summer was selected for this study.

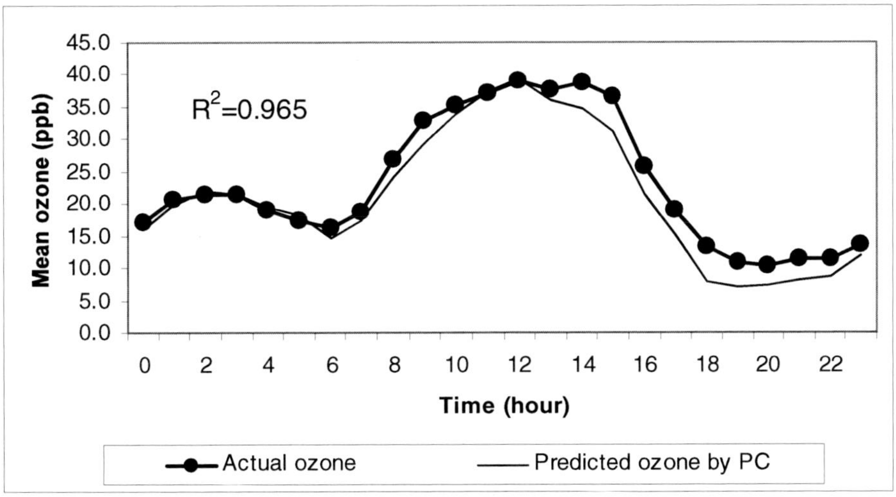

Figure 11. Predicted concentrations of ozone by principal component regression and their corresponding actual values according to the hours of the day.

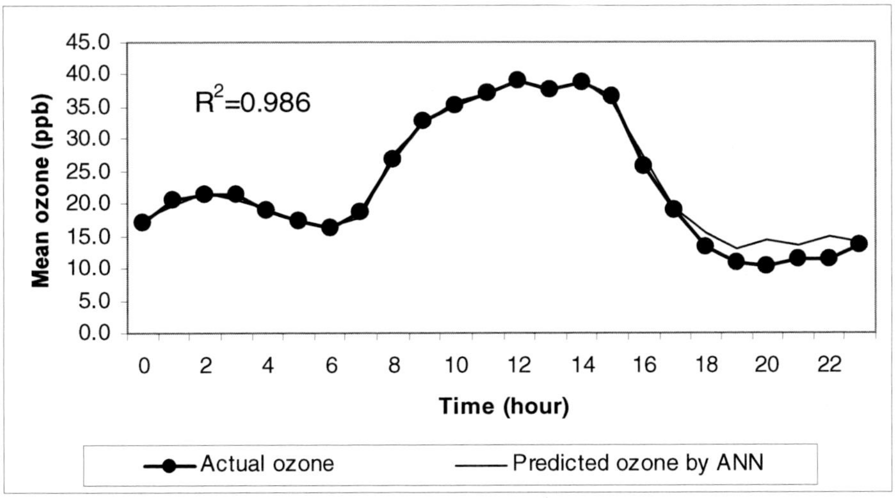

Figure 12. Predicted concentrations of ozone by an artificial neural network and their corresponding actual values according to the hours of the day.

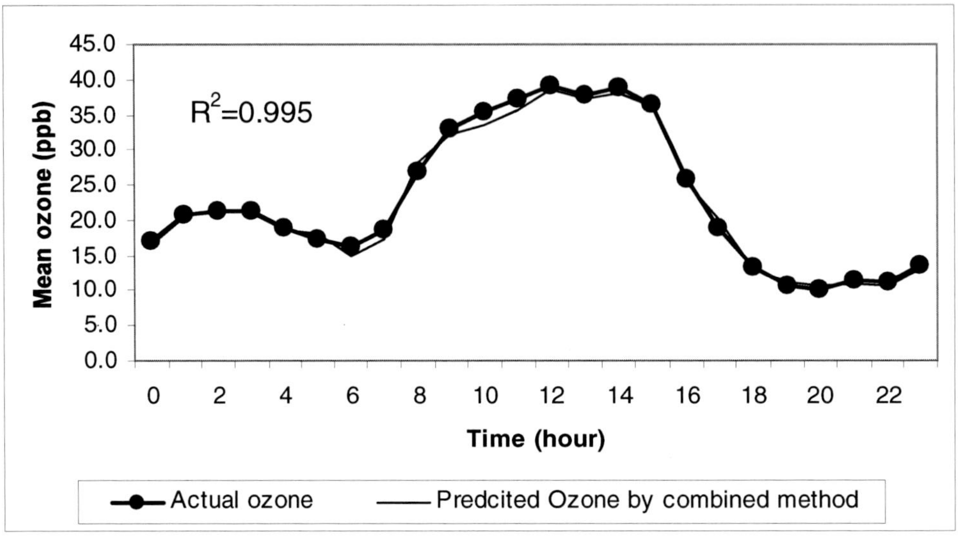

Figure 13. Predicted concentrations of ozone by the combined method and their corresponding actual values according to the hours of the day.

3.3.2. Modeling Ozone Concentrations

Figure 11 shows the results of applying the PCR method to model ozone concentration. While the adjusted coefficient of variations for this model are quite satisfactory (0.965), Figure 11 clearly shows an underestimation of ozone concentration levels during part of the morning hours, and more so from the afternoon hours to midnight. When the data were fitted using ANN, there was a significant improvement, yielding an R^2 value of about 0.986, as shown by Figure 12. Nevertheless, the fit resulted in an over-estimation of the ozone concentration levels during the evening hours. The application of ANN on the residuals from the PCR model resulted in an almost perfect fit of the ozone concentrations, with $R^2 = 0.995$ (Figure 13). Table 6 gives the actual hourly mean concentration levels and those computed by the three methods.

Figure 14 shows the graphs of the observed and fitted values of hourly mean ozone concentration as produced by the combined model for a typical working day, represented here by a Saturday, and for a weekend, represented by a Friday. Although there appears to be some discernable systematic under- and over-estimations, the overall fits are adequate, with very high coefficients of variation. While the ozone concentrations are highest between 8:00 AM and 4:00 PM on weekdays, during the weekend the concentrations tend to increase after 1:00 PM to a peak between 6:00 and 7:00 PM. There is usually low ozone concentration during the morning hours on weekends and at night for all days of the week. This indicates low traffic levels at these times.

Figures 15-17 are the scatter plots of the logarithms of predicted ozone concentrations of each of the three methods against the corresponding logarithms of the observed values. The slopes of the regression lines of the scatter plots, 0.823, 0.992 and 0.979, for the PCR, ANN and the combined models, respectively, are not significantly different from unity, the slope of a perfect fit. However, the R^2 value, 0.818, for the PCR method is the lowest of the three. The R^2 value for the ANN model is 0.845 while the one for the combined model is 0.970. These results yield correlation coefficients, with the logarithms of the actual ozone concentrations, namely, 0.904 for PCR, 0.919 for ANN, and 0.985 for the combined model. An added advantage of the combined model is that the principal component application identifies the predictor variables that best explain the variation in the ozone concentration levels and the nature of the variation. In this case, the chief predictor variables, as selected by the principal component method, were the pollutant variables, [NO] and [SO_2], and the meteorological variables of temperature and solar radiation, during day time, while only the polynomial function of the nitrogen compounds [NO] and [NO2] explained much of the variation at night time (Abdul-Wahab et al., 2005).

Table 6. Actual and predicted mean hourly ozone concentrations by the different forecasting methods

		Mean (ppb)		
Time (hour)	Actual ozone	Predicted ozone by ANN	Predicted ozone by PC	Predicted Ozone by combined method
0	17.13	17.52	16.10	16.63
1	20.67	19.71	19.89	20.51
2	21.39	21.75	21.86	21.60
3	21.34	20.55	21.47	20.99
4	19.02	19.37	19.65	18.73
5	17.26	17.21	18.26	18.16
6	16.19	16.61	14.75	14.84
7	18.62	17.84	17.29	17.36
8	26.85	27.67	24.24	28.10
9	32.92	32.55	29.28	32.30
10	35.32	35.77	33.91	33.57
11	37.26	37.18	37.33	35.78
12	39.09	39.09	39.19	38.73
13	37.74	37.54	36.03	37.37
14	38.80	39.11	34.58	38.17
15	36.46	35.72	31.05	36.16
16	25.81	27.04	21.53	25.17
17	18.94	19.22	15.09	20.13
18	13.24	15.58	7.80	13.06
19	10.72	13.01	7.08	11.09
20	10.24	14.44	7.24	10.72
21	11.50	13.47	8.08	11.04
22	11.29	14.94	8.62	10.76
23	13.50	14.17	11.89	13.06
Total	23.08	23.63	21.02	22.76

Zero hour represent 12:00 midnight.

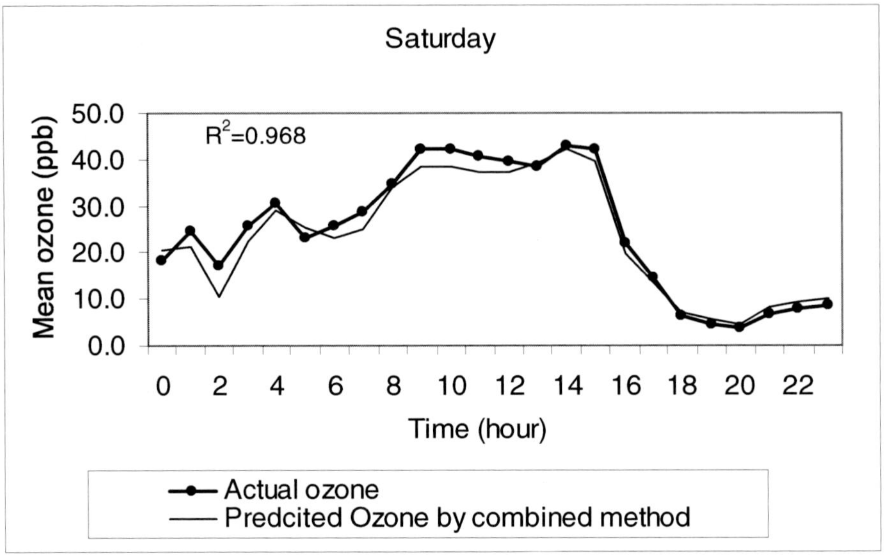

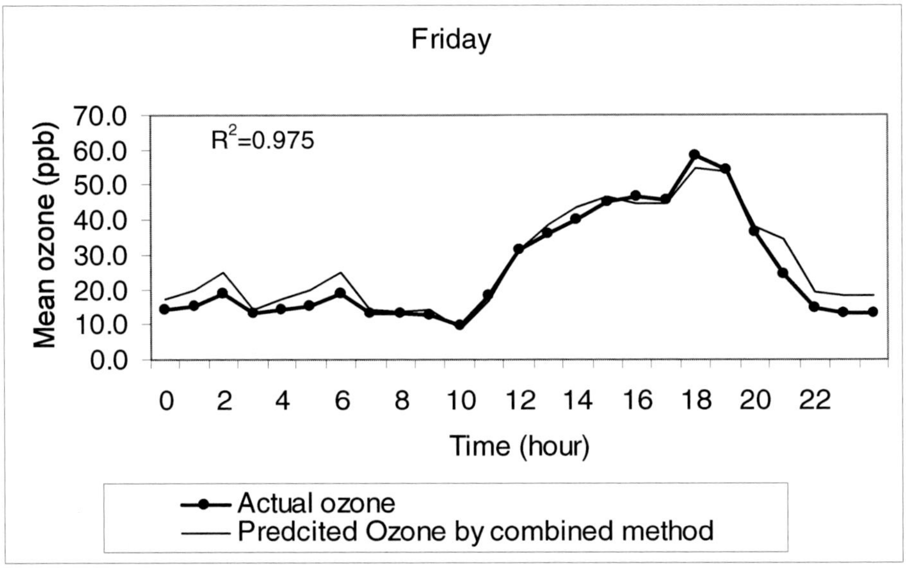

Figure 14. Predicted concentrations of ozone by the combined method and their corresponding actual values as produced for Saturday (e.g., working day) and Friday (e.g., weekend).

In summary, the case study demonstrated that a more accurate prediction model for ozone can be obtained by using PCR as a variable selection technique to identify the most appropriate explanatory variable subset for a regression model of ozone concentration, followed by applying an ANN model on the resulting residuals. Although the ANN model alone provided better fit than the PCR model, the combined PCR and ANN significantly improved the forecast accuracy for the ozone concentration levels.

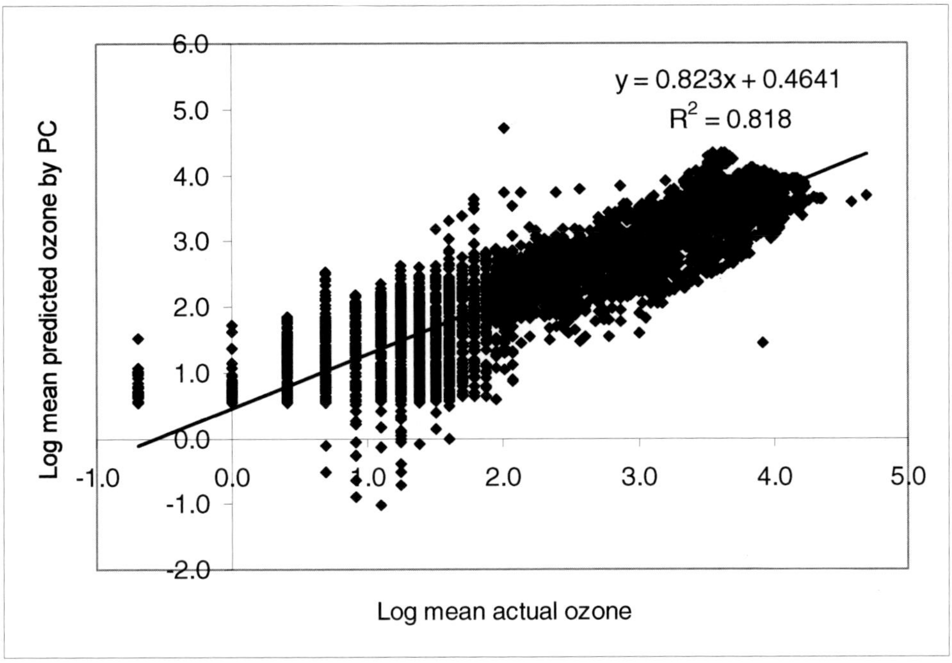

Figure 15. Predicted values of log ozone by principal component regression vs the log actual ozone values.

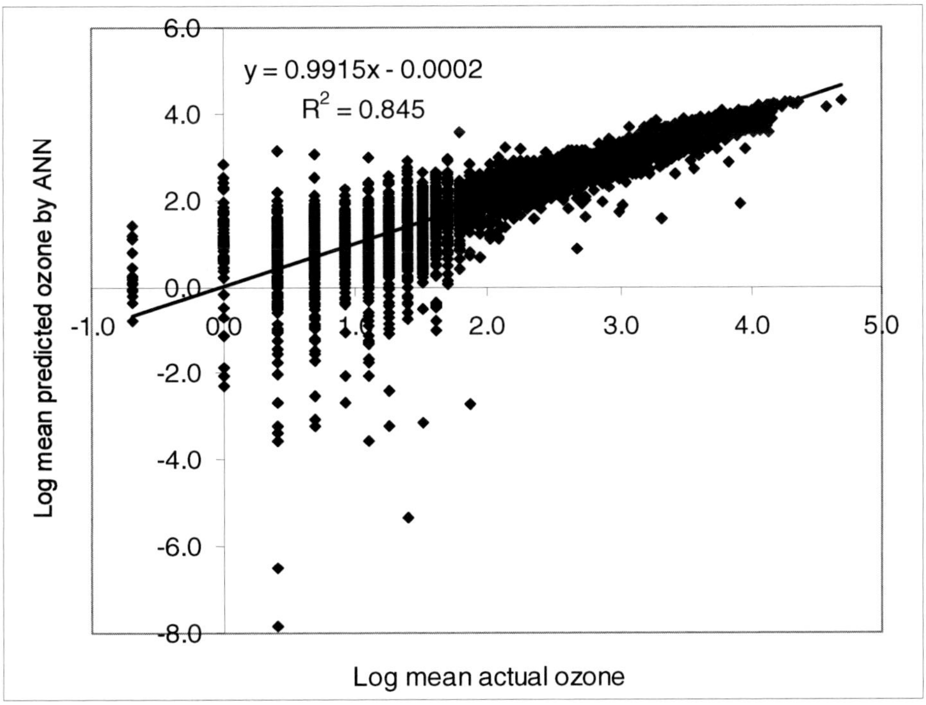

Figure 16. Predicted values of log ozone by an artificial neural network vs the log actual ozone values.

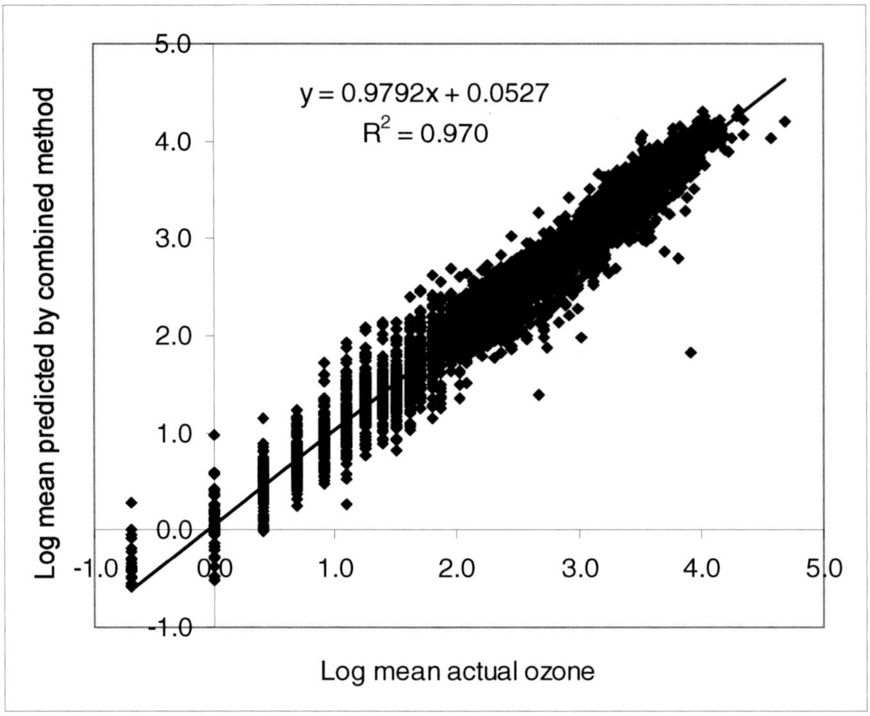

Figure 17. Predicted values of log ozone by the combined method vs the log actual ozone values.

4. Conclusions

In this chapter, a background and introduction on the use of artificial neural networks in predicting pollutants concentrations was covered. The chapter illustrated different algorithms used for training ANNs. It presented several examples that illustrated the predictive performance of the prepared network models when compared against linear and non-linear regression models, commercial simulators, and measured data. Three differenct case studies illustrated that the use of artificial neural networks (ANNs) is a promising tool for atmospheric pollution forecasting of pollutants. For instance, ozone in the lower atmosphere is a complex non-linear process. Therefore, ANNs is a well-suited method for modeling this process since it allows for non-linear relationships between ozone and other controlling variables. Accordingly, the application of ANNs to ozone modeling has recently become available to capture those non-linear features of the relationship that a conventional statistical technique (e.g., regression models) might overlook. Unlike other modeling approaches, ANNs by their unique structure possess the ability to learn non-linear relationships with limited prior knowledge about the process structure. The deterministic models (i.e., theoretical or detailed atmospheric diffusion models), for example, would require resolving the underlying chemical and physical equations that control pollutant concentrations. They are highly sophisticated because they require a high level of human resources and intense data input such as detailed emission data and meteorological conditions for the region of interest. There are generally severe limitations in both spatial and temporal accuracy of such data. In addition, some input data are not easily acquired by environmental protection agencies or local industries. Therefore, it is much more practical and cost-effective to rely on ANNs for atmospheric pollution modeling of pollutants.

Combining ANN with principal component analysis is yet another advantage that can be utilized to improve the predictions accuracy. In the third case study considered in this chapter, predicting ozone concentrations was presented by using such a a combined approach.. Although the ANN model alone provided better fit than the PCR model, the combined PCR and ANN significantly improved the forecast accuracy for the ozone concentrations. The R^2 values were 0.965, 0.986, and 0.995 for PCR, ANN, and the combined model prediction, respectively. It was also found that combining the predictions from the PCR and ANN models reduced the root mean square errors of ozone concentrations, compared to either PCR or ANN forecasts. The results of the study proved that combining predictions generated by different methods could improve the accuracy and provide a prediction that is superior to a single model prediction.

REFERENCES

Abdul-Wahab,A., Bouhamra,W., Ettouney,H., Sowerby,B. and Crittenden,B.D., 1996. Predicting ozone levels: A statitical model for predicting ozone levels in the Shuaiba Industrial Area, Kuwait. *Environ.Sci. & Pollut. Res.* 3(4), 195-204.

Abdul-Wahab, S.A., Al-Alawi, S.M., 2002. Assessment and prediction of tropospheric ozone concentration levels using artificial neural networks. *Environmental Modelling & Software* 17, 219-228.

Abdul-Wahab, S.A., Bakheit, C.S., Al-Alawi, S.M., 2005. Principal component and multiple regression analysis in modelling of ground-level ozone and factors affecting its concentrations. *Environmental Modelling & Software* 20, 1263-1271.

Al-Alawi, S., Abdul-Wahab, S.A., bakheit, C., 2008. Combining principal component regression and artificial neural networks for more accurate predictions of ground-level ozone. *Environmental Modelling & Software* 23, 396-403.

Baughman,D.R. and Liu, Y.A., 1990. *Neural Networks in Bioprocessing and Chemical Engineering*, Academic Press , New York.

Boznar, M., Lesjak,M. and Mslakar,P.,1993. A network-based method for short-term predictions of ambient SO_2 concentrations in highly polluted industrial areas of complex terrain. *Atmos.Environ.,* 27B, 221-230.

Elkamel, A.,1998. An artificial neural network for predicting and optimizing immiscible flood performance in heterogeneous reservoirs. *Computers Chem. Engng.* 22, 1699-1709.

A. Elkamel, S. Abdul-Wahab, W. Bouhamra, and E. Alper, "Measurement, Prediction, and Modeling of Ozone Levels Nearby an Industrial Area", *Advances in Environmental Research,* 5 (1), 47-59, (2001).

E. Fatehifar, A. Elkamel, M. Taheri, W. A. Anderson, and S. A. Abdul-Wahab, "Modeling and Simulation of Multi-Pollutants Dispersion from a Network of Refinery Stacks Using a Multiple Cell Approach", *Environmental Engineering Science*, Volume 24, Number 6, p 795-811, 2007.

Fausett, L.V., 1998. Neural Network approaches to problems in ground water contamination. 1998 Advanced Simulation Technologies Conference (ASTC'98), *Proceedings of International XV,* Edited by Ades,M. and Griebenow,R., pp.29-34.

Garson, G.D., 1991. Interpreting Neural-Networks Connection Weights. AI Expert, 6(7):47.

Gery, M.W. ,Whitten,G.Z, Killus,J.P and Dodge,M.C., 1989. A photochemical kinetics mechanism for urban and regional scale computer modeling, *Journal of Geophysical Research* Vol.94, No.D10, pp. 12 925-12 956.

Kinosian, J.R.,1982. Ozone-precursor relationship from EKMA diagrams, *Environ. Sci. Technol.* 16(22), 880-883.

Kuntasal,G. and Chang, T.Y.,1987. Trends and relationships of O_3, NO_x and HC in the south coast air basin of California, *J.Air Poll.Contr.Assocn.,*37, 1158-1163.

Mayora-Ibarra, O., Motta, L. and Curatelli, F.,1998. Forecasting of atmospheric pollution levels with neural networks. *Neural Networks & Their Applications*, Marseilles, March-11-13, pp.323-328.

Mohesen, J., 2000. *Measurement and Modeling of Air Pollution Nearby a Cross-Road Intersection,* M.S. Thesis, Kuwait University.

Passini, A. and Potesta, S.,1996. Neural Network Modeling: Perspectives of Application for Monitoring and Forecasting Physical-Chemical Variables in the Boundary Layer, pp.329-340. *NATO ASI Series, Partnership Sub-Series, 2. Environment-Vol.8, Urban Air Pollution.* Edited by Allegrini, I. And De Santis, F., Springer-Verlag berlin,Heidelberg.

Rumelhart, D.E., Hinton,G.E. and Williams,R.J., 1986. Learning representations by back-propogating errors. *Nature* 323, 533-536.

Rege, M.A. and Tock, R.W., 1996. Simple neural network for estimating emission rates of hydrogen sulfide and ammonia from single point sources , *Journal of the Air & Waste Management Association*. Vol. 46, no. 10, pp. 953-962.

Sarle,W./S., 1997. *Why use a bias input?* ftp.sas.com/pub/neural/FAQ2.html#A_bias.

Rege, M.A. and Tock, R.W., 1996. A Simple Neural Network for Estimating Emission Rates of Hydrogen Sulfide and Ammonia from Single Point Sources. *Journal of the Air & Waste Management Association,* 46, pp. 953-962.

Schumuller, J., 1988. Neural Networks and Environmental Applications. *Expert Systems for Environmental Applications*. Edited bt Hushou,H. ACS Symposium Series 431, Miami Beach, Florida.

Seinfeld, J.H. and Pandis, S. N., 1998. *Atmospheric Chemistry and Physics, from Air Pollution to Climate Change,* John Wiley & Sons, New York,

Tambe, S.S. Kulkarni, B.D. and Desphande, P.B.,1996. *Elements of artificial networks with selected applications in Chemical Engineering, and Chemical and Biological Sciences*. Simulation and Advanced Controls Ltd., Lousville-KY, U.S.A.

Utojo, U. and Bakshi, B.R., 1995. *Connection between Neural Networks and Multivariate Statitical Methods: An Overview.* (In "Neural Networks in Bioprocessing and Chemical Engineering" by Baughman,D.R. and Liu, Y.A., 1990, Academic Press , New York, pp-435-459).

Zannetti, P.,1990. *Air Pollution Modeling: Theories, Computational Methods and Available Software,* Van Nostrand Reinhold, New York.

Leone, J.A. and Seinfeld,J.H., 1986. Comparitive analysis of chemical reaction mechanisms for phtochemical smog, *Atmospheric Environment,* 19, 487.

Whitten, G.Z., Hogo,H. and Killus, J.P., 1980. The carbon-bond mechanism: A condensed kinetic mechanism for photochemical smog. *Environ.Sci.Technol.,*14, 690.

Kinosian, J.R.,1982. Ozone-precursor relationship from EKMA diagrams, *Environ. Sci. Technol.* 16(22), 880-883.

Kuntasal,G. and Chang, T.Y.,1987. Trends and relationships of O_3, NO_x and HC in the south coast air basin of California, *J.Air Poll.Contr.Assocn.,*37, 1158-1163.

Zhang, G.P, Patuwo, E.B., Hu, M.Y., 1998. Forecasting with artificial neural networks: the state of the art. *Int. J. Forecasting* 14, 35-62.

Zhang, G. P., 2003. Time series forecasting using a hybrid ARIMA and neural network model. *Neurocomputing* 50, 159-175.

In: Environmental Chemistry Research Progress
Editors: P. Robinson and R. Gallo
ISBN 978-1-60741-055-3

Chapter 4

ENVIRONMENTAL FRIENDLY METHODS FOR THE EXTRACTION AND DETERMINATION OF ORGANOPHOSPHORUS AND ORGANOCHLORINE PESTICIDES

J. J. Santana Rodríguez**, *Z. Sosa Ferrera and D. Vega Moreno
Department of Chemistry, Faculty of Marine Sciences, University of Las Palmas de Gran Canaria, 35017, Las Palmas de Gran Canaria, Spain

ABSTRACT

Organochlorine pesticides have been used during decades as insecticides and for the control of vector-borne diseases. In 1980s these pesticides were banned in developed countries and were replaced by other kind of compounds like organophosphorus pesticides; however, they are usually more toxic than organochlorine pesticides. Nowadays these compounds are present in the environment; thus the long-term persistence and their metabolites are presented in water, sediments, soils and vegetables.

The methods usually used for the extraction of these compounds from environmental samples are often too time consuming, involve multi-step procedures that are prone to loss of analytes and use high amounts of toxic organic solvents.

Modern strategies in analytical chemistry have been developed towards the simplification of sample's preparation, as well as the minimization of organic solvent used on the extraction process. An alternative to organic solvents is the use of micellar systems as extractants of organic pollutants in liquid and solid environmental samples. Micellar media facilitate the solubilization of different compounds present in different kind of samples. Benefits obtained when micellar media are used, are lesser toxicity and cost than organic solvents.

Cloud-point extraction (CPE) process in liquid samples and microwave assisted micellar extraction (MAME) in solid samples, have been developed for the extraction of pesticides prior to their determination by high performance liquid chromatography. Moreover, during extraction process, many interfering components are usually co-extracted from solid samples together with target analytes. For increasing sensitivity and selectivity, MAME procedure can be coupled with other extraction and preconcentration techniques, introducing a clean-up step for removing substances that could interfere with the final determination of target analytes in HPLC system. This coupling can be done with solid phase microextraction (SPME), called MAME-SPME, and with solid phase extraction (SPE), MAME-SPE, producing also an intensification of the analytes signals by means of extract clean-up and preconcentration.

In this study an overview of the use of micellar systems well alone, or well combined with the aforementioned extraction/preconcentration techniques, for the determination of organochlorine and organophosphorus pesticides in liquid and solid samples is presented. Fundaments, experimental, variables optimization, quantitative procedures and applications are described and discussed, making emphasis on the advantages of these green methodologies with regard to traditional extraction techniques.

ABBREVIATIONS

ASE:	Acelerated Solvent Extraction
CMC:	Critical Micellar Concentration
CPE:	Cloud Point Extraction
DDT:	Dichlorodiphenyltrichloroethane
EPA:	Environmental Protection Agency
HPLC:	High Pressure Liquid Chromatography
LLE:	Liquid-Liquid extraction
LOD:	Limit of detection
MAE:	Microwave Assisted Extraction
MAME:	Microwave Assisted Micellar Extraction
PLE:	Pressurized Liquid Extraction
POLE:	Polyoxyethylene 10 Lauryl Ether
RSD:	Relative Standard Desviation
SFE:	Supercritical Fluid Extraction
SPE:	Solid Phase Extraction
SPME:	Solid Phase Microextraction

INTRODUCTION

Organochlorine and Organophophorus Pesticides

Organochlorine pesticides have been used during decades in agricultural as insecticide and for the control of vector-borne diseases. Although these pesticides have banned in developed countries since 1970s and 1980s, some of them have continued to use DDT, especially for public health applications such as malaria control [1, 2]. Despite not being used by many countries for some years they are present in the environment; in this way, the long-term persistence of DDT and its metabolites in soil have been reported [3-6].

The lipophilic nature, hydrophobicity and low chemical and biological degradation rates of organochlorine pesticides have led their accumulation in biological tissues and subsequent magnification of concentrations in organisms progressing up the food chain with consequences for the humans [7, 8]; for these reasons are listed as US Environmental Protection Agency (EPA) priority pollutants [9].

Organophosphorus pesticides, which may be considered as second generation of synthetic pesticides, after the use of organochlorine pesticides, have become more popular, as they are generally of low persistence, biodegradable and do not tend to accumulate in the food chain. However they are generally more toxic (potent cholinesterase inhibitors) than organochlorine or carbamate compounds [10].

Over recent decades, research into developing new pesticides and analytical methodology has been severely curtailed in favour of alternate means of pest control. Nevertheless, the use of pesticides continues to be our major weapon for controlling pests. Large amounts of pesticide residues remain in agricultural and forest soils after their application [11]. They leach into groundwater and are also found in run-off water. Therefore, the development of new and improved techniques for analysing these agents and their breakdown products is mandatory.

Tradional techniques for analysis of organochlorine and organophosphorus pesticides are based in gas-chromatographic multiresidue procedures [12-14]. However, these methods are often too time consuming, due to the slow chromatographic elution and the intensive pre-analytical clean-up procedures [15-17] followed, in some cases, by a derivatization [17, 18].

They usually involve multi-step procedures, are prone to loss of analytes, and use toxic organic solvents [19, 20]. Consequently, most of the analysis time is spent on the sample preparation steps alone. Furthermore, some pesticides are thermolabile and therefore difficult to volatilize without decomposition [21].

New methods therefore have been developed in order to supplement or even replace existing gas-chromatographic methods. These altenative methodologies are based in a high performance liquid chromatography determination combined with a previous micellar extraction: cloud point extraction (CPE) procedure for liquid samples and microwave assisted micellar extraction (MAME) for solid samples. Specifically, the organochlorine and organophosphorus pesticides listed in Table 1 have been studied and determined using these green methods.

Table 1. Organochlorine and Organophosphorus pesticides studied

Organochlorine Pesticides	Organophosphorus Pesticides
	Dimethoate
4,4'-Dichlorodiphenyldichloroethane (4,4'-DDD)	Methidation
Dieldrin	Parathion methyl
4,4'-Dichlorodiphenyltrichloroethane (4,4'-DDT)	Malathion
2,4'-Dichlorodiphenyltrichloroethane (2,4'-DDT)	Ethoprophos
4,4'- Dichlorodiphenyldichloroethylene (4,4'-DDE)	Parathion ethyl
Aldrin	Diazinon
	Chlorpyrifos

Main Conventional Extraction Techniques for Liquid and Solid Samples

The need to determine low concentrations of pollutants in environmental samples, as well as the matrix complexity, means that analytical methods with high sensitivity, selectivity and resolution have to be applied to soil, sediment, water and biological samples. Elaborate sample preparation involving analyte isolation and enrichment is generally necessary before the final analysis, which is usually performed using gas or liquid chromatography.

The most frecuently used methods for the extraction and preconcentration of organochlorine and organophosphorus pesticides in water samples are liquid-liquid extraction (LLE) and solid-phase extraction (SPE) [22,23]. Solvent extraction methods have the disadvantage of poor recoveries, which vary from compound to compound and use a large amount of organic solvent. There is therefore a trend to replace solvent extraction procedure in order to minimised sample manipulation, analyte losses and the use of toxic solvents [24]. So there is an increasing tendency to replace LLE by solid-phase extraction (SPE). SPE is widely accepted as an alternative extraction/cleanup method to LLE in order to determine pollutants in liquid samples [25]. However, SPE have certain disadvantages for water analysis: the cross-sectional area is small, sample processing rates are slow, the particle blockage tolerance and adsorbed matrix components is low and channelling reduces the capacity to retain analytes. Another drawback in SPE, which is the same for LLE, is the considerable amount of time needed and the manual operations involved [26].

Other methodologies have been developed with a view to eliminating or, at least minimising, the use of organic solvents. Solid-phase microextraction (SPME) is a recent technique for the extraction of organic compounds from environmental samples. SPME is based on the partition equilibrium of analytes between a polymeric stationary phase, and the sample matrix. The desorption step is usually carried out by placing the fiber in a hit injector of a gas chromatograph [27] or it can also be performed in an HPLC system by introducing an interface [28]. An inherent disadvantage of SPME is that quantitative work is still rather laborious because severe carry-over may occur between samples. Moreover, a relatively long equilibration time for the absorption of analytes is needed.

An alternative extraction technique for liquid samples is cloud point methodology or cloud point extraction (CPE) [29]. It lets extraction and preconcentration steps based on phase separation using non ionic surfactants as extractants. It offers a convenient alternative to more conventional extraction procedures. From an analytical point of view, one of the most important properties of surfactants is their good capacity to solubilise solutes of different type and nature due to their organised structure and their anphipatic character.

Table 2 gives a description of the aforementioned techniques and lists their advantages and their disadvantages for liquid samples.

The extraction of organochlorine and organophosphorus pesticides from solid samples requires the use of organic solvent, which competes in the releasing of the analytes retained owing to the high activity of the matrix. Traditional methods employ large volumes of solvent under aggressive shaking and/or temperature conditions. The most frequently used method for the extraction of organic compound from soils is Soxhlet extraction or the use of an ultrasonic bath. Soxhlet extraction, the most conventional of all methods, is particularly suitable for organic pollutants strongly absorbed in soil matrices but it requires long extraction times (from 24 to 48 hours) and the use of large volumes of frequently toxic organic solvents (300 to 500 ml per sample) [30, 31].

Table 2. Extraction techniques for liquid samples [from ref. 54]

	LLE	SPE	SPME	CPE
Brief description	Analyte is partitioned between two immiscible solvents	Analyte retained on a solid adsorbent	Partition equilibrium of analytes between a polymeric stationary phase and the sample matrix	Analyte is portioned between two isotropic phase 'surfactant-rich phase' and 'aqueous phase'
Extraction Time	Up to 24 h	20-30 min	Up to 60 min	20 min
Solvent Usage	Organic Solvent	Organic Solvent	None	Surfactant solution
Cost	Low	Relatively low	Relatively low	Low
Ease of operation	Easy	Relatively easy	Relatively easy	Easy
Disadvantages	Large consumption of solvent and concentration of sample required after extraction	Insufficient retention of very polar compounds and limited selectivity	Choice of fiber and optimisation of procedure required	Optimisation of operating conditions

Sonication extraction is faster than Soxhlet extraction (30 to 60 min per sample) and allows the extraction of large amounts of sample, but it still uses about as much solvent as the Soxhlet extraction [32].

The newer extraction techniques such as supercritical fluid extraction (SFE), based in the extraction of target compounds with solvents near supercritical conditions [33]; accelerated solvent extraction or pressured liquid extraction (ASE or PLE), which extracts target compounds using solvents at high pressures and high temperatures [34] and microwave assisted solvent extraction (MAE), based in the extraction by microwave energy and temperature [35,36]. All of them are very attractive because they are faster, use much smaller amounts of solvents and are more environmentally friendly techniques than traditional ones [37].

In the last few years, the number of procedures using extraction of organic compounds from environmental matrices by microwave energy has increased [38, 39]. Microwave assisted extraction (MAE) allows rapid extraction of solutes from solid matrices by employing microwave energy as a heat source, with an extraction efficiency comparable to that of the classical techniques. It exhibits many substantial improvements in analytical sample preparation, as it requires much lower volume of organic solvent (30 ml in MAE versus 300 - 500 ml in Soxhlet extraction), reduces extraction time (20 to 30 min per sample) and lets prepare multiple samples in one step [40-42]. Optimisation of MAE conditions is rather easy, as there are few parameters (nature of solvent, time and power) and it is cheaper than other modern techniques, such as supercritical fluid extraction. The main drawbacks of this system are: extraction solvent has to be able to absorb microwaves, a clean-up procedure may be required before analysis and the vessels need to be cooled to room temperature after extraction and before they can be opened, thus increasing the overall extraction time. This methodology has been applied to the extraction of different organic compounds, such as organochlorine insecticides [43], polynuclear aromatic hydrocarbons, [44], arochlors [45], etc. with good results. In all these studies, the extractant has always been an organic solvent.

Table 3 provides a summarised description of these techniques and show their advantages and disadvantages for solid samples.

Surfactants

An alternative to the use of organic solvents would be the use of micellar systems (surfactants) as extractants.

Organic solvents are toxic compounds which are dangerous to researchers who work with them and also for environment. For this reason it is necessary look for new extraction solvents that will be less toxic for environment and work's atmosphere. In this sense, surfactants can be used as extractants in the analysis of organic pollutants like organochlorine and organophosphorus pesticides in different kind of matrices.

Table 3. Extraction techniques for solid samples [from ref. 54]

	Soxhlet	**Sonication**	**SFE**	**PLE**	**MAE**
Brief description	Sample is placed in an extraction thimble and leached with hot solvent	Sample is placed in a ultrasonic bath with an organic solvent, with shaking	Sample is placed in a high-pressure cartridge and extracted with a supercritical fluid	Sample is placed in an extraction vessel and pressurized with a solvent heated	Sample is placed immersed in solvent, and heated with microwave energy
Extraction Time	Up to 24-48 h	30-60 min	30-60 min	15 min	20-30 min
Solvent usage	Organic solvent	Organic solvent	Solvent in its supercritical state	Organic solvent at high temperature and pressure	Organic solvent heated with microwave energy
Cost	Low	Low	High	High	Moderate
Ease of operation	Easy	Easy	Relatively easy	Relatively easy	Relatively easy
Dis-advantages	Large amount of solvent and evaporation of extract are neccesary	Large amount of solvent and filtration are required	Limited sample size, and dependence of matrix and analyte type	Dependence on matrix	Clean-up procedure is required

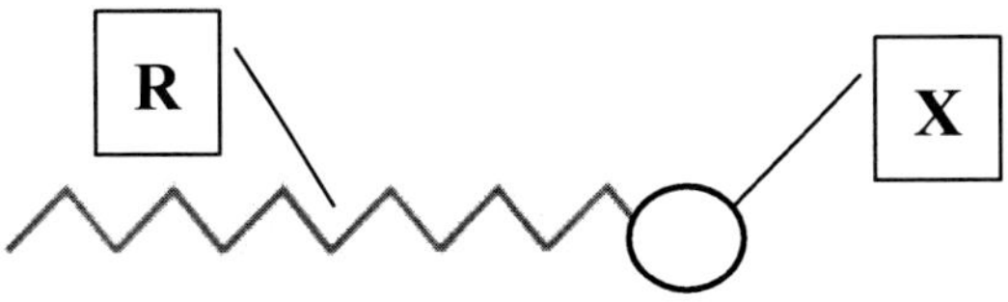

Figure 1. Scheme of surfactant's monomer molecular structure.

Surfactants are known as tensoactives, whose main characteristic is their anphipatic structure; the presence in a same molecule of two or more groups with anthagonic properties respect the solubility of them [46]. A monomer of surfactant has an structure R-X where R is a hydrocarbon chain between 8 and 18 carbon of hydrophobic characteristics and X is a polar group of ionic or non-ionic nature (Figure 1).

Depending on the nature of hydrophilic head, surfactants are classified in four groups: Cationic, Anionic, Zwitterionic (positive and negative charge) and Non-ionic [47]:

Due to the anphipatic character of surfactants, their molecules in presence of water are absorbed in interfaces, decreasing interfacial pressure and forming organized structures, called 'micelles' when surfactant concentration are higher than critical micellar concentration (c.m.c.).

Main property of surfactants, related with their capacity to micelle formation, it is analytes solubilization. Micelles, formed in aqueous solution (direct micelle) or in non aqueous solution (reverse micelle) can solubilize polar and non polar molecules. In next scheme it is showed a micellar aggregated depending on their nature, polar, semipolar or non polar (Figure 2).

Solubilization is located in different points depending on the nature of solubilized substance, but also depending on surfactant. In non-ionic surfactants the solubilization is generally in hydrophilic heads [47]. This capacity for solubilization produces that micellar solutions have important applications in fields like chemistry, pharmacological, cosmetics, etc. [48]

From the analytical point of view, this capacity to solubise solutes of different types and nature as well as their compatibility with the mobile phase in HPLC are the most important properties. Moreover, other advantages of surfactants are safety, low cost and no toxicological effects due to the biodegradability [49, 50].

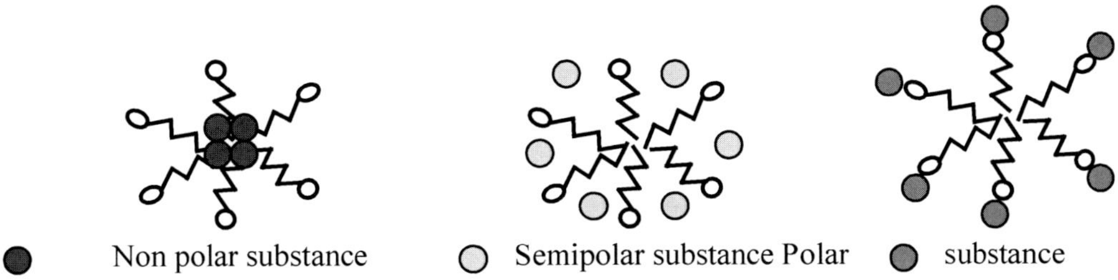

Figure 2. Micellar aggregated depending on substance nature.

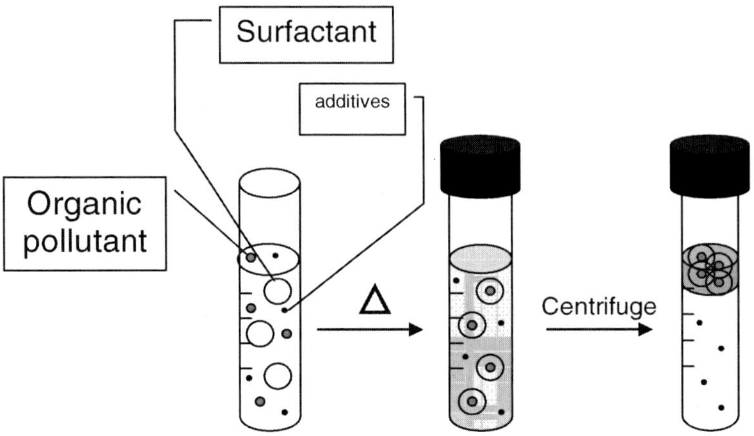

Figure 3. Squeme of the Cloud-Point process.

For liquid samples, the use of extraction and preconcentration steps based on phase separation using the cloud point methodology offers a convenient alternative to more conventional extraction systems. Aqueous solutions of some surfactants have been used in cloud point extraction (CPE) of different species prior to their determination by several techniques [51-53]. The small volume of the surfactant-rich phase obtained in CPE process allows to extract and preconcentrate the analytes in one step, prior to gas or liquid chromatographic analysis (Figure 3).

For solid samples, using the combination of the surfactant capacity to solubilise different species and the energy supplied by microwave assisted extraction, hereinafter microwave assisted micellar extraction (MAME), it can be applied to the extraction of different compounds present in solid samples [54], which always offers the benefit of low cost, easy handling, reduced toxic effects and compatibility with aqueous-organic mobile phase in HPLC (Figure 4).

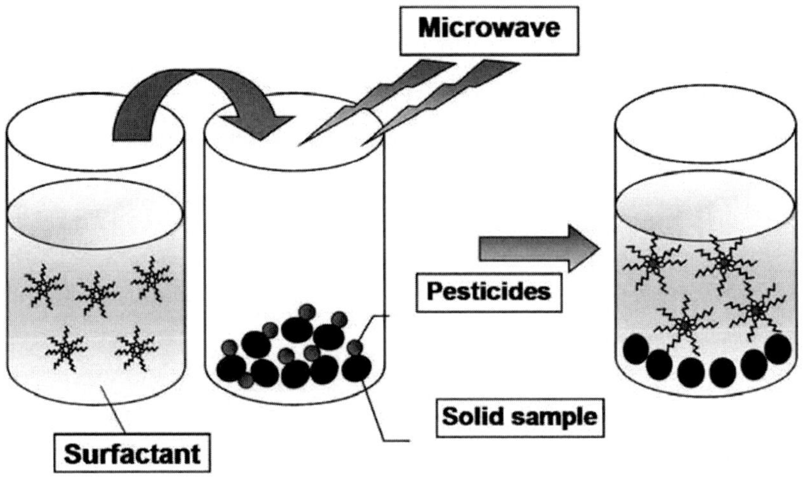

Figure 4. Squeme of Microwave Assisted Micellar Extraction process.

The MAME procedure have been applied to the extraction of several compounds as phenols [55], organochlorine and organophosphorous pesticides [56,57] present in different solid environmental samples such as marine sediments, soils and biological samples.

Determination of Pesticides by Using Cloud Point Extraction (CPE)

Introduction

When a micellar solution of a non-ionic surfactant is heated, it become turbid over a narrow temperature range, which is referred to as its cloud point temperature [58]. Above the cloud point temperature, such solutions separate into two isotropic phases, one of which will contain a surfactant-rich phase with a small amount of water, "surfactant rich phase", which is separated from the bulk aqueous solution, "aqueous phase", in which the surfactant concentration will be approximately equal to the critical micelle concentration of the non-ionic surfactant present. The phenomenon is reversible and, upon cooling, a single phase is again obtained. Table 4 shows the main characteristics of the non-ionic surfactants most used in CPE.

The solubilisation of organic material in micelles and subsequent cloud point extraction technique thus offers a convenient alternative to conventional liquid-liquid extraction that uses organic solvents. The extraction and preconcentation of organic compounds by CPE prior to their analysis by gas or liquid chromatography has been used for different authors [51-53].

Table 4. Characteristics of non-ionic surfactants [from ref. 54]

Surfactant	**Commercial name**	**CMC**[a]	**Cloud point (°C)**[b]
Polyoxyethylene 4 lauryl ether	Brij 30 ($C_{12}E_4$)	0.064	2.0
Polyoxyethylene 6 lauryl ether	$C_{12}E_6$	0.068	50.3
Polyoxyethylene 9 lauryl ether	Polidocanol ($C_{12}E_9$)	0.069	86.6
Polyoxyethylene 10 lauryl ether	Pole ($C_{12}E_{10}$)	0.090	77.0
Polyoxyethylene 8 dodecanol	Genapol X80 ($C_{12}E_8$)	0.050	75.0
Polyoxyethylene 7.5 nonylphenyl ether	Ponpe 7.5	0.085	5.0
Polyoxyethylene 9.5 octylphenyl ether	Triton X-100	0.240	64.0
Polyoxyethylene 10 cetyl ether	(Brij 56) $C_{16}E_{10}$	0.001	69.0
Polyoxyethylene 10 stearyl ether	Brij 97 ($C_{18}E_{10}$)	0.002	72.0

a) Critical micellar concentration (CMC) values given in mM/L.

b) Values given are the critical temperatures for micellar solutions in which the surfactants concentration is 1% (w/v).

The cloud point temperature depends on the structure of the surfactant and on its concentration. For a homologous series of polyoxyethylene non-ionic surfactant, the cloud point temperature increases with decreasing length of the hydrocarbon chain or increasing length of the ethylene oxide moiety. This temperature can be modified by the presence of salts, alkalis, acids, polymers, urea and other surfactants [58]. The ionic strength can alter the cloud point temperature as well as facilitate the separation of the two phases by altering the density of the bulk aqueous phase. It has been shown, for example, that for Genapol X-080, the cloud point temperature obtained when NaCl is added in a 5% w/v concentration, is lesser than the temperature obtained in the absence of this salt [59].

In CPE, extraction needs to be carried out under conditions in which the preconcentration factor will be the maximum or the extraction yield will be 100%. The preconcentration factor is defined by the expression $F_c= C_s/C_{in}$, where C_s is the analyte concentration in the surfactant-rich phase after phase separation, and C_{in} is the analyte concentration in the initial solution, before the preconcentration step. This depends on the phase relationship, on the distribution constant of the analyte between the phase and on the surfactant concentration used.

Moreover, the ratio between the volume of aqueous solution to be preconcentrated and the surfactant-rich phase volume (V_w/V_s) increases with the decrease in the surfactant concentration. This shows that the smaller the surfactant concentration, the higher the preconcentration factor; but when the volume of surfactant-rich phase is small, the extraction process become more difficult, and the accuracy and reproducibility probably suffer [29,60]. In this way, a compromise must be reached so that the surfactant concentration will allow a high phase ratio and a manageable volume of surfactant-rich phase.

The extraction process can be also altered by different factors such as equilibration time, pH, concentration and nature of the surfactant and addition of salt [61]. The effect of these factors on the percentage extraction of the analytes studied therefore needs to be established.

Optimization of Variables

Surfactant Concentration

The effect of the surfactant concentration is related with the surfactant capacity for the extraction of target pesticides [62]. It depends on the kind of surfactant used and on the concentration of pesticides in the sample. The range of surfactant concentration can vary between 1-10% (v/v) [56,62].

As an example, for a mixture of organophosphorus pesticides, it was determined by measuring a series of aqueous solutions containing the mixture of target pesticides (listed in Table 5) and different surfactant concentrations ranging from 2 to 7% (v/v) for each of the surfactants under study, POLE and Genapol (X-080).

Figure 5 shows how the recovery percentages with POLE remain practically constant up to 4% (v/v), from which point onwards they drop sharply, before stabilising again round 5% (v/v). This indicates that pesticides are not extracted as efficiently at high surfactant concentrations. The Genapol X-080 concentration can therefore be considered practically not to influence on the extraction of these analytes in the range studied.

Table 5. Organophosphorus Pesticides Studied

Identification Number	*Compound*
1	Dimethoate
2	Methidation
3	Parathion methyl
4	Malathion
5	Ethoprophos
6	Parathion ethyl
7	Diazinon
8	Chlorpyrifos

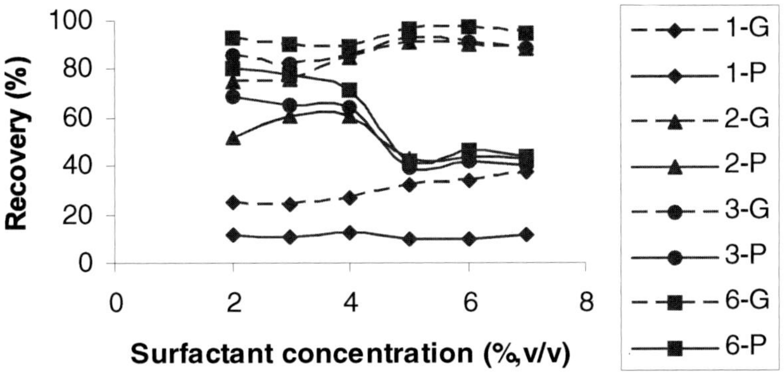

Figure 5. Effect of surfactant concentration on recovery. Dashed line, Genapol X-080 (G); continuos line, POLE (P). CPE conditions: 95°C, 15 min. Identification number in Table 5. [from ref. 56]

In view of these results and taking into account that the surfactant concentration has a proportionally inverse impact on the preconcentration factor [63], the lowest possible surfactant concentrations must be used, but within a range that allowed the samples to be adequately handled.

Equilibration Time

The recovery percentage depends on the time that the analytes have to interact with the micelles and get into their core. It has been reported that longer equilibration times (more than 30 min) do not have any significant effect on the extraction parameters and that an equilibration time of 20 min is sufficient to obtain a good extraction [59,60]. It has likewise been demonstrated that the analyte preconcentration factor and percent recovery in CPE process increases as the equilibration temperature for phase separation is progressively increased above that of the cloud point temperature of the system [64]. Likewise, an increase in the cloud point temperature leads to a slight decrease in the volume of the surfactant-rich phase [61]. As the temperature, or the equilibration time, increases, the amount of water in a surfactant-rich phase decreases and hence the volume of that phase decreases. This can be interpreted in terms of the fact that as the temperature is increased, the hydrogen bonds are disrupted and dehydratation occurs.

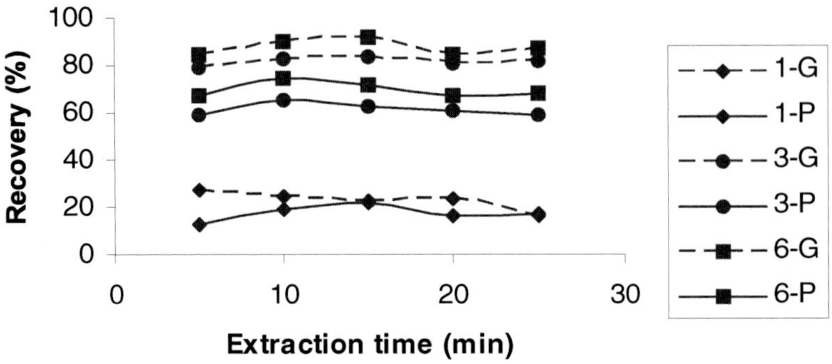

Figure 6. Effect of the extraction time on the recovery when using 2% (v/v) Genapol X-080 and 3% (v/v) POLE with 3 and 4% (w/v) of NaCl, respectively. Dashed line, Genapol X-080 (G); continous line, POLE (P). CPE conditions: 95°C, 15 min. Identification number in Table 5. [from ref. 56]

Equilibration time studies normally range between 5 and 25 min, higher extraction times than 20 min generally don't produce better recoveries.

Figure 6 shows the results obtain for the optimization of this variable on the recovery of a mixture of organophosphorus pesticides. For this case, this variable practically does not affect the recovery of the analytes, in the case of POLE and Genapol X-080 surfactants [56].

pH

The ionic form of a neutral molecule normally does not interact with the micellar aggregate as strongly as its neutral form does and a lesser amount of the analyte is therefore extracted [61, 65]. Thus, the pH should be adjusted to ensure that the neutral molecular form of the analytes is present prior to performing the CPE step. The influence of the pH on the recovery percentages is not important for those compounds that do not present ionic forms, such as organochlorine or organophosphorus pesticides. However, in the case of the phenolic derivatives (can appear as degradation products of some pesticides like 2,4-D), when working with pH modified by the addition of 1% acetic acid, the percentage extraction is better for most solutes and is the most noticeable change for more polar compounds [59].

Ionic Strength

The addition of an inert salt can facilitate the phase separation process for some non-ionic surfactant systems since it increases the density of the bulk aqueous phase [58,66]. When the salt concentration is increased, the micelle size and the aggregation number are increased and the critical micellar concentration remains constant [67]. In addition, non-polar analytes may become less soluble in the solution at higher salt concentrations and thus contribute to higher recoveries. The results obtained indicate that the addition of salt produces an increase in the extraction of the more polar solutes while the recoveries of the less polar compounds is not affected [65,68].

In the case of organophosphorus pesticides the maximum extraction for surfactant POLE is obtained with 4% (w/v) NaCl, while with Genapol X-80 practically this parameter does not affect on the recovery. It can be observed in Figure 7.

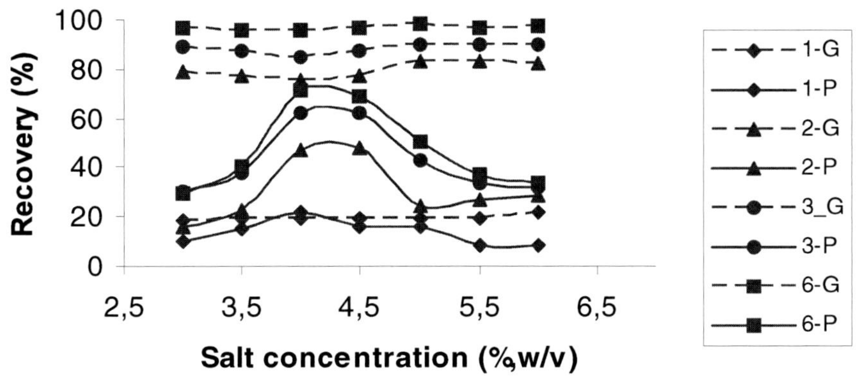

Figure 7. Effect of salt addition on the recovery when using 2% (v/v) Genapol X-080 and 3% (v/v) POLE. Dashed line, Genapol X-080 (G); continous line, POLE (P). CPE conditions: 95°C, 15 min. Identification number in Table 5. [from ref. 56]

Analytical Parameters and Applications

Most analytical applications of CPE methodology for the extraction of organic compounds make use of reversed-phase high performance liquid chromatography (HPLC). The surfactant-rich phase obtained in the extraction process is compatible with the hydro-organic phase usually employed in this chromatographic mode. From this, another important step is the optimization of the chromatographic conditions. There are two important factors to be taken into account in liquid chromatography separations: the analysis time and a good separation of the analytes.

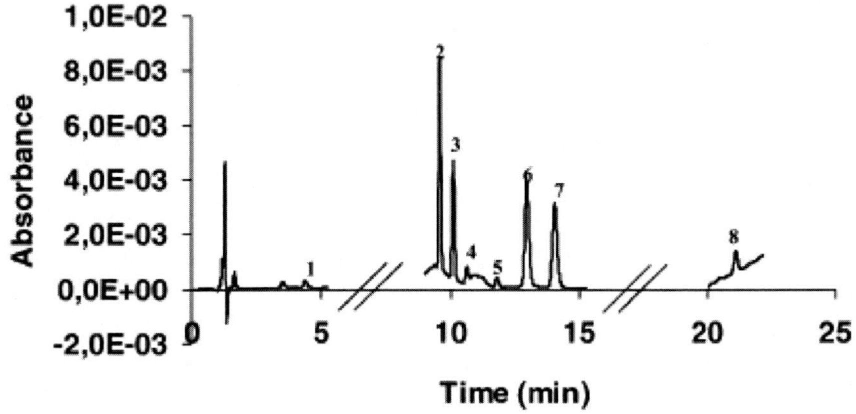

Figure 8. Chromatogram of the organophosphorus pesticides in the presence of 5% (v/v) Genapol X-080. Pesticide concentration: 2mg L^{-1}; Detection wavelength: 247 nm. Identification number in Table 5. [from ref. 56]

Table 6. Analytical parameters for the cloud point extraction method of organophosphorous pesticides [from ref. 56]

Compounds	POLE		Genapol X-80	
	R.S.D[a] (%)	L.O.D.[b]($ngml^{-1}$)	R.S.D[a] (%)	L.O.D.[b]($ngml^{-1}$)
Dimethoate	1.3	1.86	0.6	0.61
Methidathion	2.5	1.65	0.8	1.37
Parathion methyl	3.2	1.86	2.1	0.03
Malation	0.5	28.45	0.7	1.37
Ethoprophos	5.5	0.88	1.4	1.84
Parathion ethyl	1.6	2.03	2.6	0.36
Diazinon	2.8	2.96	1.0	0.47
Chlorpyrifos	2.4	3.54	2.2	2.81

a) Relative standard desviation (n = 6).
b) Limit of detection (s/n = 3).

Table 7. Recoveries obtained on the application of the CPE method using Genapol X-080 to a certified standard pesticide mixture [from ref. 56]

Compound	Recovery (%)[a]	
	Added amount (10 ng·ml^{-1})	Added amount (100 ng·ml^{-1})
Parathion methyl	81.7 ± 2.3	80.5 ± 1.1
Malation	75.0 ± 0.0	66.0 ± 0.7
Parathion ethyl	80.4 ± 1.6	84.5 ± 1.1

a) Mean ± S.D (n = 3).

The CPE methodology has been successfully exploited for the preconcentration of pesticides as a step prior to their determination by HPLC using non-ionic surfactants [56]. Figure 8, shows a chromatogram obtained for the mixture of the organophosphrus pesticides in Genapol X-080 solution, after CPE process was applied.

Table 8. Recoveries obtained on the application of the CPE method using Genapol X-080 to pesticides added to real aqueous samples [from ref. 56]

Compound	Added amount (ng ml^{-1})	Recovery (%)a	
		Depurated waste water	Groundwater
Dimethoate	10	27 ± 0.5	29 ± 1.5
Methidathion	10	86 ± 2.2	85 ± 3.1
Parathion methyl	6	92 ± 2.0	83 ± 1.2
Malation	10	87 ± 1.1	81 ± 0.6
Ethoprophos	30	43 ± 1.8	47 ± 2.3
Parathion ethyl	6	105 ± 1.2	100 ± 1.6
Diazinon	6	89 ± 1.0	84 ± 0.0
Chlorpyrifos	30	99 ± 3.6	99 ± 5.0

a) Mean ± S.D (n = 3).

Relative standard deviation values are listed in Table 6, where values ≤5.5 and 2.6% were obtained for POLE and Genapol X-080 respectively, in the case of organophosphorus pesticides extraction. The limits of detection were also calculated for each analyte using a signal-to-noise ratio (s/n = 3) [69] once the CPE method was fully applied. The results obtained are also listed in Table 6.

In order to corroborate its efficiency, the method was applied to a certified standard containing the pesticides: Parathion-methyl, Parathion-ethyl, Malathion and Diazinon (Pesticide Mix 1, Dr. Ehrenstorfer). The results (Table 7) show the satisfactory recoveries obtained using the optimised CPE procedure.

The proposed method can be applied to real samples for the determination of pesticide's concentration in them. As example, it was applied to various spiked aqueous samples from different locations in Gran Canaria island (Canary Islands, Spain). Samples of depurated waste water and groundwater were employed in this study. After checking the absence of pesticides in the samples, they were spiked with the analytes under study (6–30 $ng.ml^{-1}$) and the established method with Genapol X-080 was applied to them. The results obtained for all the samples analysed are set out in Table 8. Recoveries between 70 and 100% were obtained for the majority of cases. Only Dimethoate and Ethoprophos were extracted with recoveries <50%.

Determination of Pesticides by Using Microwave Assisted Micellar Extraction (MAME)

Introduction

In recent years, the use of microwave technology for sample preparation has been widely accepted. The reasons for the large dissemination of microwave-assisted extraction (MAE) relate to its clear advantages over more traditional technology:

- a shorter digestion time
- a supposed better recovery of volatile elements and compounds
- lower contamination levels
- minimal volumes of reagent being required
- more reproducible procedures
- better working environment

Ganzler et al. [70] were the first researchers ever to use a microwave oven to extract organic compounds from contamined soil. Since then microwave digestion methods have been developed for different types of matrices to extract organic contaminants, such as PAHs [23, 71], PCBs [72,73], pesticides [73,74], phenols [75, 76], etc. using an organic solvent as extractant.

It has been reported that the combination of MAE with micellar media as extractants (MAME) allows the extraction of different organic compounds from solid samples [77-80]. The extraction process starts with the loading of the sample into the extraction vessel, followed by surfactant addition and closing of the vessel. Microwave radiation is applied and an extraction step is initiated in order to heat the extractant to set values. When the extraction is concluded the samples are allowed to cool down to a temperature that is reasonable to handle. Prior to the analysis, a clean-up step might be needed. In most applications, simple filtration of the extract using a membrane syringe filter is enough.

Optimization of MAME Process

There are different parameters that can influence the extraction process when MAME is used. The most commonly studied parameters are: nature and volume of surfactant, extraction temperature or extraction power and duration of extraction. Factorial, central composite and orthogonal array can be used to find optimal conditions.

Nature and Volume of Surfactant

A correct choice of kind of surfactant is fundamental for obtaining an optimal extraction process. When selecting the extractant, consideration should be given to its interaction with the matrix, as well as the solubility of the analyte. Another important aspect is the compatibility of the extraction surfactant with the analytical method used for the final analysis step. The non-ionic surfactant solutions should be effective in desorbing and extracting organic compounds from solid matrices due to their power of solubilisation. In addition, the surfactant phase is compatible with the hydro-organic mobile phases usually employed in HPLC.

In some cases, extractant solvent volume may be an important parameter for efficient extraction. The surfactant volume must be sufficient to ensure that the entire sample is immersed, especially when having a matrix that will swell during the extraction process. Research has led to the conclusion that the proportion of sample in the extraction solution should not exceed 30-34% (w/v) sediment/solvent ratio. In conventional extraction techniques, a higher volume of solvent will generally increase the recovery, but in the case of MAME, a higher surfactant volume does not influence the extraction efficiency. Therefore, if the sample is full immersed in the extractant, an increase of the surfactant volume does not seem to have any influence on the extraction percentages [77,81]. Normally at extraction volumes lower than 5 ml irreproducible data are obtained (due to the insufficient covering of the extractant) and at volumes higher than 20 ml some evaporation losses take place (due to the high temperatures reached). This last effect maybe due to the high capacity of the aqueous surfactant solutions in absorbing the microwave radiation and transforming it into heat [82].

As example, for MAME extraction of organophosphorus pesticides using POLE and Genapol X-080 surfactants, the optimum extraction volume is around 10 ml [78] and for organochlorine pesticides using POLE, 8 ml [79], a very similar value in both cases.

Surfactant Concentration

Surfactant concentration is one of the most important parameters that needs to be optimized in MAME process. Surfactant concentration have to secure that the amount of micelles is enough for a efficient interaction with analytes and a correct solubilization, but it should be not too high due to possible viscosity problems.

Optimum surfactant concentration can be determined using sequential optimization steps or applying factorial designs.

For example, to study the effect of surfactant concentration on the recovery percentage of organophosphorus pesticides, several samples containing different POLE and Genapol X-080 concentrations, 1, 3, 5 and 7% (v/v), were analyzed using sequential mode. The pesticides from enriched samples were extracted with both surfactant solutions in the microwave system at 300 W during 5 min [78]. The recoveries obtained for some of the compounds studied can be seen in Figure 9.

For the extraction of organochlorine pesticides, the results of the experiment design applied and the correlation between variables with the recovery and each other show that the surfactant concentration is the variable which mostly affect to the recovery and is correlated with the microwave power. The results obtained for the case of 4,4'-DDT using a mixture of surfactants POLE and Polyoxyethylene 10 Stearyl Ether (Stearyl mixture) are shown in Figure 10 [79].

Extraction Temperature and Extraction Power

The most researched MAME parameter is the extraction temperature or extraction power, as an increase in the power involves a corresponding increase in heat. When MAME is conducted in closed vessels, the temperature may rise well above the boiling point of the solvent at atmospheric pressure, thus enhancing both extraction speed and efficiency. These elevated temperatures result in improved extraction efficiencies, since desertion of analytes from active sites in the matrix will increase. However, the effect of the temperature depend on the type of analyte as well as the type of matrix.

As in other extraction techniques, time is another parameter whose influence needs to be taken in account. Extraction times in MAME are very short compared to conventional techniques where extraction procedures require as much as 24 hours. 30 minutes are often sufficient to obtain good results in MAME. With thermolabile compounds, long extraction times may result in degradation, obtaining lower recovery percentages when increase extraction time [80].

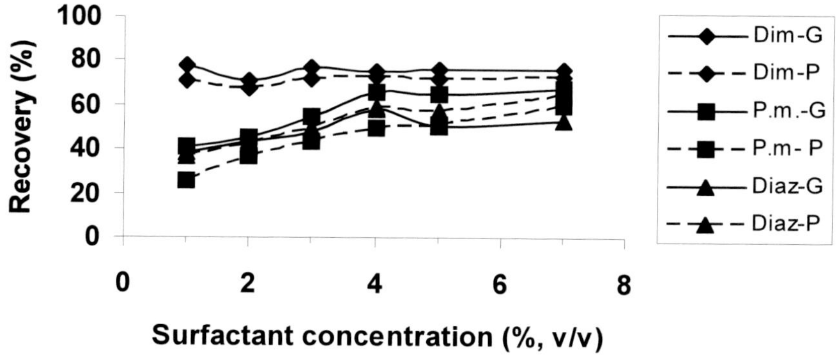

Figure 9. Effect of surfactant concentration on the recovery of the pesticides under study. Continous line (—): Genapol X-080 (G); dashed line (- - -): POLE (P). Dim: Dimethoate, P.m.:parathion-methyl and Diaz:diazinon [from ref. 78].

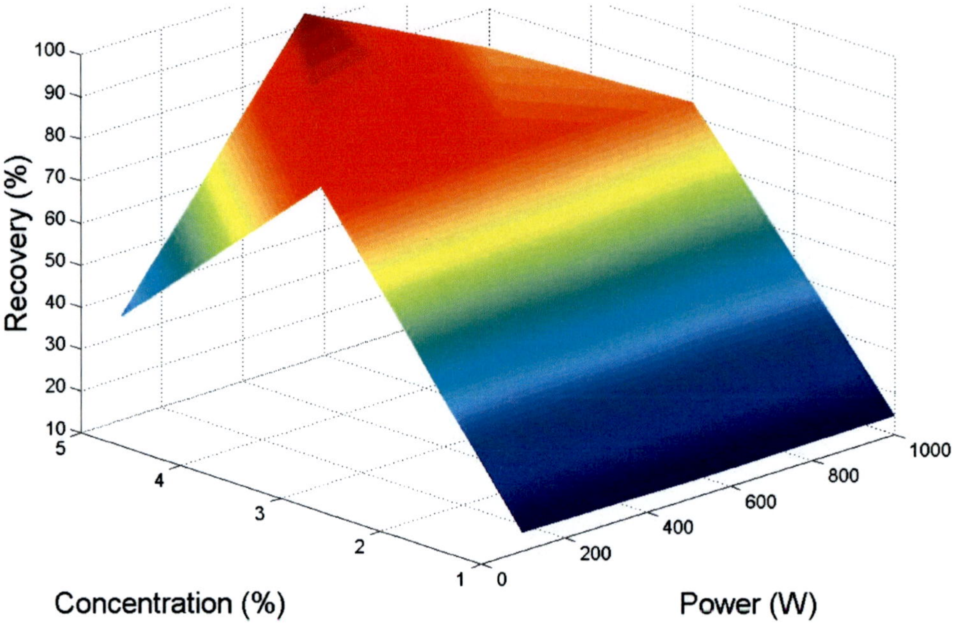

Figure 10. Response surface for the effect of surfactant concentration and power on the extraction of 4,4'-DDT using Stearyl mixture [from ref. 79].

A central composite design can be used in order to study the effect on the recovery, it allows the direct evaluation of the considered variables [83]. The values normally vary between 200 to 800 W during 2 to 14 min [78]. The optimization of both variables can be done using a 3^2 factorial design, where the recovery is the dependent variable. The experiment involves 10 run by duplicate and other variables involved in the extraction process were kept constant in their optimized values.

Figure 11a shows the response surface for 4,4'-DDT using Cetyl mixture (a mixture of POLE and Polyoxyethylene 10 Cetyl ether) as extractant, where we can observe that from 750 W for microwave power and during 14 minutes the best recoveries are obtained; for Stearyl mixture (Figure 11b) the optimum conditions are 775 W and 8 minutes .

Matrix Characteristic

The nature of the matrix in which the analytes of interest are bound can have a profound effect on the recoveries of the compounds. The different interactions that the organic pollutants present with the organic and inorganic matter of the matrix can play different roles in their solubility or diffusivity into the micellar medium.

The pH, conductivity and organic content of the sample can vary the recoveries obtained. Moreover, the texture of a soil is extremely important in the sorption process. When the particles are small they present a high superficial area thus increasing the adsorption, although this adsorption process depends on the nature of the analytes. In this sense, a high variability can be observed in the recoveries obtained with the different granulometry of the soil samples. In general, it can be said that the proposed method is applicable in the sense that it enables the extraction of the pesticides under study even if the recovery efficiency depends on the characteristics of the soil as well as the surfactant to be used.

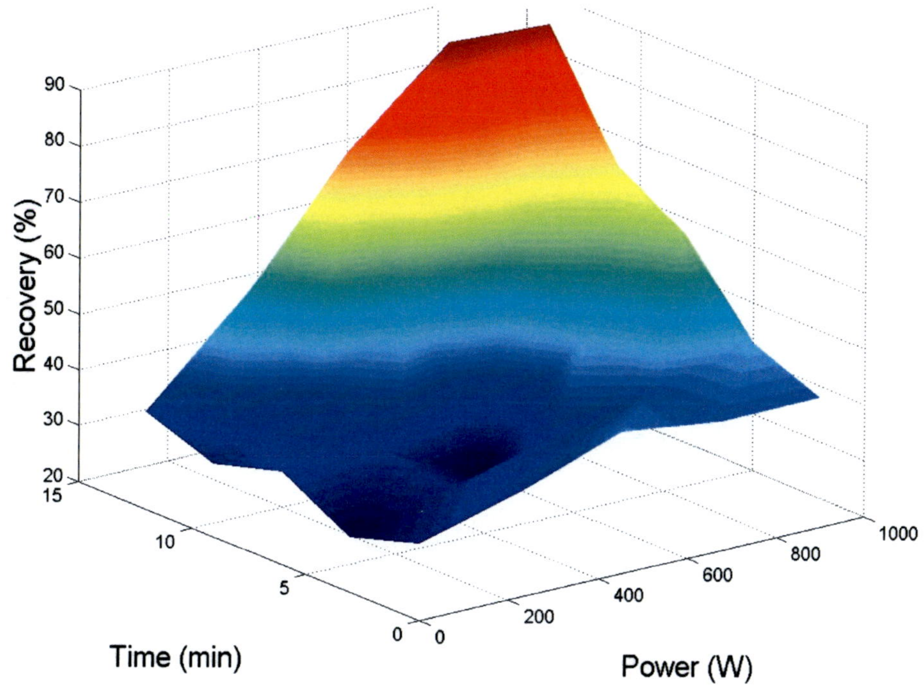

(a)

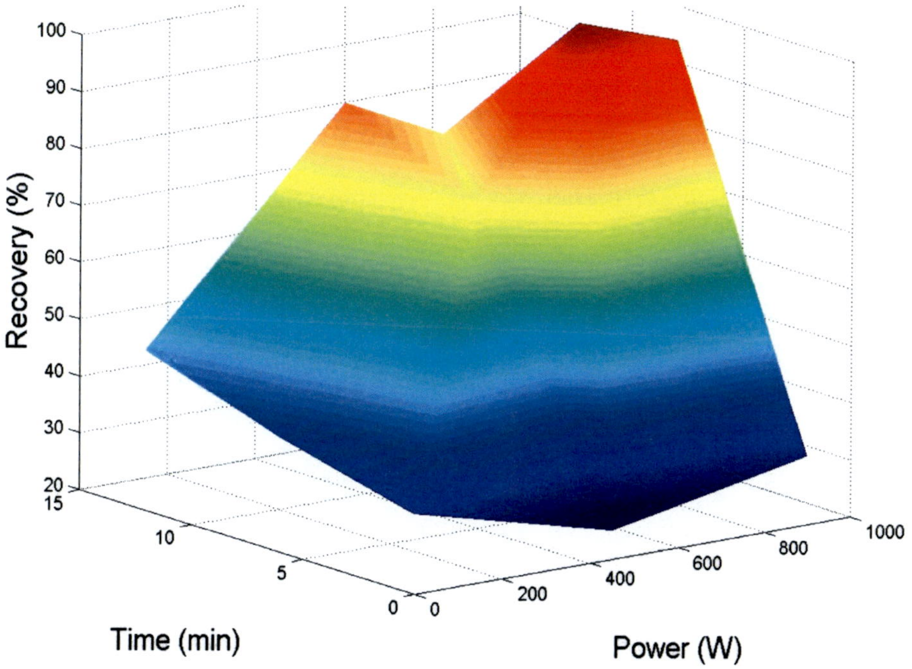

(b)

Figure 11. Response surface for the effect of power and time on the extraction of 4,4'-DDT using Cetyl mixture (a-left) and Stearyl mixture (b-right) [from ref. 79].

Table 9. Soil characteristics (from ref. [79])

Soil	Agricultural Uses	Particle size (%)				pH	Organic matter (%)	Conductivity (μS/cm)
		0.3 mm	0.2 mm	0.15mm	≤0.1 mm			
Tafira	Garden	44.4	17.2	14.6	23.8	8.3	4.8	292
San Roque	Bananas	52.1	20.0	13.2	14.6	7.86	12.54	3266
Sta Brigida	Pine Forest	56.5	17.8	12.8	12.8	5.9	3.9	483
Valleseco I	Potatoes	36.33	16.43	15.19	32.06	4.84	4.4	202
Valleseco II	Potatoes	25.16	13.17	14.38	47.29	3.91	6.21	100

Table 10a. Recoveries obtained after MAME procedure for organochlorine pesticides in five different soils using Cetyl mixture* [from ref. 79]

Soil	4,4'-DDD	Dieldrin	4,4'-DDT	2,4'-DDT	4,4'-DDE	Aldrin
Tafira	99.1	66.2	67.7	58.6	65.0	63.0
San Roque	69.4	44.7	78.0	59.4	67.1	52.6
Sta Brigida	96.8	80.7	84.0	75.1	73.1	63.8
Valleseco I	78.3	73.3	76.5	66.3	65.3	64.4
Valleseco II	65.2	72.3	61.7	59.4	56.1	60.2

* Mean of three determinations.

Table 10b. Recoveries obtained after MAME procedure for organochlorine pesticides in five different soils using Stearyl mixture* [from ref. 79]

Soil	4,4'-DDD	Dieldrin	4,4'-DDT	2,4'-DDT	4,4'-DDE	Aldrin
Tafira	103.6	78.6	101.8	62.8	71.8	79.4
San Roque	101.3	46.1	94.4	86.2	76.0	80.0
Sta Brigida	84.7	80.8	78.8	62.4	66.0	65.6
Valleseco I	55.2	56.3	100.7	83.5	44.8	57.5
Valleseco II	54.3	50.2	98.3	68.4	49.2	72.8

* Mean of three determinations.

Appling the method to different soil samples containing a mixture of organochlorine pesticides can be observed that recoveries vary depending on the soil characteristics (Table 9) and the surfactant used as extractant [79]. The results obtained are shown in Tables 10a and 10b.

The soils with high organic matter have good recoveries because the surfactant can also extract humic substances which are linked to the pesticides [84]. These recoveries decreases when the temperature is too high (observed in the most extremely conditions in the experimental design).

As we commented before, texture of a soil is quite important in the sorption process [85]. Clay is by far the most adsorbent of the three main soil textures (clay, silt and sand) due to its small particle size, high surface area, and high surface charge [86]. This has a relevant effect on the recovery; the higher clay percent a soil has, the lower the recovery is so; in this case, the Valleseco II soil has a higher amount of particles with smaller size and therefore, lower recoveries are obtained.

The effect of ageing the samples on the recovery of the analytes is a known phenomenon [87]. Decreasing recoveries resulting from aging of matrices is frecuent [88,89]. The analytes present in recent soil samples are more easily extracted than those that have had a longer contant time. This can be explained according to whether the analytes are incorporated by adsorption (short periods) [90] or by sequestration (long periods). The former phenomenon occurs at the early stages of sorption, where H-bonding and Van der Waals forces prevail. On the other hand, sequestration involves sorption at remote microsites within the soil matrix [91]. As the contact time increases between the analyte and the matrix the general recuperation of all the analytes under study was reduced which could be explained by the greater interaction between them [92]. This effect could also be caused by a process of degradation because as is well known the degradability of this family of compounds is far greater than that of organochlorine pesticides. Among the different pathways of organophosphates decomposition (hydrolysis, photolytic oxidation, microbial transformations, etc.) [93,94], hydrolysis is the most common degradation process which can occur at several reactive centres in a given organophosphorus pesticide molecule [95], but also can occur different degradation mechanisms including the presence of humic acids, metal oxides and the soil structure [94-97].

In the case of the extraction of organochlorine pesticides, MAME procedure was applied to four soils for different time periods after conditioning. Figure 12a shows the recoveries obtained for the different pesticides in the case of Valleseco I soil using Cetyl mixture with the time aging and Figure 12b shows the results obtained in the case of San Roque soil using Stearyl mixture with the same aging time.

It can be seen that the recoveries decrease with the aging time for all compounds. Only in the case of dieldrin using Stearyl mixture, the recovery remains practically constant with the time. Moreover, none of the metabolites of DDT increases with the time then we can assert that degrading processes were not happen in these aging time. These results indicate that the developed method leads determine the presence of pesticides in real samples after a contact time and show the viability of the method as alternative to the conventional ones.

Analytical Parameters and Validation

Separation and detection of pesticides have been developed using MAME procedure coupled with HPLC. The calibration curves can be obtained by injecting standard solutions containing a known concentration of the pesticide mixture and the surfactant chosen into the chromatographic system. This method provides optimum detection limits and reproducibility in soil matrices. For organophosphorus pesticides, on the application of this process a linear relationship is obtained in the interval 100–2500 ng ml^{-1}, when POLE or Genapol X-80 are used as extractants, with high correlation coefficients (0.999) for all pesticides.

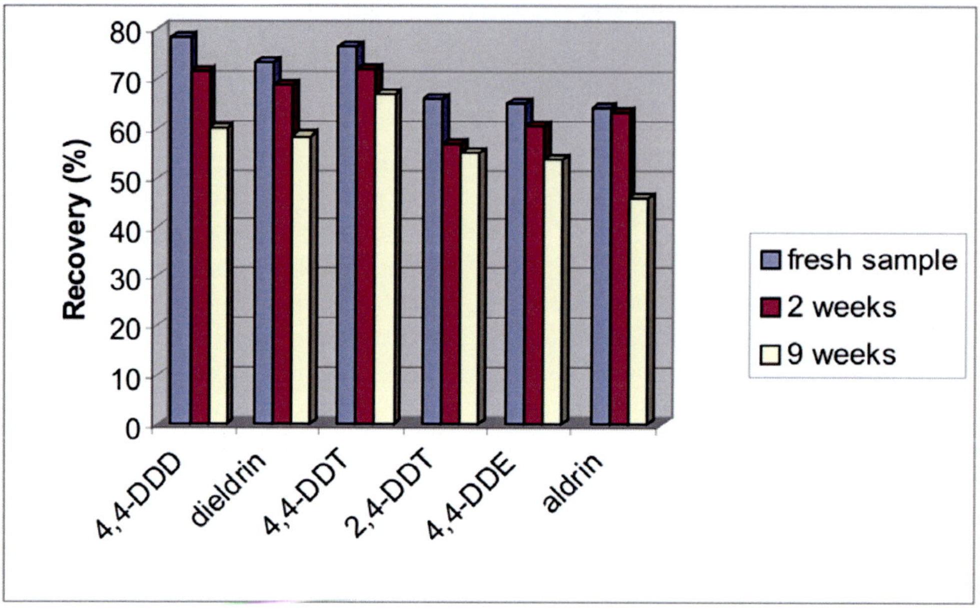

(a)

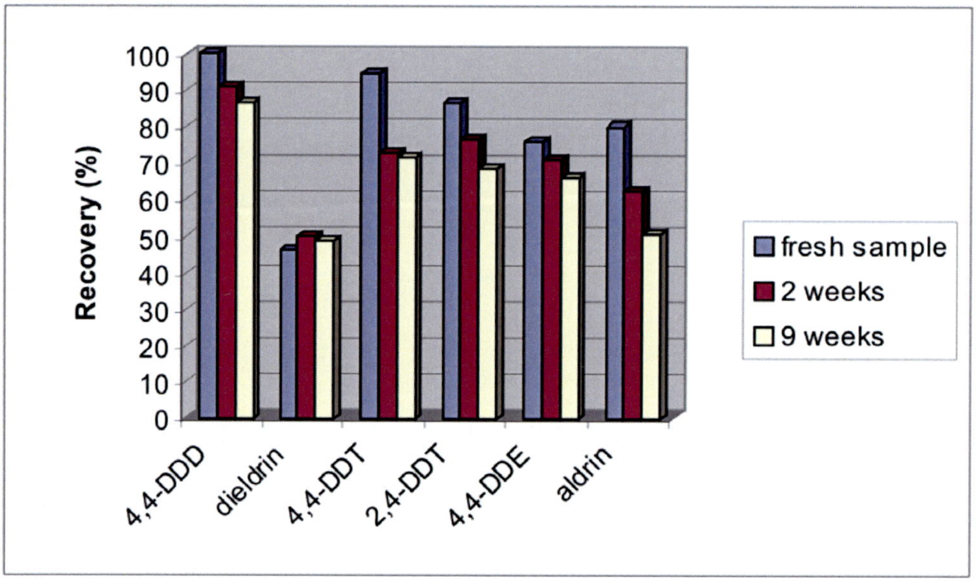

(b)

Figure 12. Recoveries obtained after MAME procedure for organochlorine pesticides for three different aging times for Valleseco I soil using Cetyl mixture (a-left) and San Roque soil using Stearyl mixture (b-right) [from ref. 79].

Relative standard deviation lower than 2.1 and 2.6% for Genapol-X080 and POLE repectively and LODs lower than 0.2 in both cases were obtained (Table 11). Figure 13 shows the chromatogram of the pesticides mixture extracted from a soil sample under the optimized conditions for POLE.

Table 11. Analytical parameters for the determination of organophosphorus pesticides under study using MAME procedure [from ref. 78]

Compound	*POLE*		*Genapol X-080*	
	RSD[a] (%)	**LOD[b] (ng·ml^{-1})**	**RSD[a] (%)**	**LOD[b] (ng·ml^{-1})**
Dimethoate	0.4	0.2	1.8	2.4
Methidathion	2.2	4.5	2.1	3.0
Parathion/methyl	1.1	2.1	1.2	4.8
Malathion	2.2	14.3	1.1	11.3
Ethoprophos	2.6	95.0	1.0	0.8
Parathion-ethyl	1.6	0.7	1.6	1.8
Diazinon	1.2	2.5	0.8	0.9
Chlorpyrifos	0.3	6.7	1.1	0.2

a) Relative standard deviation (n = 6).

b) Limit of detection.

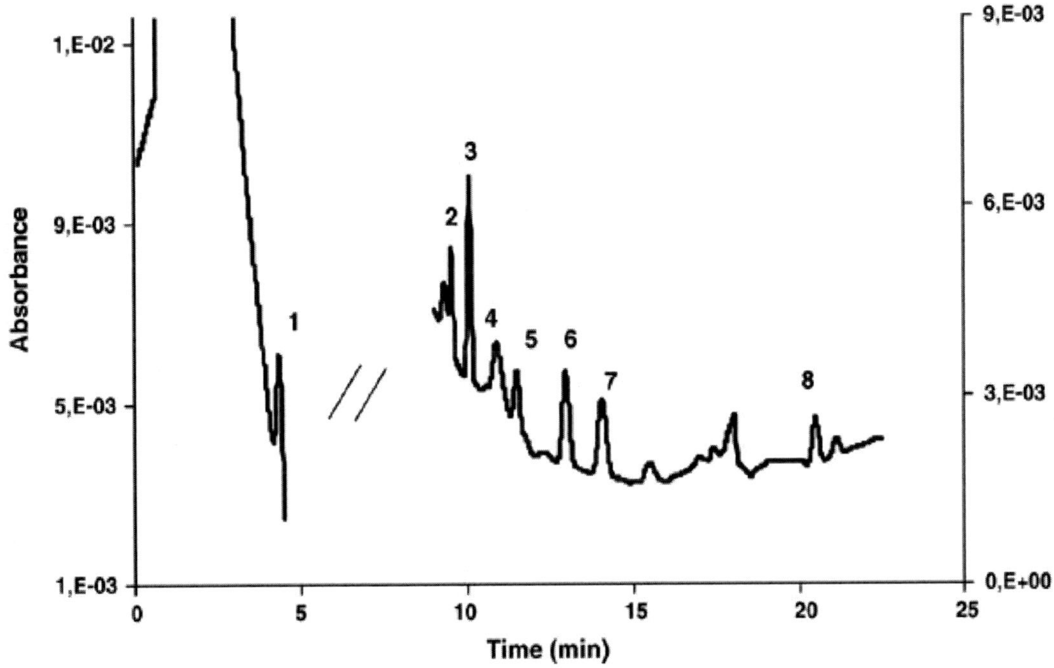

Figure 13. Chromatogram of the organophosphorous pesticide mixture extracted from soil sample n° 1 with POLE at the optimized conditions. Identification number in Table 5 [from ref. 78].

In the case of organochlorine pesticides, a good success was obtained appling MAME-HPLC method with POLE-Cetyl and POLE-Stearyl mixtures as extractants, with RSDs values between 5.2-8.4% for POLE-Cetyl and 5.2-9.7 for POLE-Stearyl and LODs of 86.4-806.4 ng·ml^{-1} and 150.8-734.0 ng·ml^{-1}, respectively [79]. In Figure 14 we can observe the chromatogram obtained for the mixture of organochlorine pesticides extracted of a spiked soil using Steary mixture as extractant.

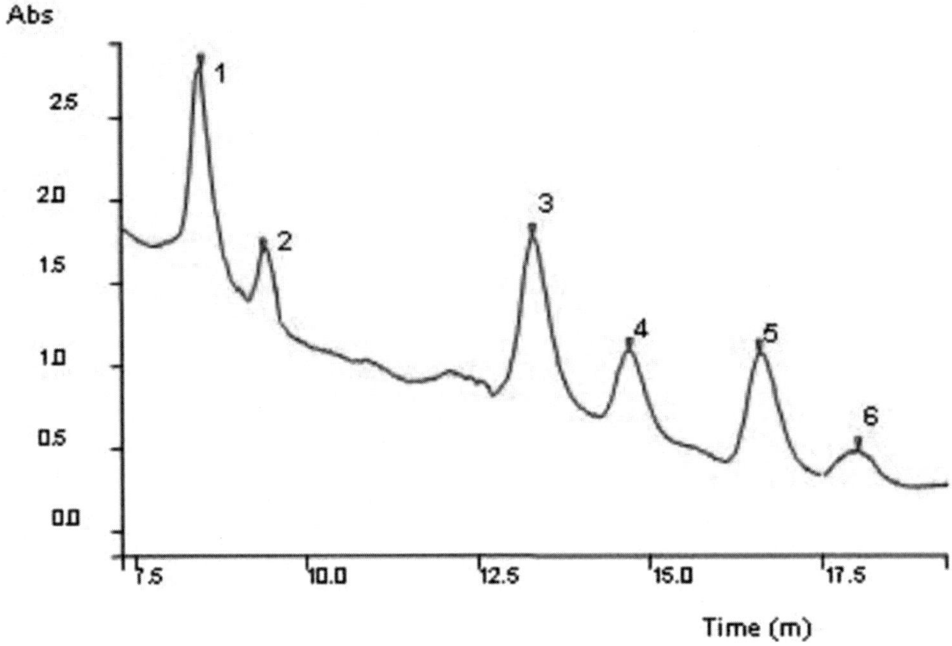

Figure 14. Chromatogram of an extract of a spiked Tafira soil after MAME procedure (*1*- 4,4'-DDD; *2*- Dieldrin; *3*- 4,4'-DDT; *4*- 2,4'-DDT; *5*- 4,4'-DDE; *6*- Aldrin) [from ref. 79].

Recoveries obtained with spiked compounds may not be representative of those found with native compounds. Spiked analytes are generally lightly coated on the surface of the matrix whereas native ones can be strongly adsorbed inside the porous matrix. This can be explained by the diffusional and the kinetic limitations of the sorption process, and the several interactions which may have been simultaneously established between native analyte and the matrix [98]. This is the reason why it is necessary to validate the extraction procedure with certified reference matrices.

Due to the lack of a certified material available containing target organophosphorus pesticides, the developed extraction method was applied in a soil sample enriched with a certified mixture, Pesticide Mix 1, EPA Method 914, and the results obtained are shown in Table 12.These results were compared with those found using the traditional Soxhlet extraction procedure as proposed by the EPA in the 3540 Method [82], finding an important similarity of results in both methods.

For organochlorine pesticides, a contaminated certified reference soil from an agricultural region of the Western United States (CRM804-050) was used. Table 13 reports the pesticides present in the certified soil, the corresponding certificated values, the confidence intervals and the concentration of each analyte that was obtained using the MAME procedure with Stearyl mixture. The concentrations obtained, with the proposed method, falling within the certified range for all compounds analyzed, indicate that the proposed extraction procedure is suitable for this kind of matrix.

Table 12. Application of MAME procedure and Soxhlet extraction to a soil sample containing a certified mixture of organophosphorous pesticides, (Pesticide Mix 1, EPA Method 914)* [from ref. 78]

	MAME		*Soxhlet*	
Compound	**Added (μg g^{-1})**	**Found (μg g^{-1})**	**Added (μg g^{-1})**	**Found (μg g^{-1})**
Parathion/methyl	1.50	1.39 ± 0.02	2.00	2.04 ± 0.02
Malathion	1.50	1.27 ± 0.04	2.00	1.89 ± 0.11
Parathion-ethyl	1.50	1.49 ± 0.12	2.00	1.85 ± 0.02
Diazinon	1.50	1.46 ± 0.02	2.00	1.64 ± 0.05

* Mean of three determinations.

Table 13. Extraction of organochlorine pesticides compounds using MAME with Stearyl mixture in a certified reference soil[a] [from ref. 79]

Compound	**Reference Value[b]**	**RSD (%)**	**Confidence Interval**	**Concentration found**
4,4'-DDD	1.531	0.476	1.294 – 1.767	1.733 ± 0.009
Dieldrin	1.863	0.655	1.539 – 2.186	2.242 ± 0.084
4,4'-DDT	1.060	0.275	0.926 – 1.195	1.079 ± 0.015
4,4'-DDE	1.520	0.410	1.325 – 1.715	1.502 ± 0.022

[a] All values are expresed in μg·g^{-1}.

[b] The pesticides values in the sample were certified by USEPA SW846 (3rd edition). Extraction Methods 3540A/3541 (soxhlet), 3550 (sonication) and Analysis Method 8081.

Determination of Pesticides by Using Coupling MAME-SPME and MAME-SPE

Coupling MAME with Solid Phase Microextraction (MAME-SPME)

Solid phase microextraction (SPME) is a new variation of extraction techniques which has been used mainly for the analysis of pollutants in environmental water samples [99], but this methodology can be coupled with other techniques, as the microwave extraction, for the analysis of soil samples [100]. SPME allows the simultaneous extraction and preconcentration of analytes from a sample. The principle of SPME is equilibration of the analytes between the sample matrix and the organic polymeric phase usually coating a fused-silica fibre. The amount of the analyte absorbed by the fibre is proportional to the initial concentration. Figure 15 shows the scheme of SPME process.

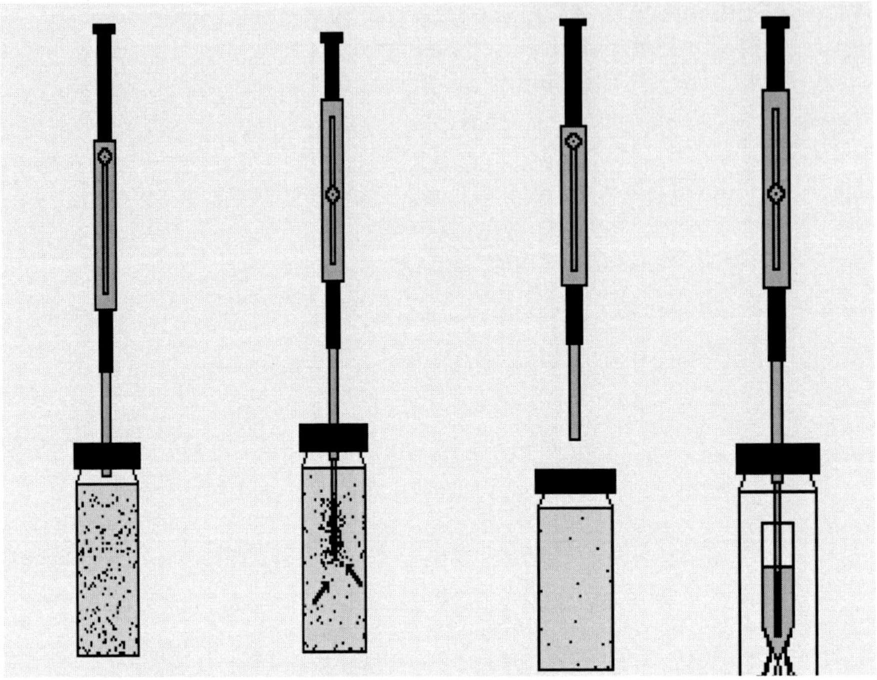

Figure 15. Scheme of the solid phase microextraction (SPME) process.

Normally the application of SPME has been performed in combination with gas chromatography, however, it is possible coupled with HPLC [101]. The advantage using HPLC is that compounds with low volatility or thermal labile compounds can be analysed. An organic solvent (static desorption) or the mobile phase (dynamic mode) is used to disorb the analytes from the SPME fiber.

One application of solid phase microextraction (SPME) procedure can be a clean-up and preconcentration steps applied to extracts obtained with other extraction systems such as microwave assisted micellar extraction (MAME). Using HPLC-UV as determination system, the coupling MAME-SPME has been used for the analysis of five organochlorine pesticides (4,4'-DDD, 4,4'-DDT, 2,4'-DDT, 4,4'-DDE, Dieldrin and Aldrin) in agricultural soil samples [57,102]. A non ionic surfactant, POLE, was used, and the different variables for the optimization of MAME and SPME procedures were studied. This method was successfully applied to the determination of these pesticides in several kinds of agricultural soil samples, obtaining lower detection limits than determining MAME extract directly.

Microwave Assisted Micellar Extraction

Once the soil sample is transferred to the vessel, the optimized volume of surfactant is added and the solid sample is subjected to the MAME process [15, 78]. The vessels are closed for avoiding the loss of extract during the extraction process at microwave power and time optimized; after this, the vessels are allowed to cool first before being opened.

Optimization of SPME Procedure

For MAME-SPME extraction, a fraction of MAME extract is diluted with water and introduced in a vial for the SPME extraction under the optimized conditions [103]. The desorption of the compounds is done in a glass vial with a small volume of desorption solvent (100 μl of methanol) during several minutes.

The efficiency of analytes extraction by SPME can vary widely depending upon the matrix nature, the time period of absorption, temperature, salt addition to the sample and pH [104,105]. Figure 16 shows different variables optimized for the analytes 4,4'-DDT, 4,4'-DDD and 4,4'-DDE as example [57]. One of the most important steps is the determination of the time needed until it reaches equilibrium between the sample matrix and the coating of the fiber. Most of organochlorine pesticides reach the equilibrium after about 40 min; with this time, it is obtained a good reproducibility and an acceptable analysis time (Figure 16a).

Desorption is done normally using static mode by immersion of the fiber into the vial during enough time for producing the efficienct analytes desorption. This time normally is studied between 1 and 15 min, longer times don't give better recoveries. In the studied case, it was proved that a time of 4 min was enough for obtaining a desorption of target analytes (Figure 16b).

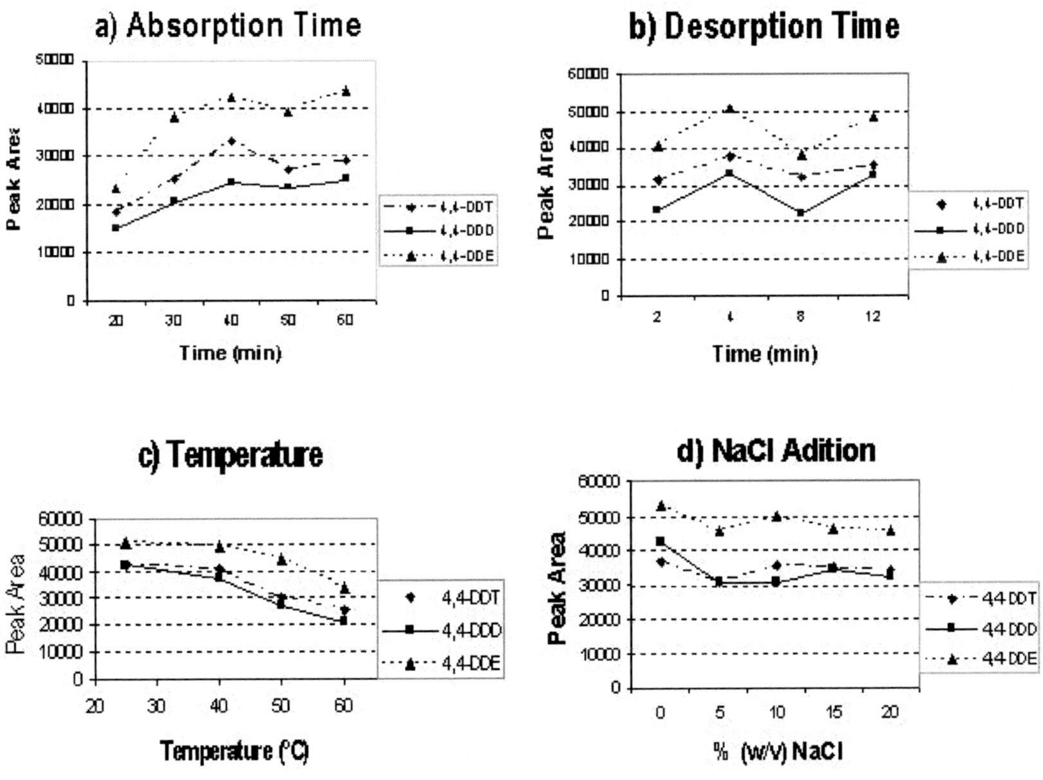

Figure 16. Optimized variables for SPME extraction using POLE coupling with PDMS/DVB fiber [from ref. 57].

Extraction temperature can play an important role on the absorption of analytes, because it influences the mass transfer rates and the partition coefficients of the analytes. A high temperature could increase the diffusion rate of the analytes, however, absorption is generally an exothermic process and the amount of analytes absorbed could decrease with a temperature increase [106]. For that, it is necessary to search the optimum extraction temperature. In the case of organochlorine pesticides, for temperatures lower to 40 °C extraction efficiency is constant, but to higher temperatures, it decrease due to the exothermic process (Figure 16c).

Another extraction parameter which has a well-establish effect in the extraction procedure is the ionic strength of the sample by addition of salt. An increase of ionic strength can improve extraction efficency due to the reduction of the solubility of least polar analytes. However, an increase of salt concentration can generate damage on the fiber by salt deposition. Figure 16d shows the effect of ionic strength on the peak area of target analytes, between 0% and 20% (w/v) NaCl concentrations. It can be seen that for these compounds salt addition doesn't improve the analyte absorption.

Analytical Parameters and Validation

Liquid chromatography with UV detection can be used in the analysis of organochlorine pesticides after their MAME-SPME extraction. Limits of detection reached using this process are in the range 56-96 $ng \cdot g^{-1}$ for all compounds studied and relative standard desviations are lower than 9% in all cases (Table 14).

When this process MAME-SPME is applied to the determination of a mixture of pesticides 4,4'-DDD, 4,4'-DDT, 4,4'-DDE and Dieldrin present in a certified soil sample (CRM804-050), recoveries obtained for all compounds analyzed, falling within the certificated range.

Applications

Soil characteristics could also have a great influence on the recovery. However, if we applied the optimized method MAME-SPME-HPLC to the extraction and determination of organochlorine pesticides mixture in different types of agricultural soils, with different characterisitics, extraction efficiency is satisfactory for most compounds in study, independently of the characteristics of the matrix (Figure 17).

Table 14. Analytical parameters for the determination of organochlorine pesticides using MAME-SPME procedure [from ref. 57]

Compound	RSD (%)*	LOD ($ng \cdot g^{-1}$)
4,4'-DDD	7.4	68
Dieldrin	5.5	96
4,4'DDT	7.3	72
2,4'DDT	8.4	84
4,4'-DDE	7.9	56

* n=6

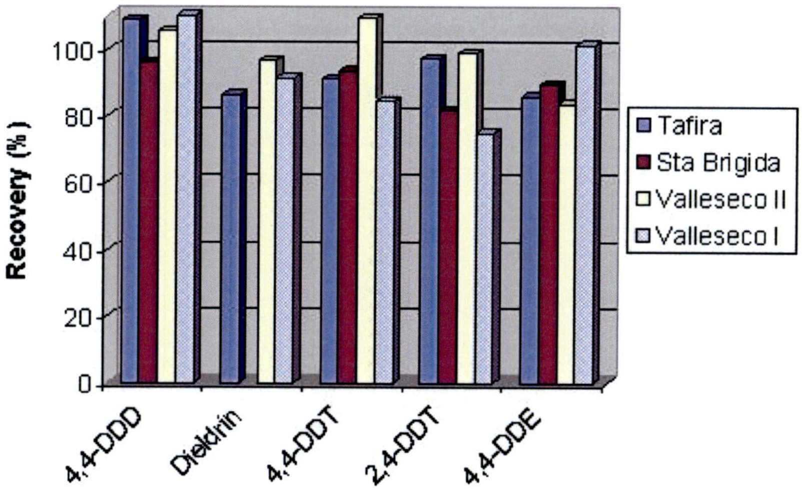

Figure 17. Recoveries obtained after MAME-SPME procedure for organochlorine pesticides for four different soils using POLE as extractant [from ref. 57].

This method has been also applied to different mud samples with satisfactories results [107]. Therefore, the versatility and less complex sample's preparation make MAME-SPME a suitable process for solid samples.

Coupling MAME with Solid Phase Extraction (MAME-SPE)

During extraction step, many interfering components are co-extracted from solid samples together with target analytes which could interfere on the final quantification. Removal of interfering substances can be made by SPME [100] as we studied before, and by SPE [108] process, producing also an intensification of the analytes signal by means of extract clean-up and preconcentration.

With the aim to obtain cleaner extracts and lower detection limits, it can be applied the coupling between MAME procedure and solid phase extraction (SPE). This coupling has a simply handling and allows simultaneous analysis. MAME-SPE procedure has been applied to the extraction and determination of organochlorine pesticides from seaweed [109] and mollusc samples [110] with very satisfactories results.

MAME-SPE extraction can be divided in two steps: microwave assisted micellar extraction (MAME) to obtain MAME extract under the optimum conditions (optimized before) and the application of solid phase extraction (SPE) clean up procedure to this extract.

Solid phase extraction (SPE) is widely used as preparation process based on the selective retention of interest's compounds in a solid sorbent and then eluted with an organic solvent. But also can be coupled with other extraction process as clean-up step (see Figure 18).

In SPE procedure, before the extraction it is necessary conditioning the cartridge. After absorption, a wash step is done for reducing surfactant remains with ultra-pure water. The retained analytes are eluted with methanol. Finally, 50 μl of extract were injected into chromatographic system.

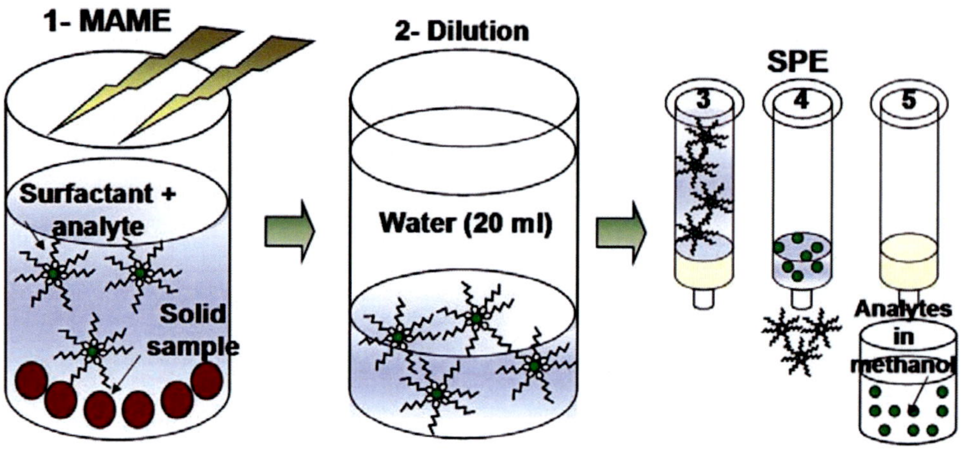

Figure 18. Scheme of the coupling MAME-SPE (1- MAME extract, 2- Extract's Dilution, 3- SPE Absorption, 4- SPE Clean step, 5- SPE Desorption) [from ref. 110].

There are different parameters that may influence in the clean-up and preconcentration step such as type of sorbent, sample volume, composition and volume of washing solution and type and volume of desorption solution.

Optimization of MAME-SPE Procedure

It is necessary to optimize and compare different SPE cartridges for obtaining the best extraction efficiencies [111]. Several papers reported on the evaluation of a number of stationary phases that can be used based on compound's characteristics. For example, in the case of organochlorine pesticides, results obtained (as peak area) for different SPE cartridges show that the best recoveries were obtained with C_{18} and Envirelut-Pesticides cartridges. However, the last one was selected because it gives a better baseline in the chromatogram.

For MAME-SPE procedure it is necessary to study the amount and concentration of MAME extract which are going to be use in SPE step [112]. Sample must be diluted with water to facility the pass of the sample through the solid phase and minimize the competition for analytes between surfactant and cartridge sorbent.

Figure 19 shows a comparison between relative peak areas obtained using 5 ml of MAME extract (5%, v/v of surfactant) without and with dilution with different volumes of ultra-pure water. It can be observed that, results were quite similar for each analyte using different final volume samples. However, from a volume of 25 ml the ratio signal-noise obtained in the chromatogram was better and this volume was selected as the optimum for further experiments.

In order to increase preconcentration factor in MAME-SPE procedure, methanol on the desorption volume can be reduced until a value which would allow high sensitivity but with an acceptable reproducibility of the method [109,110].

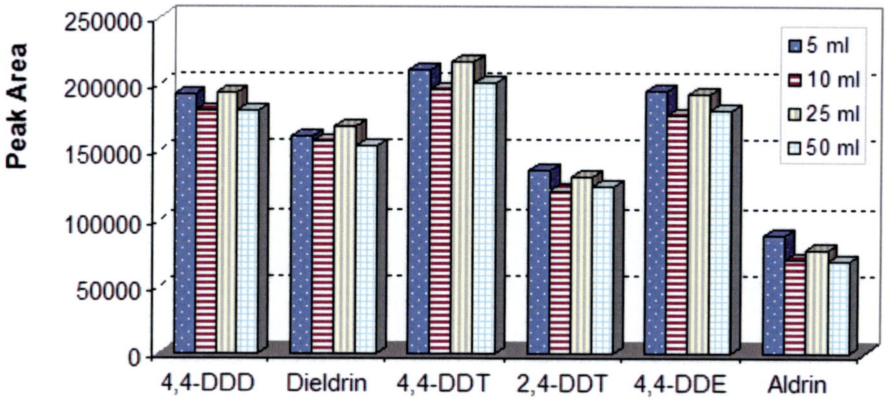

Figure 19. Study of absorption volume, represented as final volume, used in MAME-SPE process [from ref. 109].

Analytical Parameters

After MAME-SPE extraction, analytes can be determined by HPLC. This procedure has been applied to the determination of organochlorine pesticides in seaweeds [109] and molluscs [110]. In both cases, average recoveries were in the range between 78.8-107.7% for seaweeds and 68.2-100.4% for molluscs. Precision was determined by reproducibility studies, expressed by relative standard desviation (RSD). Reproducibility was measured by analysis of six spiked seaweed samples under optimum conditions. RSD values were below 10% and limits of detection of all studied compounds were calculated [69], obtaining values ranged from 2 to 38 $ng{\cdot}g^{-1}$ for seaweeds and from 6 to 38 $ng{\cdot}g^{-1}$ for molluscs. Results obtained for seaweeds are reported in Table 15.

Characteristic chromatograms for MAME-SPE of spiked (4 $\mu g{\cdot}g^{-1}$ for 4,4'-DDD, 4,4'-DDT, 2,4'-DDT and 4,4'-DDE and 8 $\mu g{\cdot}g^{-1}$ for Aldrin and Dieldrin) (a) and non-spiked (b) seaweed samples are shown in Figure 20.

Table 15. Analytical parameters in the determination of organochlorine pesticides using MAME-SPE [from ref. 109]

Compound	Recovery (%)	RSD[a] (%)	LOD[b] ($ng{\cdot} g^{-1}$)
4,4'-DDD	92.9	3.5	12
Dieldrin	100.5	3.3	22
4,4'DDT	92.6	2.8	2
2,4'DDT	91.2	3.6	6
4,4'-DDE	107.7	5.3	2
Aldrin	78.8	4.6	38

a) Relative Standard Desviation (n=6).

b) Limit of Detection.

c) Concentration of spiked seaweed: 4 μg/g for 4,4'-DDD; 4,4'-DDT; 2,4'-DDT and 4,4'-DDE, and 8 μg/g for Aldrin and Dieldrin.

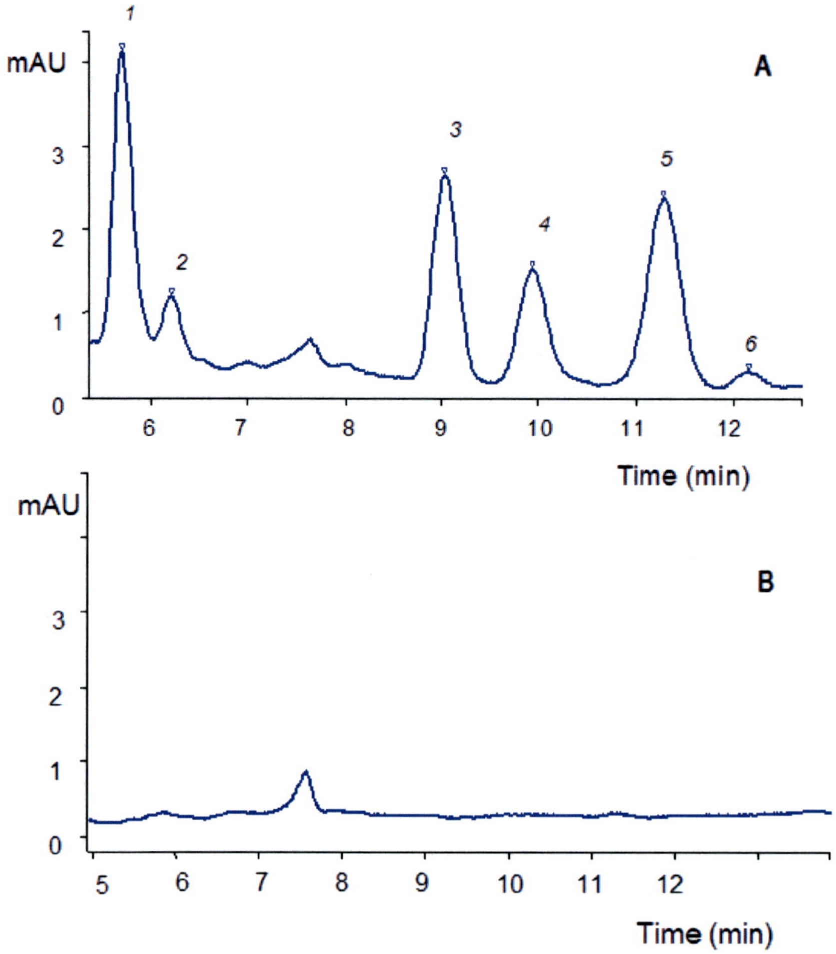

Figure 20. Chromatogram of an extract of a spiked (a) and non-spiked (b) Ulva seaweed sample after MAME-SPE procedure. λ=238 nm. (*1*- 4,4'-DDD; *2*- Dieldrin; *3*- 4,4'-DDT; *4*- 2,4'-DDT; *5*- 4,4'-DDE; *6*- Aldrin) [from ref. 109].

Applications and Validation

When this method is applied to several seaweed samples or different kind of molluscs, recoveries obtained in all cases were equal or higher than 73.9 and 68.3%, respectively. It is demonstrated that the proposed method MAME-SPE-HPLC can be an alternative to the determination of organochlorine pesticides in complex solid samples [109,110].

If we compared traditional Soxhlet extraction [113] with MAME-SPE process developed, results obtained by both methods are comparable (Table 16). A clean-up of Soxhlet extracted was not done because it was only used to validate our method, and with direct injection was enough for obtaining recovery's results but with bad baseline and higher detection limits than in MAME-SPE. Moreover, MAME-SPE offers as advantages that is a faster process and it needs much less organic solvent.

Table 16. Application of MAME-SPE procedure and Soxhlet extraction to a seaweed sample containing a certified mixture of pesticides[a,b] [from ref. 109]

Compound	MAME-SPE ($\mu g \cdot g^{-1}$)	Soxhlet ($\mu g \cdot g^{-1}$)
4,4'-DDD	3.72 ± 0.08	4.12 ± 0.24
Dieldrin	4.04 ± 0.14	3.94 ± 0.30
4,4'DDT	3.82 ± 0.10	4.04 ± 0.18
2,4'DDT	3.52 ± 0.18	3.44 ± 0.22
4,4'-DDE	3.78 ± 0.14	3.76 ± 0.15

a) Concentration of spiked seaweed: 4 μg/g for all compounds.
b) Mean of two determinations.

General Conclusions and Future Trends

New analytical methods are continually being developed for sample preparation. Although LLE continues to be widely used, CPE methodology is a good alternative extraction technique for liquid samples and offers a series of highly interesting advantages from an analytical point of view, such as the possibility of extracting and preconcentrating the analytes of different polarities in only one step.

The pre-concentration factor can be optimised by modifying the type and the concentration of surfactant as well as the experimental conditions under which extraction and phase separation are carried out. Surfactants are lesser toxic and cheaper than the extractants used in LLE. The most commonly used surfactants are commercially available and, since it is not necessary to evaporate off the solvents, no analyte is lost in the process. The experimental operations involved in CPE methodology are very simple and the surfactant-rich phase is compatible with the mobile phases used in HPLC.

MAME procedure could replace the conventional Soxhlet extraction, offering several advantages over traditional extraction techniques applied to solid samples. The major benefits are:

- a reduction in extraction time (typically 10–30 min)
- a reduction in solvent consumption (around 25–50 ml, reduced by 5–10-fold compared to traditional extractions); and,
- the opportunity to perform multiple extractions.

MAME procedure is easy to use and is cheaper than other modern techniques. Furthermore, the use of micellar medium reduces exposure to solvent vapours and wastes. When considering economical and practical aspects, the use of micellar systems for the extraction of organic compounds in solid matrices offers strong competition to other extraction techniques.

This competitiveness may be improved in the future in favour of two-directional micellar extraction. It may also be applied to a great number of types of solid real samples (plants, marine organisms, etc.), using other types of surfactants, as well as a mixture of them as extractants. In addition, combining the use of surfactants with other extraction techniques and with highly sensitive chromatographic techniques, such as LC–MS and GC–MS or GC-ECD, may be of great interest for the extraction and the determination of organochlorine and organophosphorus pesticides that, in many occasions, are found in the environment in concentrations too small to be detected using conventional techniques.

Finally, the combination of microwave assisted micellar extraction (MAME) with SPME or SPE as a clean-up and preconcentration step of MAME extract, provided sensitive and selective methods for the extraction of organochlorine and organophosphorus pesticides residues in environmental samples. Specifically, the results obtained for MAME-SPE-HPLC indicate that this technique is more sensitive for the extraction and determination of this kind of compounds where the LODs are from 10 to 100 times lower than those obtained using MAME-SPME-HPLC.

REFERENCES

[1] Kusvuran, E. & Erbatur, O. (2004) Degradation of aldrin in adsorbed system using advanced oxidation processes: comparison of the treatment methods. *Journal of Hazardous Materials,* 106B, 115-125.

[2] Stockholm Convention on Pops: *A Global Public Health Treaty.* (2001) Sweden.

[3] Shivaramaiah, H.M., Odeh, I.O.A., Kennedy, I.V. & Skerritt, J.H. (2002) Mapping the distribution of DDT residues as DDE in the soils of the irrigated regions of Northern New South Wales, Australia using ELISA and GIS. *Journal of Agricultural and Food Chemistry,* 50, 5360-5367.

[4] Chen, W., Zhang, L., Xu, L., Wang, X., Hong, L. & Hong, H. (2002) Residue levels of HCHs, DDTs and PCBs in shellfish from coastal areas of east Xiamen Island and Minjiang Estuary, China. *Marine Pollution Bulletin, 4*5, 385-390.

[5] Nawab, A., Aleem, A. & Malik, A. (2003) Determination of organochlorine pesticides in agricultural soil with special reference to gamma-HCH degratation by Pseudomonas strains. *Biosesource Technology,* 88, 41-46.

[6] Megharaj, M., Kantachote, D., Singleton, I. & Naidu, R. (2000) Effects of long-term contamination of DDT on soil microflora with special reference to soil algae and algal transformation of DDT. *Environmental Pollution,* 109, 35-42.

[7] Mwevura, H., Othman, O. & Mhehe, G.L. (2002) Organochlorine pesticide residues in sediments and biota from the coastal area of Dar es Salaam city, *Tanzania. Marine Pollution Bulletin,* 45, 262-267.

[8] Maroni, M., Colosio, C., Ferioli, A. & Fait, A. (2000) Organochlorine pesticides. *Toxicology,* 143, 61-75.

[9] U.S. Environmental Protection Agency. *Toxic Substance Control Act,* US EPA. (1979) Washington DC.

[10] Anson Moye, H. (Ed.) *Analysis of Pesticide Residues* (1981) vol 58. John Wiley & Sons, New York/Chichester/Brisbane/Toronto. EEUU.

[11] Woodwell, G.M., Craig, P.P. & Johnson, H.A. (1971) DDT in the Biosphere: Where Does It Go?. *Science,* 174, 1101-1107.

[12] Hernández, F., Pitarch, E., Beltran, J. & López, F.J. (2002) Headspace solid-phase microextraction in combination with gas chromatography and tandem mass

spectrometry for the determination of organochlorine and organophosphorus pesticides in whole human blood. *Journal of Chromatography B,* 769, 65-77.

[13] Zulin, Z., Huasheng, H., Xinhong, W., Jianqing, L., Weiqui, C. & Li, X. (2002) Determination and load of organophosphorus and organochlorine pesticides at water from Jiulong River Estuary, China. *Marine Pollution Bulletin*, 45, 397-402.

[14] Viñas, P., Campillo, N., López-García, I., Aguinaga, N. & Hernández-Córdoba, M. (2002) Determination of pesticides in waters by capillary gas chromatography with atomic emission detection. *Journal of Chromatography* A, 978, 249-256.

[15] Torres, C.M. & Mañes, Y.P.J. (1996) Determination of pesticide residues in fruit and vegetables. *Journal of Chromatography* A, 754, 301-331.

[16] Tadeo, J.L., Sánchez-Brunete, C., Pérez, R.A. & Fernández, M.D. (2000) Analysis of herbicide residues in cereals, fruits and vegetables. *Journal of Chromatography* A, 882, 175-191.

[17] Carabias-Martínez, R., Rodríguez-Gonzalo, E., Paniagua-Marcos, P.H. & Hernández-Méndez, J. (2000) Analysis of pesticide residues in matrices with high lipid contents by membrane separation coupled on-line to a high-performance liquid chromatography system. *Journal of Chromatography* A, 869, 427-439.

[18] Caballo-López, A. & Luque de Castro, M.D. (2003) Continuous ultrasound-assisted extraction coupled to on line filtration–solid-phase extraction–column liquid chromatography–post column derivatisation–fluorescence detection for the determination of N-methylcarbamates in soil and food. *Journal of Chromatography* A, 998, 51-59.

[19] Rastrelli, L., Totaro, K. & De Simone, F. (2002) Determination of organophosphorus pesticide residues in Cilento (Campania, Italy) virgin olive oil by capillary gas chromatography. *Food Chemistry,* 79, 303-305.

[20] Russo, M.V., Campanella, L. & Avino, P. (2002) Determination of organophosphorus pesticide residues in human tissues by capillary gas chromatography-negative chemical ionization mass spectrometry analysis. *Journal of Chromatography* B, 780, 431-441.

[21] Jeong, M.L. & Chesney, D.J. (1999) Determination of *N*-methylcarbamate pesticides from spiked matrices using supercritical fluid extraction. *Analytica Chimica Acta,* 389, 53-57.

[22] Hennion, M.C., Cau-Dit-Coumes, C. & Pichon, V. (1998) Trace analysis of polar organic pollutants in aqueous samples: Tools for the rapid prediction and optimisation of the solid-phase extraction parameters. *Journal of Chromatography* A, 823, 147-191.

[23] Hennion, M.C. (1999) Solid-phase extraction: method development, sorbents, and coupling with liquid chromatography. *Journal of Chromatography* A, 856, 3-54.

[24] Chiron, S., Fernández Alba, A. & Barceló, D. (1993) Comparison of on-line solid-phase disk extraction to liquid-liquid extraction for monitoring selected pesticides in environmental waters. *Environmental Science Technology,* 27, 2352-2359.

[25] Ahmed, F.E. (2001) Analyses of pesticides and their metabolites in foods and drinks. *Trends in Analytical Chemistry,* 20, 649-661.

[26] Rodriguez, I., Llompart, M.P. & Cela, R. (2000) Solid-phase extraction of phenols. *Journal of Chromatography* A, 885, 291-304.

[27] Lores, M., Llopart, M., González-García, R., González-Barreiro, C. & Cela, R. (2002) On-fibre photodegradation studies of polychlorinated biphenyls using SPME-GC-MS-MS: a new approach. *Chemosphere,* 47, 607-615.

[28] Eisert, R. & Pawliszyn, J. (1997) Automated In-Tube Solid-Phase Microextraction Coupled to High-Performance Liquid Chromatography. *Analytical Chemistry,* 69, 3140-3147.

[29] Moreno Cordero, B., Pérez Pavón, J.L., García Pinto, C. & Fernández Laespada, E. (1993) Cloud point methodology: A new approach for preconcentration and separation in hydrodynamic systems of analysis. *Talanta,* 40, 1703-1710.

[30] Heemke, O.P., Norbert, T. & Wenclawiak, B.W. (1997) Comparison of ASE and SFE with Soxhlet, Sonication, and Methanolic Saponification Extractions for the Determination of Organic Micropollutants in Marine Particulate Matter. *Analytical Chemistry*, 69, 2171-2180.

[31] Snyder, J., Grob, R., McNally, M. & Gostdyk, T. (1992) Comparison of supercritical fluid extraction with classical sonication and Soxhlet extractions for selected pesticides. *Analytical Chemistry,* 64, 1940-1946.

[32] Fernández, P., Grifoll, M., Solanas, A.M., Bayona, J. & Albaigés, J. (1992) Bioassay-directed chemical analysis of genotoxic components in coastal sediments. *Environmental Science Technology,* 26, 817-829.

[33] Punín Crespo, M. & Lage Yusty, M. (2006) Comparison of supercritical fluid extraction and Soxhlet extraction for the determination of aliphatic hydrocarbons in seaweed samples. *Ecotoxicology and Environmental Safety,* 64, 400-405.

[34] Li, K.; Landrialult, M.; Fingas, M. & Llompart, M. (2003) Accelerated solvent extraction (ASE) of environmental organic compounds in soils using a modified supercritical fluid extractor. *Journal of Hazardous Materials,* 102, 93-104.

[35] Zuloaga, O., Etxebarria, N., Fernández, L.A. & Madariaga, J.M. (2000) Fresenius *Journal Analytical Chemistry,* 367, 733-737.

[36] Sanghi, R. & Kannamkumarath, S. (2004) Comparison of extraction methods by soxhlet, sonicator, and microwave in the screening of pesticides residues from solid matrices. *Journal of Analytical Chemistry,* 59, 1032-1036.

[37] Sporring, S.; Bowadt, S.; Svensmark, B. & Björklund, E. (2005) Comprehensive comparison of classic Soxhlet extraction with Soxtec extraction, ultrasonication extraction, supercritical fluid extraction, microwave assisted extraction and accelerated solvent extraction for the determination of polychlorinated biphenyls in soil. *Journal of Chromatography* A, 1090, 1-9.

[38] Pastor, A., Vazquez, E., Ciscar, R. & de la Guardia, M. (1997) Efficiency of the microwave-assisted extraction of hydrocarbons and pesticides from sediments. *Analytica Chimica Acta,* 344, 241-249.

[39] Shu, Y.Y., Tey, S.Y. & Wu, D.K.S. (2003) Analysis of polycyclic aromatic hydrocarbons in airborne particles using open-vessel focused microwave-assisted extraction. *Analytica Chimica Acta,* 495, 99-108.

[40] López-Ávila, V. & Young, R. (1994) Microwave-assisted extraction of organic compounds from standard reference soils and sediments. *Analytical Chemistry,* 66, 1097-1106.

[41] Esteve-Turrillas, F.A., Aman, C.S., Pastor, A. & de la Guardia, M. (2004) Microwave-assisted extraction of pyrethroid insecticides from soil. *Analytica Chimica Acta,* 522, 73-78.

[42] Barriada-Pereira, M., Concha-Graña, E., González-Castro, M.J., Muniategui-Lorenzo, S., López-Mahía, P., Prada-Rodríguez, D. & Fernández-Fernández, E. (2003)

Microwave-assisted extraction versus Soxhlet extraction in the analysis of 21 organochlorine pesticides in plants. *Journal of Chromatography A,* 1008, 115-122.

[43] Onuska, F.I. & Terry, K.A. (1993) Extraction of pesticides from sediments using a microwave technique. *Chromatographia,* 36, 191-194.

[44] Letellier, M., Budzinski, H. & Garrigues, P. (1999) Focused microwave-assisted extraction of polycyclic aromatic hydrocarbons. *LC-GC International,* 12, 222-225.

[45] López-Avila, V., Young, R., Benedicto, J., Ho, P., Kim, R. & Beckert, W.F. (1995) Extraction of organic pollutants from soild samples using microwave energy. *Analytical Chemistry,* 67, 2096.

[46] Rosen, M.J. (1986) *Surfactants and Interfacial phenomena,* John Wiley & sons. USA.

[47] Pelizzetti, E. & Pramauro, E. (1985) Analytical applications of organized molecular assemblies. *Analytica Chimica Acta,* 169, 1-29.

[48] Uhrich, K.E. (1999) Chemical Review, 99, 3181-3198.

[49] Goto, T., Ito, Y., Oka, H., Saito, I., Matsumoto, H. & Nakazawa, H. (2003) Simple and rapid determination of N-methylcarbamate pesticides in citrus fruits by electrospray ionization tandem mass spectrometry. *Analytica Chimica Acta,* 487, 201-209.

[50] Namera, A., Yashiki, M., Liu, J., Okajima, K., Hara, K., Imamura, T. & Kojima, T. (2000) Simple and simultaneous analysis of fenfluramine, amphetamine and methamphetamine in whole blood by gas chromatography–mass spectrometry after headspace–solid phase microextraction and derivatization. *Forensic Science International,* 109, 215-223.

[51] Paleologos, E. K., Giokas, D. L. & Karayannis, M. I. (2005) Micelle-mediated separation and cloud point extraction. *Trends in Analytical Chemistry,* 24, 426-436.

[52] Faria, A. M., Dardengo, R. P. Lima, C. F., Neves, A. & Queiroz, M. E. (2007) Determination of disulfoton in surface water samples by cloud-point extraction and gas chromatography. *International Journal of Environmental Analytical Chemistry,* 87, 249-258.

[53] Shen, J. & Shao, X. (2006) Determination of tobacco alkaloids by gas chromatography-mass spectrometry using cloud point extraction as a preconcentration step. *Analytica Chimica Acta,* 561, 83-87.

[54] Sosa Ferrera, Z., Padrón Sanz, C., Mahugo Santana, C. & Santana Rodríguez, J.J. (2004) The use of micellar systems in the extraction and pre-concentration of organic pollutants in environmental samples. *Trends in Analytical Chemistry,* 23 (7), 469-479.

[55] Mahugo Santana, C., Sosa Ferrera, Z. & Santana Rodríguez, J.J. (2004) Use of polyoxyethlylene-6-lauryl ether and microwave-assisted extraction for the determination of chlorophenols in marine sediments. *Analytica Chimica Acta,* 524, 133-139.

[56] Padrón Sanz, C., Halko, R., Sosa Ferrera, Z. & Santana Rodríguez, J.J. (2004) Micellar extraction of organophosphorus pesticides and their determination by liquid chromatography. *Analytica Chimica Acta,* 524, 265-270.

[57] Vega Moreno, D., Sosa Ferrera, Z. & Santana Rodríguez, J.J. (2006) Microwave assisted micellar extraction coupled with solid phase microextraction for the determination of organochlorine pesticides in soil simples. *Analytica Chimica Acta,* 571, 51-57.

[58] Hinze, W.L. & Pramauro, E. (1993) A critical review of surfactant-mediated phase separations (cloud point extractions): theory and applications. *Critical Reviews in Analytical Chemistry,* 24(2), 133-177.

[59] Mahugo Santana, C., Sosa Ferrera, Z. & Santana Rodríguez, J.J., (2002) Use of non-ionic surfactant solutions for the extraction and preconcentration of phenolic compounds in water prior to their HPLC-UV detection. *The Analyst,* 127, 1031-1037.

[60] Eiguren Fernández, A., Sosa Ferrera, Z. & Santana Rodríguez, J.J., (1997) *Química Analítica,* 16(2), 283.

[61] Frankewich, R.P. & Hinze, W. (1994) Evaluation and Optimization of the Factors Affecting Nonionic Surfactant-Mediated Phase Separations. *Analytical Chemistry,* 66, 944-954.

[62] Carabias-Martínez, R., Rodríguez-Gonzalo, E., Domínguez-Álvarez, J., García Pinto, C., Hernández-Méndez, J. (2003) Prediction of the behaviour of organic pollutants using cloud point extraction. *Journal of Chromatography* A, 1005 (1-2), 23-34

[63] Carabias, R., Rodríguez, E., Moreno, B., Pérez, J.L., García, C. & Fernández, E. (2000) Surfactant cloud point extraction and preconcentration of organic compounds prior to chromatography and capillary electrophoresis. *Journal of Chromatography* A, 902, 251-265.

[64] Silva, M.F., Fernández, L., Olsina, R.A. & Stacchiola, D. (1997) Cloud point extraction, preconcentration and spectrophotometric determination of erbium(III)-2-(3,5-dichloro-2-pyridylazo)-5-dimethylaminophenol. *Analytica Chimica Acta,* 342, 229-238.

[65] Eiguren Fernández, A., Sosa Ferrera, Z. & Santana Rodríguez, J.J. (1998) Determination of polychlorinated biphenyls by liquid chromatography following cloud-point extraction. *Analytica Chimica Acta,* 358, 145-155.

[66] Faria, A.M., Dardengo, R.P., Lima, C.F., Neves, A.A., Queiroz, M.E.L.R. (2007) Determination of disulfoton in surface water samples by cloud-point extraction and gas chromatography. *International Journal of Environmental Analytical Chemistry* 87 (4), 249-258

[67] Froschl, B., Stangl, G. & Niessner, R. (1997) Combination of micellar extraction and GC-ECD for the determination of polychlorinated biphenyls (PCBs) in water. *Fresenius Journal Analytical Chemistry,* 357, 743-746.

[68] Eiguren Fernández, A., Sosa Ferrera, Z. & Santana Rodríguez, J.J. (1999) Application of cloud-point methodology to the determination of polychlorinated dibenzofurans in seawater by high-performance liquid chromatography. *Analyst,* 24, 487-491.

[69] Lindsay, S. (1992) *High Performance Liquid Chromatography,* John Wiley and Sons: New York, pp. 337.

[70] Ganzler, K., Salgo, A. & Lebensm-Unter, Z. (1987) *Forsch,* 184, 274

[71] Pineiro Iglesias, M., López Mahía, P., Vázquez Blanco, E., Muniategui Lorenzo, S., Prada Rodríguez, D. & Fernández Fernández, E. (2000) Microwave assisted extraction of polycyclic aromatic hydrocarbons from atmospheric particulate samples. *Fresenius Journal Analytical Chemistry,* 367, 29-34.

[72] López-Avila, V., Benedicto, J., Charan, C. & Young, R. (1995) Determination of PCBs in soils/sediments by microwave-assisted extraction and GC/ECD or ELISA. *Environmental Science Technology,* 29, 2709-2712.

[73] Hummert, K., Vetter, W. & Luckas, B. (1996) Fast and effective sample preparation for determination of organochlorine compounds in fatty tissue of marine mammals using microwave extraction. *Chromatographia,* 42, 300-304.

[74] Pastor, A., Vázquez, E., Ciscar, R. & de la Guardia, M. (1997) Efficiency of the microwave-assisted extraction of hydrocarbons and pesticides from sediments. *Analytica Chimica Acta,* 344, 241-249.

[75] Egizabal, A., Zuloaga, O., Etxebarria, N., Fernández, L.A. & Madariaga, J.M. (1998) Comparison of microwave-assisted extraction and Soxhlet extraction for phenols in soil samples using experimental designs. *The Analyst,* 123, 1679-1684.

[76] Llompart, M.P., Lorenzo, R.A., Cela, R., Li, K., Belanger, J.M.R. & Pare, J.R.J. (1997) Evaluation of supercritical fluid extraction, microwave-assisted extraction and sonication in the determination of some phenolic compounds from various soil matrices. *Journal of Chromatography* A, 774, 243-251.

[77] Padrón Sanz, C., Sosa Ferrera, Z. & Santana Rodríguez, J.J. (2004) Extraction of PCDDs from marine sediments using polyoxyethylene 10 lauryl ether and oligoethylene glycol monoalkyl ether surfactants. *Analytical Letters,* 37, 1385-1399.

[78] Padron Sanz, C., Halko, R., Sosa Ferrera, Z. & Santana Rodríguez, J.J. (2005) Combination of microwave assisted micellar extraction and liquid chromatography for the determination of organophosphorous pesticides in soil simples. *Journal of Chromatography* A, 1078, 13-21

[79] Vega Moreno, D., Sosa Ferrera, Z. & Santana Rodríguez, J.J. (2006) Use of polyoxyethylene surfactants for the extraction of organochlorine pesticides from agricultural soils. *Journal of Chromatography* A, 1104, 11-17.

[80] Padrón Sanz, C., Eiguren Fernández, A., Sosa Ferrera, Z. & Santana Rodríguez, J.J. (2002) Determination of Organochlorinated Compounds in Marine Organisms by Microwave-Assisted Extraction with Molecular Organized Systems and Liquid Chromatography with Fluorescence Detection. *Journal of AOAC Internacional,* 85(1), 44-49.

[81] Bianco Prevot, A., Gulmini, M., Zelano, V. & Pramauro, E. (2001) Microwave-Assisted Extraction of Polycyclic Aromatic Hydrocarbons from Marine Sediments Using Nonionic Surfactant Solutions. *Analytical Chemistry,* 73, 3790-3795.

[82] U.S. Environmental Protection Agency (1996), Method 3540 C, Washington, DC.

[83] Llompart, M.P., Lorenzo, R.A. & Par´e, J.R. (1997) Optimization of a Microwave-assisted Extraction Method for Phenol and Methylphenol Isomers in Soil Samples Using a Central Composite Design. *Analyst,* 122, 133-137.

[84] Huang, W., Peng, P., Yu, Z. & Fu, J. (2003) Effects of organic matter heterogeneity on sorption and desorption of organic contaminants by soils and sediments. *Applied Geochemistry,* 18, 955-972.

[85] Thiele-Bruhn, S., Seibicke, T., Schuelten, H.R. & Leinweber, P. (2004) Sorption of Sulfonamide Pharmaceutical Antibiotics on Whole Soils and Particle-Size Fractions. *Journal of Environmental Quality,* 33, 1331-1342.

[86] Aquino, A. J., Tunega, D., Haberhauer, G., Gerzabek, M. & Lischka, H. (2003) Adsorption of Organic Substances on Broken Clay Surfaces: A Quantum Chemical Study. *Journal of Computational Chemistry*, 24(15), 1853-1863.

[87] Shu, Y.Y., Lai, T.L., Lin, H.S., Yang, T.C. & Chang, C.P. (2003) Study of factors affecting on the extraction efficiency of polycyclic aromatic hydrocarbons from soils

using open-vessel focused microwave-assisted extraction. *Chemosphere,* 52, 1667-1676.

[88] Kiflom, G., Wandiga, S.O., Ng'ang'a, P.K. & Kamau, G.N. (1999) Variation of plant p,p'-DDT uptake with age and soil type and dependence of dissipation on temperature. *Environmental Internacional,* 25(4), 479-487.

[89] Hawthorne, S.B., Björklund, E., Bowadt, S. & Mathiasson, L. (1999) Determining PCB sorption/desorption behavior on sediments using selective supercritical fluid extraction. 3. Sorption from water. *Environmental Science Technology,* 33, 3152-3159.

[90] Dec, J. & Bollag, J. (1997) Determination of covalent and noncovalent binding interactions between xenobiotic chemicals and soil. *Soil Science,* 162, 858-874.

[91] Kopinke, F., Pörschmann, J. & Stottmeister, U. (1995) Sorption of organic pollutants on anthropogenic humic matter. *Environmental Science Technology*, 29, 941-950.

[92] Stevenson, F.J. (1982) *Humus Chemistry,* John Wiley & Sons Inc., USA.

[93] Racke, K.D. in: Chambers, J.E. & Levi, P.E. (Eds.) (1992) *Organophosphates: Chemistry*, Fate, and Effects. Academic Press, San Diego, pp. 47–73.

[94] Pehkonen, S.O. & Zhang, Q. (2002) The degradation of organophosphorus pesticides in natural waters: A critical review. *Critical Review Environmental Science Technology*, 32, 17-72.

[95] Dannenberg, A. & Pehkonnen, S.O. (1998) Investigation of the Heterogeneously Catalyzed Hydrolysis of Organophosphorus Pesticidas. *Journal of Agricultural and Food Chemistry*, 46, 325-334.

[96] Dowling, K.C. & Lemley, A.T. (1995) Organophosphate insecticide degradation by non-amended and cupric ion-amended Fenton's reagent in aqueous solution. *Journal Environmental Science Health* B, 30 (5), 585-604.

[97] Noblet, J.A., Smith, L.A. & Suffet, I.H. (1996) Influence of natural dissolved organic matter, temperature, and mixing on the abiotic hydrolysis of triazine and organophosphate pesticidas. *Journal of Agricultural and Food Chemistry*, 44, 3685-3693.

[98] Dupeyron, S., Dudermel, P.M. & Counturier, D. (1997) Focused microwave assisted extraction (FMAE) of polynuclear aromatic hydrocarbons from contaminated soil: Role of acetone and water content impact on microwave efficiency. *Analusis*, 25, 286-292.

[99] Peñalver, A., Pocurull, E., Borrull, F. & Marcé, R.M. (1999) Trends in solid-phase microextraction for determining organic pollutants in environmental samples. *Trends in Analytical Chemistry*, 18(8), 557-567.

[100] Ho, W. & Hsieh, S. (2001) Solid phase microextraction associated with microwave assisted extraction of organochlorine pesticides in medicinal plants. *Analytica Chimica Acta*, 428, 111-120.

[101] Aulakh, J. & Malik, A. (2005) A review on solid phase microextraction-high performance liquid chromatography (SPME-HPLC) analysis of pesticides. *Critical Reviews in Analytical Chemistry*, 35, 71-85.

[102] Vega Moreno, D., Sosa Ferrera, Z. & Santana Rodríguez, J.J. (2006) Sample extraction method combining micellar extraction-SPME and HPLC for the determination of organochlorine pesticides in agricultural soils. *Journal of Agricultural and Food Chemistry*, 54, 7747-7752.

[103] Beltrán, J., López, F. & Hernández, F. (2000) Solid-phase microextraction in pesticides residue analysis. *Journal of Chromatography* A, 885, 389-404.

[104] Krutz, L., Senseman, S. & Sciumbato, A. (2003) Solid-phase microextraction for herbicide determination in environmental samples. *Journal of Chromatography* A, 999, 103-121.

[105] Hernández, F., Beltrán, J., López, F. & Gaspar, (2000) Use of solid-phase microextraction for the quantitative determiantion of herbicides in soil and water samples. *Journal Analytical Chemistry*, 72, 2313-2322.

[106] Arthur, C.L., Killiam, L.M., Buchholz, K.D. & Pawliszyn, J. (1992) Automation and optimization of solid-phase microextraction. *Journal Analytical Chemistry*, 64, 1960-1966.

[107] Vega Moreno D, Sosa Ferrera Z, Santana Rodríguez JJ (2008) Application of microwave assisted micellar extraction combined with solid phase microextraction and HPLC-UV for the determination of organochlorine pesticides in diferents mud samples. *International Journal on Environmental Analytical Chemistry*, in press.

[108] Mutavdzic, D., Horvat, A., Babic, S. & Kastelan-Macan, M. (2005) SPE- Microwave-assisted extraction coupled system for the extraction of pesticides from water samples. *Journal of Separation Science*, 28, 1485-1492.

[109] Vega Moreno, D., Sosa Ferrera, Z. & Santana Rodríguez, J.J. (2007) SPME and SPE comparative study for coupling with microwave-assisted micellar extraction in the analysis of organochlorine pesticides residues in seaweed samples. *Microchemical Journal*, 87, 139-146.

[110] Vega Moreno D, Sosa Ferrera Z, Santana Rodríguez JJ (2008) Analysis of organochlorine pesticides in mollusc samples by HPLC after microwave assisted micellar extraction coupled with solid phase extraction. *Journal of Liquid Chromatography & Related Technologies*, in press.

[111] Deger, A., Gremm, T. & Frimmel, F. (2000) Problems and solutions in pesticides analysis using solid-phase extraction (SPE) and gas chromatography ion-trap mass spectrometry detection (GC-MS). *Acta Hydrochim. Hydrobiol.*, 28, 292-299.

[112] Papadakis, E. & Papadopoulou-Mourkidou, E. (2006) LC-UV determination of atrazine and its prinicipal conversion products in soil after combined microwave-assisted and solid-phase extraction. *International Journal on Environmental Analytical Chemistry*, 86, 573-582.

[113] AOAC Official Method 970.52, AOAC Official Methods of Analysis, *Pesticide and Industrial.*

In: Environmental Chemistry Research Progress
Editors: P. Robinson and R. Gallo ISBN 978-1-60741-055-3

Chapter 5

NEW TECHNOLOGY TO IMPROVE PROCESSABILITY FOR RECYCLED POLY(ETHYLENE TEREPHTHALATE)

Masayuki Yamaguchi*
School of Materials Science,
Japan Advanced Institute of Science and Technology
1-1 Asahidai, Nomi, Ishikawa 923-1292, Japan

ABSTRACT

Poly(ethylene terephthalate) (PET) is one of the most eco-friendly polymeric materials, since a large amount of PET bottles, more than 65 % in Japan, is collected and consumed as recycled PET. This trend will spread worldwide in the near future because of the rapid growth of the interest in global environment as well as the excellent cost-performance. Although the material recycling system seems to be successful, it has some problems attributed to the depression of molecular weight leading to low level of melt elasticity. The lack of the melt elasticity leads to sagging, *i.e.*, downward- deformation of the extrudate by gravitational force, which prohibits extrusion processing, and inhomogeneity of the thickness of the products at film processing and blow-molding. As a result, the application of recycled PET is quite limited, such as spinning to produce fibers and injection-molding.

In this chapter, a new method employing reactive polymer blend technique is proposed to improve the processability for polyester resins by rheological modification. The technique would be applicable to recycled PET. Further, appropriate processing condition is also demonstrated for industrial application. The technique enhances the melt elasticity, *i.e.*, the most serious problem for recycled PET and thus improves the extrusion processability.

---

* Phone +81-761-51-1621; Facsimile +81-761-51-1625; E-mail m_yama@jaist.ac.jp

INTRODUCTION

Because of a rapid increase in the attention to global environment, material recycling of conventional plastics has created great interest in the plastic industry. As is well known, one of the most successful plastic products from the view point of the material recycling is soft drink bottles of poly(ethylene terephthalate) (PET). PET is produced by polycondensation reaction of terephthalic acid and ethylene glycol, and was firstly commercialized by ICI in 1948 as an engineering plastic. The beverage bottle application was patented in 1973 by DuPont and commercialized in 1977 because of its odorless nature as well as the excellent transparency. The first application was for a soy sauce bottle in Japan by Kikkoman and Yoshino Kogyo, which was the replacement of poly(vinyl chloride) bottles. Then the amount of the application has been increasing steadily all over the world.

The collection rate of PET bottles has been also increasing to a great extent in many countries. Surprisingly, 67% of PET bottles, 360,000 metric tons, were collected in Japan in 2006 and the amount is still increasing, although the collection rate in 1999 was only 23%. The immediate and successful growth would be attributed to "Containers and packaging recycling law" started in 1997 (the details are summarized in their homepage; http://www.jcpra.or.jp/eng/index.html). This trend will spread worldwide in the near future because of the excellent cost-performance of recycled PET, *i.e.*, lower than half price of virgin PET. Further, PET Container Recycling Europe, known as Petcore, which was established in 1992 to promote the collection, sorting, and recycling of post-consumer PET bottles, announced that 665,000 metric tons were collected in Europe in 2004. They also forecast that the collection will exceed one million tons only in Europe by 2010. Moreover, in USA, 570,000 metric tons, which was 23.5 % of the total volume of PET containers available for recycling, were collected in 2006 according to the National Association for PET Container Resources (NAPCOR) and the Association of Postconsumer Plastic Recyclers (APR). The amount in 2006 is a 9 % increase over 2005 in USA. These reports strongly support the rapid increase in the collection rate of PET bottles all over the world.

Furthermore, various attractive properties, such as high strength, heat resistance, gas-barrier property, and transparency, will enhance the demand of PET. However, extensive efforts are still required to widen the application of recycled PET because of the poor processability due to the lack of melt elasticity, low viscosity, and brittle nature in the solid state. As a result, recycled PET is primarily used in the fiber market, because it does not require melt elasticity for spinning. For example, more than half of the polyester carpet in USA is made from recycled PET. Similarly, more than 50% of recycled PET was used as fiber in Japan in 2006.

It is known that chain extension reaction of polyester resins, including PET, in which polyaddition or polycondensation occurs between terminal groups of polyester and a reagent having more than two reactive sites, has been employed to overcome the difficulties as summarized by Scheirs and Long (2003). There are mainly two processes for the chain extension technology: one of them is solid-state polycondensation and the other is reactive extrusion, *i.e.*, reaction in the molten state during extrusion. In the case of recycled PET, reactive extrusion will be the dominant technology because the solid-state polycondensation requires extremely high capital and production costs. End users in the polymer processing industry cannot have such an expensive plant in general. On the contrary, the reactive extrusion requires only an extruder having specific screw segments, which is not so expensive as compared with the solid-state polycondensation method. During the reactive extrusion, terminal carboxyl and/or hydroxyl groups of polyesters are employed as reactive sites with various types of functionalities, such as oxazoline, glycidyl (epoxy), isocyanate, and maleic acid, in other reactive compounds as summarized by Xanthos (1992). The details of the reaction and the chemical structure of the functionalities are illustrated in Figure 1. These reactive compounds are sometimes called as "chain-extender" or "melt-strength-enhancer".

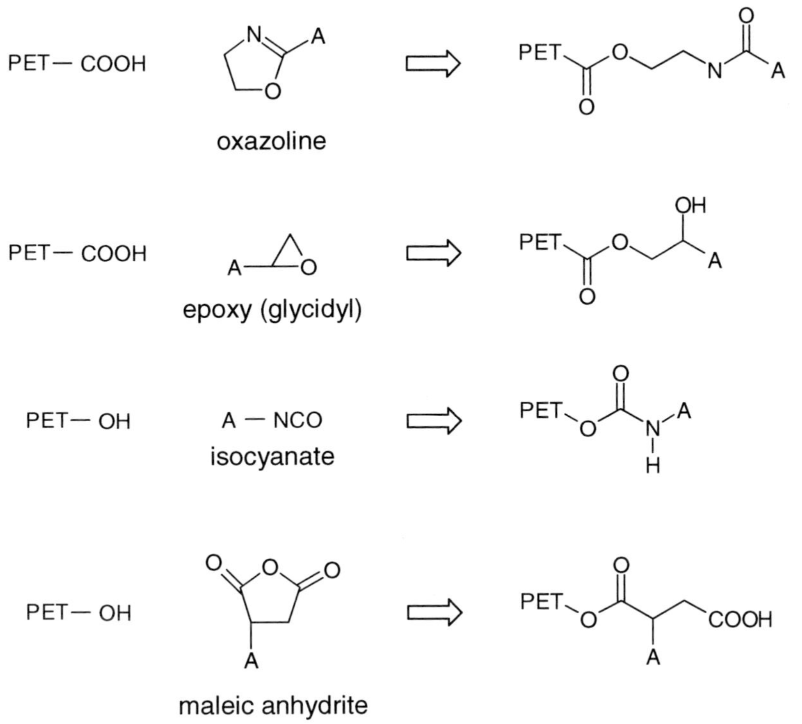

Figure 1. Chemical reaction of carboxyl and hydroxyl terminal groups in polyester with various functionalities.

REACTIVE MODIFICATION BY LOW-MOLECULAR-WEIGHT COMPOUND

Rheological modification by a reactive low-molecular-weight compound has been studied extensively to enhance the processability of PET (not only recycled PET but also virgin one). Except for the safety and cost-performance, the important technique of the modification by the reactive extrusion is to decide the appropriate compounds which react with PET for a short time in a molten state. Further, it is also important to evaluate and/or predict the processability by a small amount of materials, because direct evaluation of the processability is time-consuming and requires a large amount of materials as well as various processing machines. In general, the following three methods are performed both in the industries and academia; (1) dilute solution properties, such as intrinsic viscosity and gel permeation chromatography to evaluate the molecular weight and its distribution, (2) direct measurements of the reaction by titration of carboxyl and/or hydroxyl contents, and (3) the rheological measurements in a molten state.

Inata and Matsumura of Teijin Ltd. in Japan have performed a pioneering work in this field in the mid of 1980's, although their experiments were carried out in a flask not in an extruder. In series of their papers (Inata and Matsumura, 1985, 1986, 1987), it was clarified that various bis-heterocycle compounds, such as 2,2'-bis(2-oxazoline), 2,2'-bis(5,6-dihydro-4H-1,3-oxazine), and *N,N'*-hexamethylenebis(carbamoyl-2- oxazoline), are available as the chain extenders which are capable of addition reaction with carboxyl terminals in PET. Further, bis[5(4H)-oxazolone]s and bis(4H-3,1-benzoxazin-4-one)s were found to be the most effective chain extenders coupling with hydroxyl terminals in PET. Their characterization proved that both types of reactive modifiers enable to increase the molecular weight as a linear polymer not a branched polymer.

Loontjens et al. (1993) found that 1,2-bis(2-oxazolinyl-2) ethane is applicable to the chain extender for PET. Further, they evaluated 1,3-phenylene bis(2-oxazoline-2) and isophthaloyl biscaprolactamate as chain extenders, and found that the bisoxazorine reacts immediately with carboxyl terminals whereas the caprolactam acts as a suitable coupling agent for polymers having hydroxyl group like PET. Moreover, the pronounced change of the rheological properties is detected when both types of the chain extenders are used simultaneously (Loontjens et al. 1997).

Bikiaris and Karayannidis (1995, 1996) studied the chain extension of PET and poly(butylene terephthalate) with N,N'-bis(glycidyl ester) pyromellitimides. Although both carboxyl and hydroxyl terminals in PET are capable of the reaction with glycidyl group, they clarified that the carboxyl terminal reacts preferentially. Further, they found that long-chain branches are generated with some amounts of gel fraction, suggesting the occurrence of crosslinking reaction. The results should be notified because long-chain branches enhance the melt elasticity to a great extent (Xanthos et al., 2001; Yamaguchi, 2004). Further, it has been clarified recently that mixing a weak gel has a great impact on the enhancement of melt elasticity while remaining viscous nature (Yamaguchi, 2004). The latter technique is available to biomass-based polyesters that are lack of the melt elasticity (Arakawa et al., 2007)

The discovery of the chain extension reaction has a great impact on the industries, because it would be applicable to reactive extrusion. Cardi et al. (1993) performed the reactive modification employing recycled PET with 2,2'-bis(2-oxazoline) in a twin-screw extruder. They confirmed the addition reaction with carboxyl terminals in PET as well as thermal and hydrolytic degradations during processing.

Moreover, Haralabakopoulos et al. (1999) employed various diepoxides to modify the rheological properties of PET and investigated the chain extension reaction by intrinsic viscosity and titration of carboxyl content. They found that cyclic diepoxide compounds, such as 3,4-epoxycyclohexylmethyl-3,4-epoxycyclohexyl carboxylate, react with PET immediately, which is promoted in a nitrogen atmosphere. Further, based on the experimental results employing a laboratory scale reactor, they performed the reaction in an internal mixer and demonstrated the enhancement of shear viscosity. Their experiments revealed that the prompt reaction is suitable for the reactive mixing in a molten state because of the suppression of thermal and hydrolytic degradations.

Incarnato et al. (2000) studied the rheological properties of PET modified by pyromellitic dianhydride (PMDA), one of the most famous chain extenders for polycondensation reaction, and indicated that the reactive mixing leads to broadening of molecular weight distribution as well as the generation of long-chain branches. Further, the obtained material exhibits the pronounced shear-thinning behavior, *i.e.*, shear rate sensitivity of the steady-state viscosity, which is a considerably important rheological property at extrusion processing especially at high out-put condition.

Xanthos et al. (2000, 2001) applied the experimental results by Bikiaris and Karayannidis (1995, 1996) to reactive mixing in a molten state employing a batch-type internal mixer. They confirmed both chain extension and degradation reactions by monitoring the torque change during the reactive mixing. Further, they concluded less than 1 wt% diimidodiepoxide is good enough to enhance the melt elasticity, which is available for extrusion foaming. Their results are considerably important, because extrusion foaming requires high level of melt elasticity, which is completely different from spinning. Further, Dhavalikar et al. (2003) revealed that adding triglycidyl isocyanurate results in the generation of long-chain branches as well as the chain extension. The residence time required for the reaction with carboxyl terminals in PET is more than 5 min, which is a bit longer than the conventional residence time in an extruder. Further, various rheological data in their paper demonstrated that melt elasticity is greatly enhanced by the reactive mixing.

Reactive Modification by Reactive Polymer

Reactive modifiers composed of polymeric materials have been considered to be inappropriate for the enhancement of the viscosity and elastic nature by reactive extrusion due to prolonged reaction time. Some polymeric modifiers are, however, found to be available as a chain extender recently. Furthermore, employing a polymeric material as a reactive modifier has a great advantage because of easy handling in both solid state (pellet form) and molten state (high viscosity). Therefore, much attention has been focused on the polymeric modifiers for recycled PET these days. There have been numerous reports on the blends of PET and polymeric modifiers with reactive sites, such as polyolefin elastomers functionalized with maleic anhydride, ethylene-glycidyl methacrylate, and ethylene-ethylacrylate-glycidyl methacrylate. However, most of them have been focused on the improvement of toughness of virgin PET, and the relation with the morphology was discussed in detail. Since our interest is to control the rheological properties and thus the processability of recycled PET, the concentration is fairly low as compared with the impact modifier.

Yamaguchi and Wakabayashi (2006) studied the rheological properties and extrusion processability of a polyester resin modified by a polymeric material having reactive sites. They employed poly(ethylene terephthalate-co-ethylene isophthalate), P(ET-EI), instead of PET, in order to perform the precise rheological characterization avoiding thermal and/or hydrolytic degradations. The modifier employed is styrene-acrylate-based copolymer with glycidyl functionality, commercially available from Toagosei in Japan. The modified P(ET-EI) was prepared by the following two steps; firstly 10 wt% master batch was prepared by a co-rotating twin-screw extruder, and secondly the master batch was added into virgin P(ET-EI) by a single-screw extruder equipped with a full-flight screw, *i.e.*, a conventional screw.

Figure 2 shows the temperature dependence of dynamic tensile moduli, such as storage modulus E' and loss modulus E'', at 10 Hz for P(ET-EI), the modifier, and the master batch. Beyond the glass transition temperature, *i.e.*, 60 – 80 °C, the storage modulus, which is a measure of the rigidity, for all the materials decreases with temperature. Further, it increases around at 120 – 130 °C for P(ET-EI) and the master batch. This is attributed to the crystallization of P(ET-EI) during the measurements. The peak of the loss modulus curve indicates the glass to rubber transition. As seen in the figure, the peak of the master batch is broad, suggested by the bold arrow, as compared with that of the pure P(ET-EI). The result indicates that some level of molecular interdiffusion takes place in the blend, although P(ET-EI) is immiscible with the modifier. The block and/or graft copolymer of P(ET-EI) and the modifier, which is generated during the mixing, would act as a compatibilizer for the system. Further, the result demonstrates that the residence time in the twin-screw extruder for the preparation of the master batch is sufficient enough to react between the carboxyl terminals of P(ET-EI) and glycidyl groups of the modifier at the processing temperature

Figure 3 shows the appearance of extrudates from a single-screw extruder at 250 °C at the screw rotation speed of 30 rpm. As seen in the figure, the shape of the extrudate of P(ET-EI) is typical for polymeric materials with low viscosity. Accordingly, the extrudate exhibits prominent heat sagging, *i.e.,* downward-deformation of a molten polymer by gravitational force, which is known as one of the important problems at extrusion processing. This becomes considerably serious in the case of recycled PET, because recycled PET has lower shear viscosity and has to be processed at higher temperature due to its high melting point.

It should be noted that adding only 2 wt% of the modifier is good enough to improve the processability. The shape of the extrudate is almost the same as that of the die, demonstrating the good processability at extrusion. Further, 4 wt% of the modifier enhances the extrudate's swell, enlargement of extrudate due to the elastic nature of a molten polymer. The pronounced elastic nature shown here may be applicable to other processing operations as mentioned later.

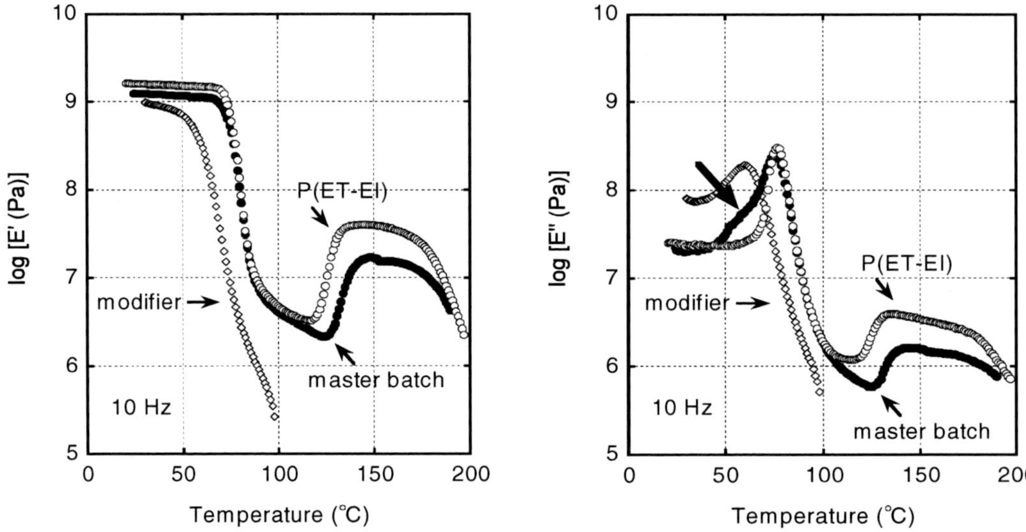

Figure 2. Temperature dependence of oscillatory tensile moduli, such as storage modulus E' (○) and loss modulus E'' (●), at 10 Hz for P(ET-EI), the modifier, and the master batch composed of P(ET-EI) containing 10 wt% of the modifier.

Figure 3. Appearance of the samples extruded from a single-screw extruder at 250 °C for (left) P(ET-EI), (middle) P(ET-EI) with 2 wt% of the modifier, (right) P(ET-EI) with 4 wt% of the modifier.

Further, the rheological properties of the extrudate are evaluated in detail. Figure 4 shows the shear rate dependence of steady-state shear viscosity at 210 °C. It is found that adding the modifier enhances the shear viscosity, which is a similar result obtained for the modified PET by low-molecular-weight reactive compounds. Furthermore, it is noted that gel permeation measurements reveal that molecular weight and its distribution are hardly changed by adding the modifier. The result indicates the followings; (1) the viscous enhancement is mainly attributed to the generation of long-chain branches and (2) thermal and hydrolytic degradations are suppressed at the processing condition.

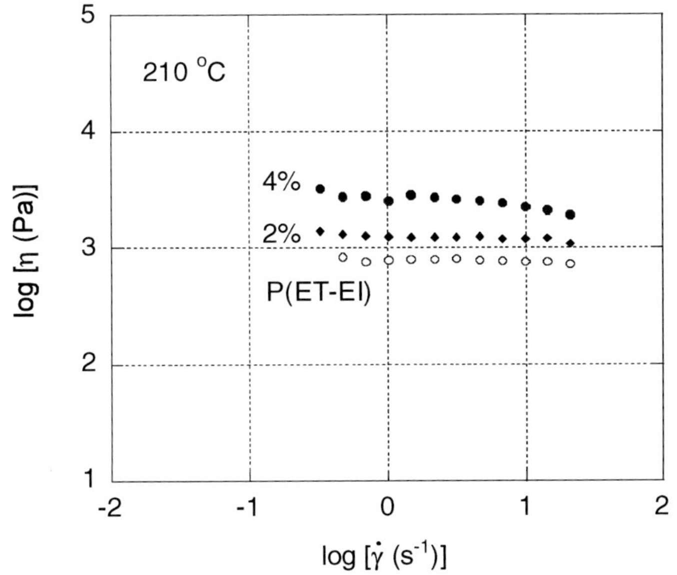

Unit of η is (Pa a)

Figure 4. Flow curves at 210 °C measured by a cone-and-plate rheometer; (open circles) P(ET-EI), (closed diamonds) P(ET-EI) with 2 wt% of the modifier, and (closed circles) P(ET-EI) with 4 wt% of the modifier.

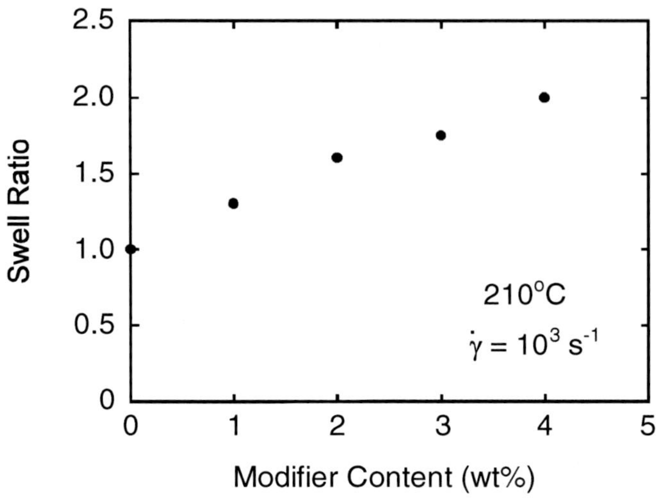

Figure 5. Swell ratio of the strands extruded from a capillary rheometer equipped with a circular die with the following dimensions: length = 10 mm, diameter = 2 mm, and entrance angle = $\pi/2$. The applied shear rate is 1000 s^{-1} and the temperature of the cylinder and the die is controlled at 210 °C.

The generation of long-chain branches affects the elastic nature, such as melt tension and extrudate's swell, which are lack of recycled PET. In particular, high level of melt tension plays a significantly important role at various processing operations such as film processing, blow molding, foaming, and thermoforming.

Figure 5 shows the relation between the modifier content and the swell ratio, defined as the ratio of the diameter of an extruded strand to that of the die, at 210 °C. The measurements were performed employing a circular die at the shear rate of 10^3 s^{-1}. The swell ratio of the pure P(ET-EI) is 1.0, showing no extrudate swell at this condition, which is a typical behavior for a low viscous material like recycled PET. Further, the swell ratio increases monotonically with the content of the modifier. The diameter of the extruded strand with 4 wt% of the modifier is twice as large as the diameter of the die.

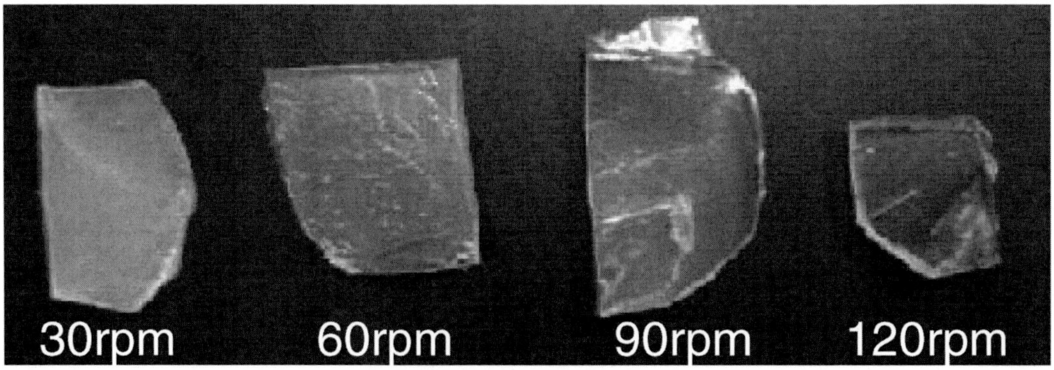

Figure 6. Appearance of the blend samples containing 4 wt% of the modifier. The samples were prepared by an internal mixer at different blade rotation speeds at 210 °C.

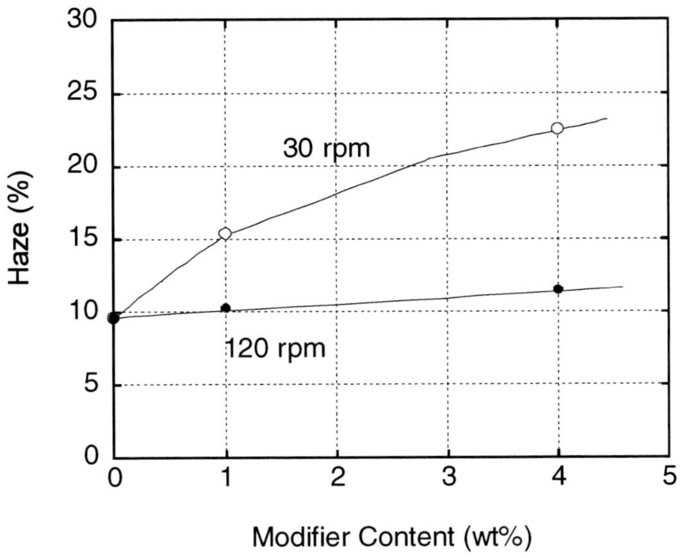

Figure 7. Haze of the blend samples prepared by an internal mixer at different blade rotation speeds at 210 °C. The thickness of the sample films is 0.3 mm.

Although the processability is significantly improved by the addition of the modifier, the optical transparency, which is one of the most important properties for such type of resins, depresses to a great extent as seen in Figure 3. The light scattering originated from the dispersed particles having different refractive index will be responsible for the opaque nature of the blends. The refractive indices of the raw materials employed are as follows; 1.575 for P(ET-EI) and 1.530 for the modifier. In general, polymers having different chemical structure are immiscible and the blend results in the phase separated morphology. In this blend system, the droplets of the modifier are dispersed in P(ET-EI) matrix. This is the problem for the polymeric modifier.

Matching of the refractive indices and the depression of the size of dispersed particles will be effective in order to avoid the problem, because the intensity of the light scattering is determined by them as demonstrated theoretically by Debye and Bueche (1949).

Yamaguchi et al. (2008) have tried to depress the size of dispersed particles by operating a mixing device at the optimum condition. They studied the relation between morphology and the mixing condition employing an internal mixer for the blends of P(ET-EI) and the pure modifier, not master batch. Figure 6 shows the photographs of the blends containing 4 wt% of the modifier obtained at the different blade rotation speed in the mixer. As clearly seen in the picture, the rotation speed, *i.e.*, mixing efficiency, greatly affect the transparency. Figure 7 shows the quantitative results by the haze value, defined as the percentage of total transmitted light passing through the specimen that is scattered from the incident beam by more than 2.5 degree; it is often employed as a measure of the turbidity of a film specimen. Therefore, low level of the haze value represents good transparency.

It should be noted that the effect of the modifier content on the transparency is quite significant when the rotation speed is low. On the other hand, the blends obtained at the high shear condition show relatively low values of haze. In particular, the haze level of the blends obtained at 120 rpm is close to that of the pure P(ET-EI) even when the modifier content is 4 wt%. Decrease in the radius of the dispersed particles is responsible for the improvement of transparency, which is proved by means of the scanning electron microscopic observation as shown in Figure 8. Although the number of dispersed particles increases with the modifier content, the size of the particles is not so different as long as the blade rotation speed is 30 rpm, suggesting that the applied shear stress plays a dominant role on the morphology. The particle size obtained at 120 rpm is approximately 100 nm for the blend with 10 wt% of the modifier, which is smaller than the wavelength of visible light. This will be the reason for the high level of transparency. The dispersed droplets of the modifier are deformed by the applied high shear stress resulting from high rotation speed, leading to the enlargement of interfacial area. In other words, intensive stress is required to overcome the interfacial tension between the polymers and/or the cohesive force of the dispersed particles. Consequently, chemical reaction between P(ET-EI) and the modifier takes place frequently. Then the generated block-copolymer, acting as a compatibilizer of the blend, makes the fine morphology.

Finally, the intensive mixing enhances the elastic nature of the modified P(ET-EI) in the molten state even at the low concentration of the modifier. This is plausible because the acceleration of the reaction leads to the generation of branched polymers to a great extent.

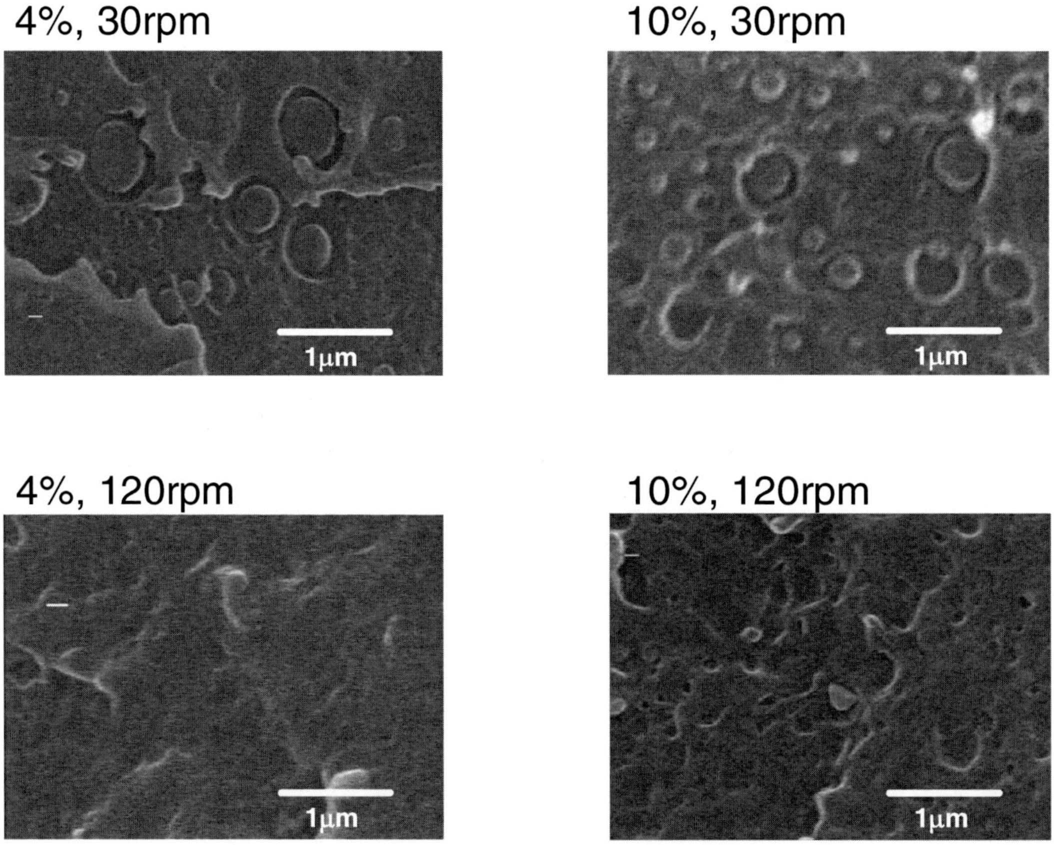

Figure 8. SEM pictures of the blend samples containing 4 and 10 wt% of the modifier. The samples were prepared by an internal mixer at different blade rotation speeds at 210 °C.

CONCLUSION

The improvement of processability for recycled PET is required intensively and it will become a serious problem in the near future, because the rapid growth of the collection rate of PET bottles all over the world as well as the excellent cost-performance. Moreover, further legislation in many countries accelerates the collection rate to a great extent. Therefore, advanced technology to improve the processability of recycled PET has to be established immediately.

Although various rheological and thus processing modifiers with low molecular weight have been proposed to enhance the viscosity and elastic nature in the past two decades, polymeric modifiers, which react with PET in a short time, will spread in the industries. This is reasonable because polymeric modifiers are available for conventional processing machines without any specific attachment such as a liquid feeding device.

Generation of long-chain branches as well as the increase in the molecular weight is responsible for the excellent processability for various processing operations including extrusion. The optical properties may be depressed by the light scattering originated from the dispersed modifier having different refractive index. This problem is avoided by intensive mixing or matching the refractive indices.

Finally, the technology shown in this chapter will be available not only for recycled PET but for various types of polyester and polyamide.

ACKNOWLEDGEMENT

The author would like to express his gratitude to Mr. Takashi Wakabayashi and Mr. Taichi Kanoh for their great efforts to accomplish the study.

REFERENCES

Arakawa K, Yokohara T, Yamaguchi M (2007) Enhancement of melt elasticity for poly(3-hydroxybutyrate-co-3-hydroxyvalerate) by addition of weak gel. *J Appl Polym Sci*: 107, 1320-1324.

Bikiaris D N, Karayannidis G P (1995) Chain extension of polyesters PET and PBT with *N,N*'-bis(glycidyl ester)pyromellitimides. I. *J Polym Sci Part A: Polym Chem*: 33, 1705-1714.

Bikiaris D N, Karayannidis G P (1996) Thermomechanical analysis of chain-extended PET and PBT. *J Appl Polym Sci* 60: 55-61.

Cardi N, Po R, Giannotta G, Occhiello E, Garbassi F, Messina G (1993) Chain extension of recycled poly(ethylene terephthalate) with 2,2'- bis(2-oxazoline). *J Appl Polym Sci* 50: 1501-1509.

Debye P, Bueche A M (1949) Scattering by an inhomogeneous solid. *J Appl Phys* 20: 518-525.

Dhavalikar R, Yamaguchi M, Xanthos M (2003) Molecular and structural analysis of a triepoxide-modified poly(ethylene terephthalate) from rheological data. *J Polym Sci Polym Chem* 41: 958-969.

Haralabakopoulos A A, Tsiourvas D, Paleos C M (1999) *J Appl Polym Sci* 71: 2121-2127.

Inata H, Matsumura S (1985) Chain extenders for polyesters. I. Addition-type chain extenders reactive with carboxyl end groups of polyesters. *J Appl Polym Sci* 30: 3325-3337.

Inata H, Matsumura S (1986) Chain extenders for polyesters. III. Addition-type nitrogen-containing chain extenders reactive with hydroxyl end groups of polyesters. *J Appl Polym Sci* 32: 4581-4594.

Inata H, Matsumura S (1987) Chain extenders for polyesters. IV. Properties of the polyesters chain-extended by 2,2'-bis(2-oxazoline). *J Appl Polym Sci* 33: 3069-3079.

Inata H, Matsumura S (1987) Chain extenders for polyesters. VI. Properties of the polyesters chain-extended by 2,2'-bis(4H-3,1-benzoxazin-4-one). *J Appl Polym Sci* 34: 2769-2776.

Incarnato L, Scarfato P, Di Maio L, Acierno D (2000) Structure and rheology of recycled PET modified by reactive extrusion. *Polymer* 41: 6825-6831.

Loontjens T, Belt W, Stanssens D, Weerts P (1993) Synthesis of 1,2-bis(2-oxazolinyl-2)ethane and its application as chain extender for poly(ethylene terephthalate). *Polym Bull* 20: 13-18.

Loontjens T, Pauwels K, Derks F, Nielen M, Sham C K, Serne M (1997) The action of chain extenders in Nylon-6, PET, and model compounds, *J Appl Polym Sci* 65: 1813-1819.

Scheirs J, Long T T (2003) *Modern polyesters*: New York, Wiley.

Xanthos M (1992) *Reactive extrusion*: Munich, Hanser.

Xanthos M, Yilmazer U, Dey S K, Quintas J (2000) Melt viscosity of polyethylene terephthalate resins for low density extrusion foaming. *Polym Eng Sci* 40: 554-566.

Xanthos M, Young M W, Karayannidis G P, Bikiaris D N (2001) Reactive modification of polyethylene terephthalate with polyepoxides. *Polym Eng Sci* 41: 643-655.

Yamaguchi M (2004) *Melt elasticity of polyolefin. in Polymeric foams, mechanism and materials*: New York, CRC press.

Yamaguchi M, Wakabayashi T (2006) Rheological properties and processability of chemically modified poly(ethylene terephthalate-co-ethylene isophthalate). *Adv Polym Technol* 25:236-241.

Yamaguchi M, Wakabayashi T, Kanoh T (2008) Effect of mixing conditions on the rheological properties and optical properties of chemically modified poly(ethylene terephthalate-co-ethylene isophthalate). *J Appl Polym Sci* 107:2665-2670.

In: Environmental Chemistry Research Progress
Editors: P. Robinson and R. Gallo
ISBN 978-1-60741-055-3

Chapter 6

ON THE NUMERICAL SOLUTION OF SOME PROBLEMS OF ENVIRONMENTAL POLLUTION

***Quang A Dang*[1], *Matthias Ehrhardt*[2], *Gia Lich Tran*[3] *and Duc Le*[4]**
[1] Institute of Information Technology, 18 Hoang Quoc Viet, Ha noi, Vietnam
[2] Institut für Mathematik, TU Berlin, Strasse des 17. Juni 136,
D–10623 Berlin, Germany
[3] Institute of Mathematics, 18 Hoang Quoc Viet, Ha noi, Vietnam
[4] Buro of Hydrology and Meteorology, Dang Thai Than, Ha noi, Vietnam

Abstract

Environmental problems are becoming more and more important for our world and their importance will even increase in the future. High pollution of air, water and soil may cause damage to plants, animals and humans. Therefore, the development of industry must be coupled with the protection of the environment. In this connection we shall consider the problems of optimal location of industrial enterprises and optimization of emissions from enterprises for ensuring sanitary environment criteria. Moreover, we study the problem of determination of the coefficients of diffusion and the coefficient of transformation of aerosols. Several numerical experiments illustrate the ability of the presented methods.

1 Introduction

At the beginning of this chapter will give a brief introduction to the mathematical formulation of air pollution models. Let G be a cylindrical domain in the three–dimensional space with the side surface Σ, the upper surface Σ_H $(z = H)$ and the bottom surface Σ_0 $(z = 0)$. In the sequel we shall use the following notations:

- $\varphi(x,y,z,t)$ denotes the concentration of aerosol pollutants
- $\mathbf{u} = u\mathbf{i} + v\mathbf{j} + w\mathbf{k}$ is the velocity vector of wind
- $\sigma \geq 0$ stands for the transformation coefficient of pollutants
- μ, ν are the horizontal and vertical coefficients of diffusion, respectively

- Σ_- and Σ_+ are the parts of the side surface Σ, where $u_n = \mathbf{u}\cdot\mathbf{n} < 0$ and $u_n \geq 0$, respectively. Here, $\mathbf{n}$ is the outward unit normal to Σ
- $w_g =$ constant > 0 denotes the falling velocity of the pollutants by gravity
- f is the power of the source.

The problem of air pollution is stated as follows: Determine the function $\varphi(x,y,z,t)$ that satisfies the differential equation

$$\frac{\partial\varphi}{\partial t} + u\frac{\partial\varphi}{\partial x} + v\frac{\partial\varphi}{\partial y} + (w - w_g)\frac{\partial\varphi}{\partial z} - \mu\left(\frac{\partial^2\varphi}{\partial x^2} + \frac{\partial^2\varphi}{\partial y^2}\right) - \frac{\partial}{\partial z}\left(\nu\frac{\partial\varphi}{\partial z}\right) + \sigma\varphi = f, \tag{1}$$

and the following initial and boundary conditions

$$\begin{aligned} \varphi = \varphi_0 &\quad \text{at} \quad t = 0, \\ \frac{\partial\varphi}{\partial z} = \alpha\varphi &\quad \text{on} \quad \Sigma_0, \\ \frac{\partial\varphi}{\partial z} = 0 &\quad \text{on} \quad \Sigma_H, \\ \varphi = \varphi_s &\quad \text{on} \quad \Sigma_-, \\ \frac{\partial\varphi}{\partial \mathbf{n}} = 0 &\quad \text{on} \quad \Sigma_+. \end{aligned} \tag{2}$$

Moreover, we assume here that (cf. [13])

$$\begin{aligned} \operatorname{div}\mathbf{u} = \frac{\partial u}{\partial x} + \frac{\partial v}{\partial y} + \frac{\partial w}{\partial z} = 0, \\ w = 0 \quad \text{at} \quad z = 0, z = H. \end{aligned} \tag{3}$$

Using the first relation in (3) and considering $w - w_g$ as the vertical component of the vector $\mathbf{u}$ with the notation $\Delta\varphi = \frac{\partial^2\varphi}{\partial x^2} + \frac{\partial^2\varphi}{\partial y^2}$ we can rewrite the equation (1) in the form

$$\frac{\partial\varphi}{\partial t} + \operatorname{div}(\mathbf{u}\varphi) - \mu\Delta\varphi - \frac{\partial}{\partial z}\left(\nu\frac{\partial\varphi}{\partial z}\right) + \sigma\varphi = f. \tag{4}$$

Instead of the two boundary conditions on two parts of Σ it is possible to pose the following one

$$\varphi = \varphi_s \quad \text{on} \quad \Sigma\,. \tag{5}$$

We shall refer to the equation (1) as the *main equation* and to the problem (1), (2) or (1), (5) as the *main problems*. It is well–known that the above problems allow for a unique solution.

The problem (1), (2) is a general 3D one. In many cases it is useful to use a 2D model that is an adequate approximation of the 3D problem:

$$\begin{aligned} \frac{\partial\varphi}{\partial t} + u\frac{\partial\varphi}{\partial x} + v\frac{\partial\varphi}{\partial y} - \mu\Delta\varphi + \sigma\varphi = f &\quad \text{in} \quad \Omega, \\ \varphi = \varphi_0 &\quad \text{at} \quad t = 0, \\ \varphi = \varphi_s &\quad \text{on} \quad \Gamma \quad \text{at} \quad u_n < 0, \\ \frac{\partial\varphi}{\partial \mathbf{n}} = 0 &\quad \text{on} \quad \Gamma \quad \text{at} \quad u_n \geq 0, \end{aligned} \tag{6}$$

where Γ is the boundary of the 2D domain Ω. Here it is assumed that the components of velocity u and v do not change with the altitude in the active layer of the substance transport and diffusion. In the above equation $\varphi(x,y,t)$ is the average of the concentration of pollutants in all heights of the cylindrical domain under consideration.

Many practical problems, for example, the substance (e.g. saline and alum) propagation in rivers (see [3], [16]), the stationary problem of air pollution generated by a point source (see [4], [15]) can even be reduced to the following 1D problem:

$$\begin{gathered}\frac{\partial\varphi}{\partial t}+u\frac{\partial\varphi}{\partial x}-\mu\frac{\partial^2\varphi}{\partial x^2}+\sigma\varphi=f,\quad -\infty<x<+\infty,\ t>0,\\ \varphi=\varphi_0\quad \text{at}\quad t=0,\\ \varphi\to 0\quad \text{as}\quad |x|\to\infty.\end{gathered}\tag{7}$$

If we limit the problem to a finite interval then we can pose the following boundary conditions

$$\begin{aligned}\varphi=\varphi_s\quad &\text{at}\quad B_-,\\ \frac{\partial\varphi}{\partial x}=0\quad &\text{at}\quad B_+,\end{aligned}$$

where B_- and B_+ denote the endpoints at which the flow comes in and comes out, respectively.

2 Adjoint Problems

In thi section we consider the main problem (1), (2) with homogeneous initial and boundary conditions, i. e. the problem

$$\begin{gathered}\frac{\partial\varphi}{\partial t}+\operatorname{div}(\mathbf{u}\varphi)-\mu\Delta\varphi-\frac{\partial}{\partial z}\left(\nu\frac{\partial\varphi}{\partial z}\right)+\sigma\varphi=f,\\ \varphi=0\quad \text{at}\quad t=0,\\ \frac{\partial\varphi}{\partial z}=\alpha\varphi\quad \text{on}\quad \Sigma_0,\\ \frac{\partial\varphi}{\partial z}=0\quad \text{on}\quad \Sigma_H,\\ \varphi=0\quad \text{on}\quad \Sigma_-,\\ \frac{\partial\varphi}{\partial\mathbf{n}}=0\quad \text{on}\quad \Sigma_+\end{gathered}\tag{8}$$

By the same way as in [13], [14] for the main problem with homogeneous boundary conditions and a periodic condition in the time variable we obtain the adjoint problem

$$\begin{aligned}
-\frac{\partial\varphi^*}{\partial t} - \operatorname{div}(\mathbf{u}\varphi^*) - \mu\Delta\varphi^* - \frac{\partial}{\partial z}\left(\nu\frac{\partial\varphi^*}{\partial z}\right) + \sigma\varphi^* = p,& \\
\varphi^* = 0 \quad \text{at} \quad t = T,& \\
\frac{\partial\varphi^*}{\partial z} = \alpha\varphi^* \quad \text{on} \quad \Sigma_0,& \\
\frac{\partial\varphi^*}{\partial z} = 0 \quad \text{on} \quad \Sigma_H,& \\
\varphi^* = 0 \quad \text{on} \quad \Sigma_-,& \\
\mu\frac{\partial\varphi^*}{\partial \mathbf{n}} + u_n\varphi^* = 0 \quad \text{on} \quad \Sigma_+.&
\end{aligned} \tag{9}$$

Here, the *complementary conditions* (3) are used. There is the following adjoint relation

$$\int_0^T dt \int_G p\varphi dG = \int_0^T dt \int_G \varphi^* f dG. \tag{10}$$

Similarly to the 3D case we consider the 2D problem with homogeneous initial and boundary conditions

$$\begin{aligned}
\frac{\partial\varphi}{\partial t} - \operatorname{div}(\mathbf{u}\varphi) - \mu\Delta\varphi + \sigma\varphi = f,& \\
\varphi = \varphi_0 \quad \text{at} \quad t = 0,& \\
\varphi = \varphi_s \quad \text{on} \quad \Gamma \quad \text{at} \quad u_n < 0,& \\
\frac{\partial\varphi}{\partial \mathbf{n}} = 0 \quad \text{on} \quad \Gamma \quad \text{at} \quad u_n \geq 0.&
\end{aligned} \tag{11}$$

The adjoint problem for the above problem reads

$$\begin{aligned}
-\frac{\partial\varphi^*}{\partial t} - \operatorname{div}(\mathbf{u}\varphi^*) - \mu\Delta\varphi^* + \sigma\varphi^* = p,& \\
\varphi^* = 0 \quad \text{at} \quad t = T,& \\
\varphi^* = \varphi_s^* \quad \text{on} \quad \Gamma \quad \text{at} \quad u_n < 0,& \\
\mu\frac{\partial\varphi^*}{\partial \mathbf{n}} + u_n\varphi^* = 0 \quad \text{on} \quad \Gamma \quad \text{at} \quad u_n \geq 0,&
\end{aligned} \tag{12}$$

and the adjoint relation has the form

$$\int_0^T \int_\Omega p\varphi dG = \int_0^T dt \int_\Omega \varphi^* f d\Omega. \tag{13}$$

In the 1D case for the main problem with homogeneous initial and boundary conditions

$$\begin{aligned}
\frac{\partial\varphi}{\partial t} + u\frac{\partial\varphi}{\partial x} - \mu\frac{\partial^2\varphi}{\partial x^2} + \sigma\varphi = f, \quad a < x < b, \ t > 0,& \\
\varphi = 0 \quad \text{at} \quad B_-,& \\
\frac{\partial\varphi}{\partial x} = 0 \quad \text{at} \quad B_+,&
\end{aligned} \tag{14}$$

the adjoint problem reads

$$\begin{aligned}
-\frac{\partial \varphi^*}{\partial t} - u\frac{\partial \varphi^*}{\partial x} - \mu\frac{\partial^2 \varphi^*}{\partial x^2} + \sigma\varphi^* = p, \quad a < x < b, \\
\varphi^* = 0 \quad \text{at} \quad t = T, \\
\varphi^* = 0 \quad \text{at} \quad B_-, \\
\mu\frac{\partial \varphi^*}{\partial x} + u\varphi^* = 0 \quad \text{at} \quad B_+.
\end{aligned} \tag{15}$$

If the 1D main problem is considered in the unbounded domain $-\infty < x < +\infty$ with the decay condition $\varphi \to 0$ for $|x| \to \infty$, then its adjoint problem is

$$\begin{aligned}
-\frac{\partial \varphi^*}{\partial t} - u\frac{\partial \varphi^*}{\partial x} - \mu\frac{\partial^2 \varphi^*}{\partial x^2} + \sigma\varphi^* = p, \quad -\infty < x < +\infty, \\
\varphi^* = 0 \quad \text{at} \quad t = T, \\
\varphi^* \to 0 \quad \text{at} \quad |x| \to \infty.
\end{aligned} \tag{16}$$

It is important to remark that all the stated adjoint problems can be reduced to the corresponding problems by making the substitution $t' = T - t$ and $\mathbf{u}' = -\mathbf{u}$. Therefore, all the qualitative results such as the uniqueness of solution for the main problems are also valid for the adjoint ones and all numerical methods for solving main problems are also applicable to adjoint problems.

3 Numerical Solution of Main and Adjoint Problems

For solving the main and adjoint problems of aerosol propagation there exists a large number of numerical schemes. Mainly, they are based on splitting methods developed by Yanenko [22] and Marchuk [13]. These schemes are stable and possess good approximation properties, but they may lead to solutions with negative values that are meaningless. Therefore, it is desired to construct difference schemes that avoid this defect. These difference schemes must ensure that if all initial and boundary conditions are nonnegative, then the solution of the corresponding problem is nonnegative, too. The difference schemes of this type are called *monotone* (or *positive*) ones (see [7]). Below, we present the method for constructing monotone difference schemes for the main problem (1), (2) developed in [6], [5]. Throughout this chapter we shall use the standard difference notations of Samarskii [17].

We begin with the consideration of the 1D parabolic problem

$$\begin{aligned}
& c(x,t)\frac{\partial \varphi}{\partial t} = L\varphi + f(x,t), \quad 0 < x < 1, \quad 0 < t \le T, \\
& \varphi(0,t) = \mu_1(t), \quad \varphi(1,t) = \mu_2(t), \\
& \varphi(x,0) = \varphi_0(x),
\end{aligned} \tag{17}$$

where

$$\begin{aligned}
& L\varphi = \frac{\partial}{\partial x}\left(k(x,t)\frac{\partial \varphi}{\partial x}\right) + r(x,t)\frac{\partial \varphi}{\partial x} - q(x,t)\varphi, \\
& 0 < c_1 \le k(x,t) \le c_2, \quad c(x,t) \ge c_1, \quad q(x,t) \ge 0.
\end{aligned} \tag{18}$$

We will construct a difference scheme for this problem on the uniform grid

$$\omega_{h\tau} = \{x_i = ih,\ t_j = j\tau,\ i = 0,1,\ldots,N;\ j = 0,1,\ldots,J;\ h = 1/N;\ \tau = T/J\}.$$

Firstly, we associate with the operator L a perturbation operator

$$\tilde{L}\varphi = \chi\frac{\partial}{\partial x}\left(k(x,t)\frac{\partial\varphi}{\partial x}\right) + r(x,t)\frac{\partial\varphi}{\partial x} - q(x,t)\varphi, \tag{19}$$

where $\chi = 1/(1+R)$, $R = 0.5h|r|/k$ and approximate the latter one by the difference operator

$$\tilde{\Lambda}y = \chi(ay_{\bar{x}})_x + b^+a^{(+1)}y_x + b^-ay_{\bar{x}} - dy, \tag{20}$$

where

$$b^\pm = \tilde{r}^\pm(x,t), \quad a = k(x-0.5h,t), \quad d = q(x,t),$$
$$r^\pm = (r \pm |r|)/2, \quad \tilde{r}^\pm = r^\pm/k, \quad a_i^{(+1)} = a_{i+1}.$$

Next, the problem (17) is replaced by the difference scheme

$$\begin{aligned} c(x,\bar{t})y_t &= \tilde{\Lambda}(\bar{t})\hat{y} + f, \quad \bar{t} = t + \tau/2,\\ y(0,t) &= \mu_1(t), \quad y(1,t) = \mu_2(t),\\ y(x,0) &= \varphi_0(x). \end{aligned} \tag{21}$$

This scheme has a truncation error of order $O(h^2+\tau)$ and is monotone. In this aspect the Crank–Nicolson difference scheme for the problem (17) is not better than (21) although it is of order $O(h^2+\tau^2)$ because the Crank–Nicolson method is monotone only if $\tau/h^2 \le 1/(2\min k + h\max|r|)$. The same conclusion holds for the so–called *optimal weighting scheme* of Wang and Lacroix [21].

If instead of a Dirichlet boundary condition there are Robin boundary conditions posed at the endpoints, for example,

$$\left(\alpha\varphi + \frac{\partial\varphi}{\partial x}\right)(1,t) = \mu_2(t),$$

then using the difference boundary condition

$$\alpha\varphi_N + \frac{y_{N+1} - y_{N-1}}{2h} = \mu_2,$$

we also obtain a monotone difference scheme for the corresponding differential problem.

Now, we consider the three–dimensional problem for the main problem (1), (2). For simplicity we suppose that the domain G is a parallelepiped $[0,X]\times[0,Y]\times[0,Z]$. In order to construct a difference scheme for (1), (2) we rewrite the equation in a convenient form

$$\frac{\partial\varphi}{\partial t} - (L_1 + L_2 + L_3)\varphi = f \quad \text{in} \quad G\times(0,T], \tag{22}$$

where

$$\begin{aligned} L_1\varphi &= \mu\frac{\partial^2\varphi}{\partial x^2} - u\frac{\partial\varphi}{\partial x}, \\ L_2\varphi &= \mu\frac{\partial^2\varphi}{\partial y^2} - v\frac{\partial\varphi}{\partial y}, \\ L_3\varphi &= \frac{\partial}{\partial z}\left(\nu\frac{\partial\varphi}{\partial z}\right) - (w-w_g) - \sigma\varphi. \end{aligned} \tag{23}$$

We employ on the domain G the uniform grid $G_h = \{x_i = ih_1, y_j = jh_2, z_k = kh_3\}$ and approximate the above differential operators (23) by the following monotone difference operators

$$\begin{aligned} \tilde{\Lambda}_1\phi &= \chi^{(x)}\mu\phi_{\bar{x}x} - u\phi_{\bar{x}}, \\ \tilde{\Lambda}_2\phi &= \chi^{(y)}\mu\phi_{\bar{y}y} - v\phi_{\bar{y}}, \\ \tilde{\Lambda}_3\phi &= \chi^{(z)}(a\phi_{\bar{z}})_z + b^+a^{(+1)}\phi_x + b^-a\phi_{\bar{x}} - \sigma\phi, \end{aligned}$$

where

$$\begin{aligned} &\chi^{(x)} = 1/(1+R^{(x)}), \quad R^{(x)} = 0.5h_1u/\mu, \\ &\chi^{(y)} = 1/(1+R^{(y)}), \quad R^{(y)} = 0.5h_2v/\mu, \\ &\chi^{(z)} = 1/(1+R^{(z)}), \quad R^{(z)} = 0.5h_3|r|/\nu, \quad r = w - w_g, \\ &a = \nu(z-0.5h_3), \quad a_i^{(+1)} = a_{i+1}, \quad b^{\pm} = r^{\pm}/\nu, \quad r^{\pm} = (r \pm |r|)/2. \end{aligned}$$

Now we write the difference scheme for the equation (22) with the boundary conditions for the case that the wind velocity is given by $\mathbf{u} = (u,0,0)$:

$$\begin{aligned} &\frac{\phi^{l+1/3} - \phi^l}{\tau} - \tilde{\Lambda}_1\phi^{l+1/3} = 0, \quad \phi_0^{l+1/3} = 0, \quad \phi_{I+1}^{l+1/3} - \phi_{I-1}^{l+1/3} = 0, \\ &\frac{\phi^{l+2/3} - \phi^{l+1/3}}{\tau} - \tilde{\Lambda}_2\phi^{l+2/3} = 0, \quad \phi_0^{l+2/3} = 0, \quad \phi_J^{l+2/3} = 0, \\ &\frac{\phi^{l+1} - \phi^{l+2/3}}{\tau} - \tilde{\Lambda}_3\phi^{l+1} = f^{l+1}, \quad \alpha\phi_0^{l+1} - \frac{\phi_1^{l+1} - \phi_{-1}^{l+1}}{2h_3} = 0, \quad \phi_K^{l+1} = 0, \\ &l = 0,1,\dots . \end{aligned} \tag{24}$$

Here, for brevity, we write only one space index for the computation direction, omitting other indexes, for example ϕ_I stands for ϕ_{Ijk}.

Due to the monotonicity of each component difference scheme in (24) it is possible to prove the positiveness of the solution of (24), its stability and the convergence order $O(h^2+\tau)$. Finally, we note that we considered in [15] another finite difference scheme based on the Abrashin alternative directions technique [1].

It should be remarked that, as many authors, we consider problems of propagation of pollutants in large bounded domains and put conditions on its boundary. These conditions indeed are artificial and close to real conditions. In the 1D case in [7] we have constructed

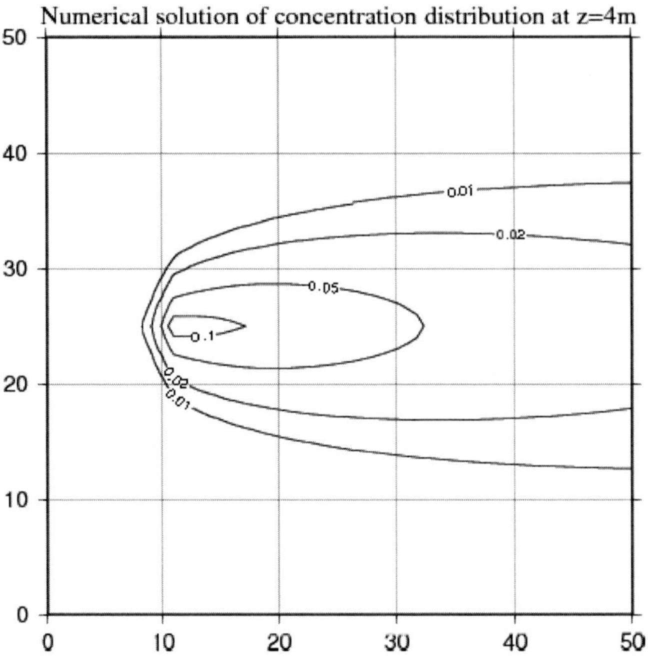

Figure 1: Concentration distribution computed by numerical solution at$z = 4$.

and studied monotone difference schemes for air pollution problems on unbounded domains using transparent boundary conditions (TBC). This TBC method was developed by Ehrhardt for 1D parabolic problems [9], [10].

Numerical Example. In order to test the above proposed difference method we consider the equation (1) with $f = \delta(x_0, y_0, H)q(t)$, $\sigma = 0$, $\mathbf{u} = (u, 0, 0)$, $w_g = 0$ and the following initial and boundary conditions

$$\begin{aligned} \varphi = 0 &\quad \text{at} \quad t = 0, \\ \varphi \to 0 &\quad \text{as} \quad x, y \to \pm\infty, \quad z \to \infty, \\ \frac{\partial \varphi}{\partial z} = 0 &\quad \text{at} \quad z = 0. \end{aligned}$$

This problem admits the exact solution [12]

$$\varphi = \int_0^t \frac{q(t)}{8\pi^{3/2}\mu\nu^{1/2}} \exp\Big(-\frac{(x - x_0 - u(t-\tau))^2}{4\mu(t-\tau)} - \frac{(y-y_0)^2}{4\mu(t-\tau)}\Big) \times \Big[\exp\Big(-\frac{(z-H)^2}{4\nu(t-\tau)}\Big) + \exp\Big(-\frac{(z+H)^2}{4\nu(t-\tau)}\Big)\Big] d\tau. \quad (25)$$

We performed the numerical computation using the difference scheme for the problem with the following data: $\mu = 2$, $\nu = 0.2$, $q(t) = 10$, $u = 2$. We limit the computational domain to $[0, 50] \times [0, 50] \times [0, 10]$ and locate a source at the point $(10, 25, 5)$. Below we present some isograms of concentration of aerosols obtained by the difference scheme and the exact solution.

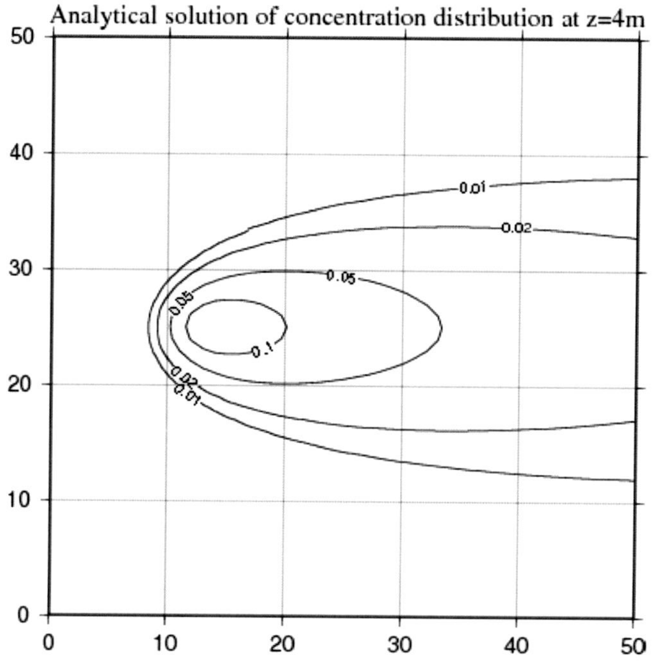

Figure 2: Concentration distribution computed by analytical solution at$z = 4$.

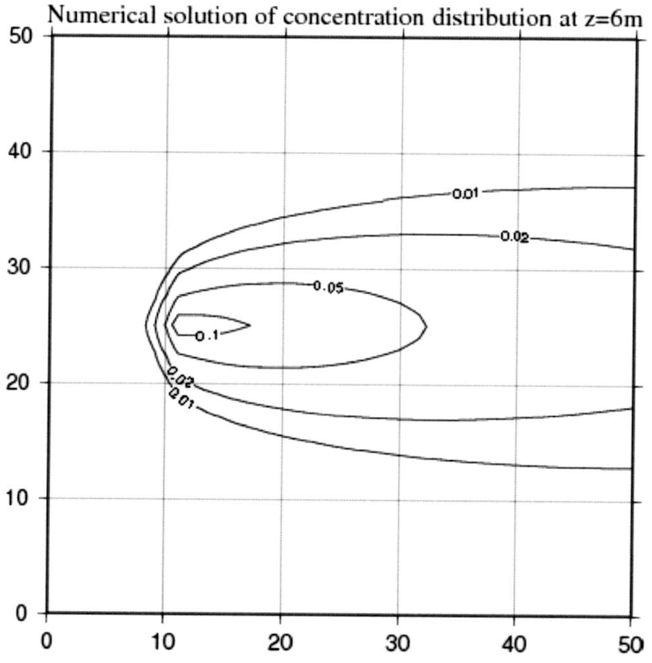

Figure 3: Concentration distribution computed by analytical solution at$z = 6$.

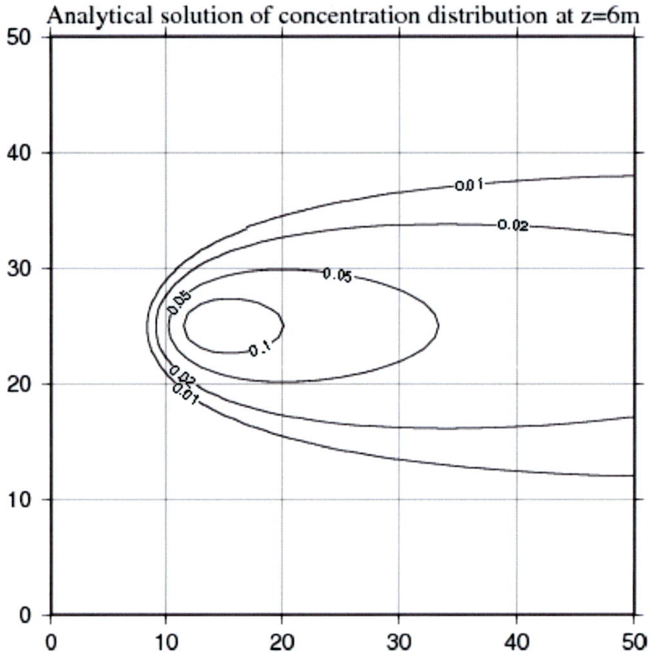

Figure 4: Concentration distribution computed by analytical solution at$z = 6$.

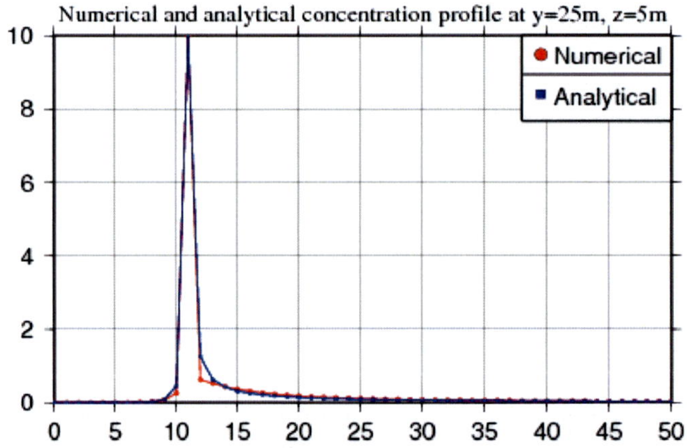

Figure 5: Concentration distribution computed by analytical solution at$y = 25$, $z = 5$.

From the Figures 1–5 one can clearly observe that the numerical solution of the problem (1) in a limited domain agrees well with the analytical solution (25) in the unbounded domain.

For the problem when the source of emission is a point and has a constant power under some assumptions on wind velocity and vertical coefficient of diffusion a monotone difference scheme for unbounded domains was constructed and a high order of accuracy was proven (see [5]).

4 Rational Location of Industrial Plants

First, as in [13], [14], we introduce a functional characterizing the level of pollution in a sensitive area Σ_k on the plane $z = 0$ in the base domain Σ_0

$$J_k = \int_0^T dt \int_{G_k} p_k \varphi dG_k, \tag{26}$$

where G_k is a unit cylinder over the sensitive area Σ_K. In Equation (26) φ denotes, as before, the concentration of pollutants,

$$p_k = \begin{cases} b_k + a_k\delta(z) & \text{in } G_k \\ 0 & \text{outside } G_k \end{cases}. \tag{27}$$

Usually we take

$$b_k = \frac{1}{T}, \quad a_k = w_g + \alpha\nu, \tag{28}$$

where $w_g =$ constant > 0 denotes the falling velocity of the pollutants by gravity. Then

$$\frac{1}{T}\int_0^T dt \int_{G_k} \varphi dG_k \tag{29}$$

is the average amount of aerosols in G_k and

$$\int_0^T dt \int_{G_k} a_k\delta(z)\varphi dG_k = a_k \int_0^T dt \int_{\Sigma_k} \varphi d\Sigma_k \tag{30}$$

is the total fallout of pollutants in the area Σ_k.

Now let there be *m sensitive areas* Σ_k, $k = 1,\dots,m$, on the plane $z = 0$ in the domain Σ_0. These may be populated areas, recreation zones, water reservoirs, etc. We consider the problem of locating a new industrial plant so that the pollution of all the m areas Σ_k does not exceed the approved standards (if such a location is possible in Σ_0 at all). If the location proves impossible on Σ_0 we formulate restrictions on the rate of pollution emissions Q which will make the location of the plant possible.

Suppose that the *approved standards of air pollution* for the sensitive area Σ_k are given by c_k. Then the problem is to choose the position $\mathbf{r}_0 = \mathbf{r}_0(x,y,H)$ of the source of aerosol emission such that the conditions

$$J_k \le c_k, \quad k = 1,2,\dots,m, \tag{31}$$

are satisfied.

Instead of this problem we can state the problem of finding a position $\mathbf{r_0}$, such that $\max_k J_k$ is minimal, i.e.

$$\min_{\mathbf{r}\in G}\max_k J_k\,. \tag{32}$$

There may be two methods for solving the above problems.

The first method consists of the following steps:

1. Choose of a position $\mathbf{r_0} \in G$,
2. Solve the main problem (1), (2) with $f = Q\delta(\mathbf{r}-\mathbf{r_0})$,
3. compute levels of pollution J_k of the areas Σ_k and verify the conditions (31).

If all the conditions are satisfied then the selected position$\mathbf{r_0}$ may be a location of the source of pollutant emission. Otherwise, choose another $\mathbf{r_0}$ and repeat the above process until a position $\mathbf{r_0}$ satisfies all the conditions.

The described method for solving the problem of plant location requires a great deal of computation due to the solution of many main problems for different test positions $\mathbf{r_0}$, especially if there is a large number of sensitive areas and industrial plants to be constructed. This method in some sense is a "blind method". Another method for the plant location problem is to use adjoint problems. Below we describe this method.

Let us represent the solution of the main equation

$$\frac{\partial\varphi}{\partial t} + L\varphi = Q\delta(\mathbf{r}-\mathbf{r_0}),$$

where for short we denote

$$L\varphi = \mathrm{div}(\mathbf{u}\varphi) - \mu\Delta\varphi - \frac{\partial}{\partial z}\left(\nu\frac{\partial\varphi}{\partial z}\right) + \sigma\varphi,$$

with the initial and boundary conditions (2) in the form of sum$\varphi = \varphi_1 + \varphi_2$, where φ_1 and φ_2 are the solutions of the problem

$$\begin{aligned}
\frac{\partial\varphi_1}{\partial t} + L\varphi_1 &= 0,\\
\varphi_1 &= \varphi_0 \quad \text{at} \quad t=0,\\
\frac{\partial\varphi_1}{\partial z} &= \alpha\varphi_1 \quad \text{on} \quad \Sigma_0,\\
\frac{\partial\varphi_1}{\partial z} &= 0 \quad \text{on} \quad \Sigma_H,\\
\varphi_1 &= \varphi_s \quad \text{on} \quad \Sigma_-,\\
\frac{\partial\varphi_1}{\partial n} &= 0 \quad \text{on} \quad \Sigma_+,
\end{aligned} \tag{33}$$

$$\begin{aligned}
&\frac{\partial \varphi_2}{\partial t} + L\varphi_2 = Q\delta(\mathbf{r}-\mathbf{r_0}),\\
&\varphi_2 = \varphi_0 \quad \text{at} \quad t=0,\\
&\frac{\partial \varphi_2}{\partial z} = \alpha\varphi_1 \quad \text{on} \quad \Sigma_0,\\
&\frac{\partial \varphi_2}{\partial z} = 0 \quad \text{on} \quad \Sigma_H,\\
&\varphi_2 = 0 \quad \text{on} \quad \Sigma_-,\\
&\frac{\partial \varphi_2}{\partial n} = 0 \quad \text{on} \quad \Sigma_+.
\end{aligned} \tag{34}$$

We remark that the solution φ_1 does not depend on the source of emission while the solution φ_2 depends only on it.

The pollution level functional for a sensitive area Σ_k now is

$$J_k = \int_0^T dt \int_G p_k\varphi dG = \int_0^T dt \int_G p_k\varphi_1 dG + \int_0^T dt \int_G p_k\varphi_2 dG. \tag{35}$$

We denote by φ_{2k}^* the solution of the problem adjoint to (34), namely, the problem

$$\begin{aligned}
&-\frac{\partial \varphi_{2k}^*}{\partial t} - \operatorname{div}(\mathbf{u}\varphi_{2k}^*) - \mu\Delta\varphi_{2k}^* - \frac{\partial}{\partial z}\left(\nu\frac{\partial \varphi_{2k}^*}{\partial z}\right) + \sigma\varphi_{2k}^* = p_k,\\
&\varphi_{2k}^* = 0 \quad \text{at} \quad t=T,\\
&\frac{\partial \varphi_{2k}^*}{\partial z} = \alpha\varphi_{2k}^* \quad \text{on} \quad \Sigma_0,\\
&\frac{\partial \varphi_{2k}^*}{\partial z} = 0 \quad \text{on} \quad \Sigma_H,\\
&\varphi_{2k}^* = 0 \quad \text{on} \quad \Sigma_-,\\
&\mu\frac{\partial \varphi_{2k}^*}{\partial n} + u_n\varphi_{2k}^* = 0 \quad \text{on} \quad \Sigma_+.
\end{aligned} \tag{36}$$

In view of the duality relation

$$\int_0^T dt \int_G p_k\varphi_2 dG = \int_0^T dt \int_G \varphi_{2k}^* Q\delta(\mathbf{r}-\mathbf{r_0}) dG, \tag{37}$$

we obtain

$$J_k = \int_0^T dt \int_G p_k\varphi_1 dG + Q\int_0^T \varphi_{2k}^*(\mathbf{r_0},t). \tag{38}$$

Thus, in the adjoint approach for calculating the functionals J_k, $k=1,\ldots,m$ we have to solve one main problem (33) and m adjoint problems (36) for m sensitive areas. Doing so, the amount of computation is significantly reduced in comparison with the direct approach. This saving of computational effort is a remarkable advantage of the proposed adjoint approach.

In order to choose a suitable position for a plant it is convenient to introduce the following function

$$J(\mathbf{r}) = \max_k J_k(\mathbf{r})$$

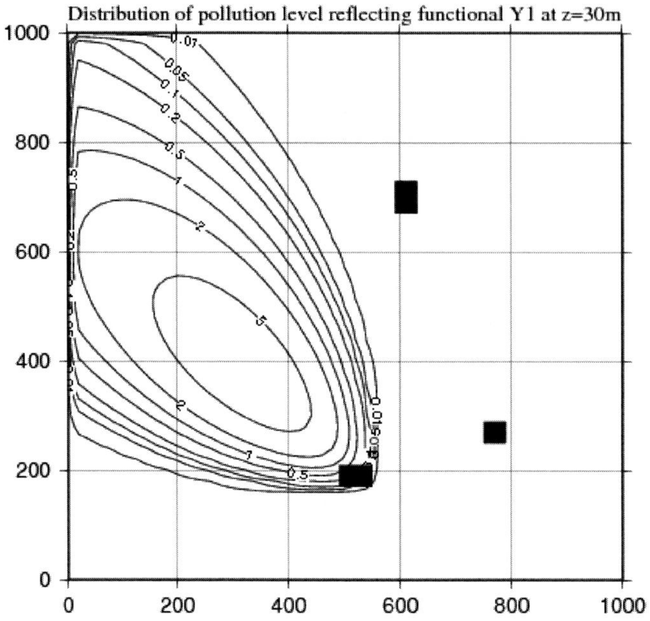

Figure 6: Isogram of pollution level functional J_1 at $z = 30$m.

and draw isograms of this function.

Numerical Example. In this example we want to study the problem of a plant location with given sensitive areas and approved standards of air pollution. We choose the following parameters:

– domain $G = [0,1000] \times [0,1000] \times [0,50]$ (in meters)

– uniformly grid with step sizes $dx = 20$m, $dy = 20$m, $dz = 5$m

– time of simulation $T = 1000$sec, time step $dt = 5$sec

– velocity vector $\mathbf{u} = (1, -1, 0)$

– falling velocity $w_g = 0.1$

– diffusion coefficients $\mu = 2, \nu = 0.2$

– coefficient of transformation $\sigma = 0$

– source of emission $Q = 50$

– three sensitive areas:

$$\Sigma_1 = [24.5, 27.5]dx \times [8.5, 10.5]dy,$$
$$\Sigma_2 = [37.5, 39.5]dx \times [12.5, 14.5]dy,$$
$$\Sigma_3 = [29.5, 31.5]dx \times [33.5, 36.5]dy$$

– approved standards of air pollution $c_k = 1$, $k = 1, 2, 3$

– initial and boundary conditions are homogeneous.

Below we present in Figures 6–13 isograms of the pollution level functionals $J_k(\mathbf{r})$ for each sensitive area and of the functional $J(\mathbf{r})$ at the heights $z = 30$m and $z = 35$m. Afterwards, we show in Figures 14 and 15 all the places where a plant cannot be located at the heights $z = 30$m and $z = 40$m for the approved standards of air pollution c_k set to unity.

Remark that in this section we consider the problem of rational location of industrial plants in 3D formulation. The 2D and 1D cases were studied by many authors, for example, in [20]its authors found places for locating plants in Halong Bay with the use of 2D model.

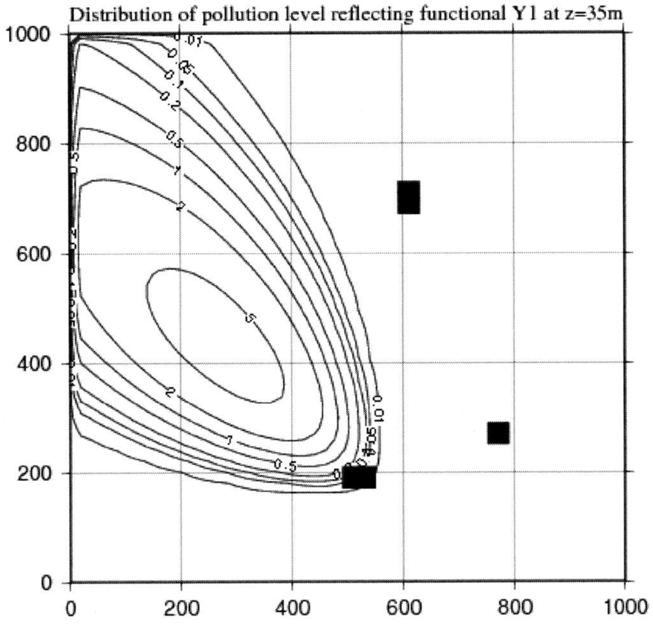

Figure 7: Isogram of pollution level functional J_1 at $z = 35$m.

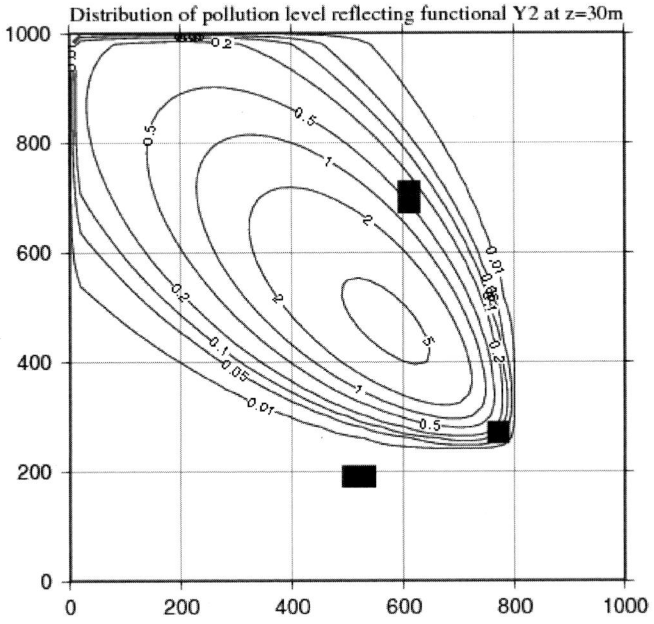

Figure 8: Isogram of pollution level functional J_2 at $z = 30$m.

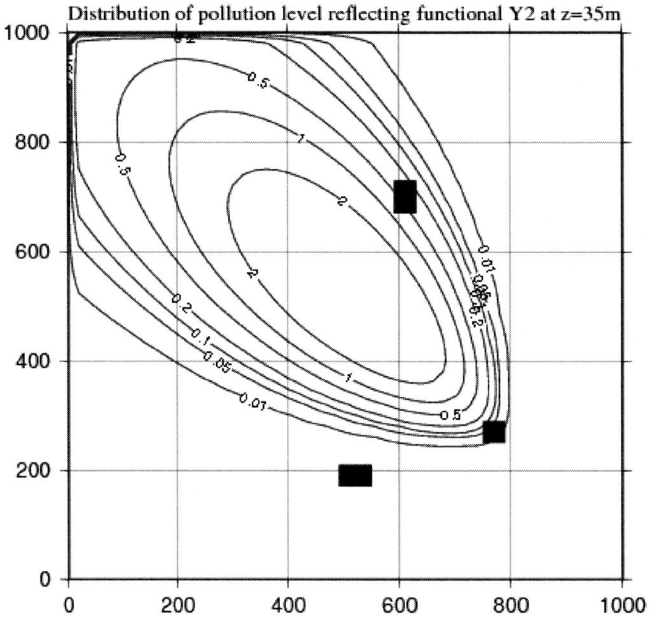

Figure 9: Isogram of pollution level functional J_2 at $z = 35$m.

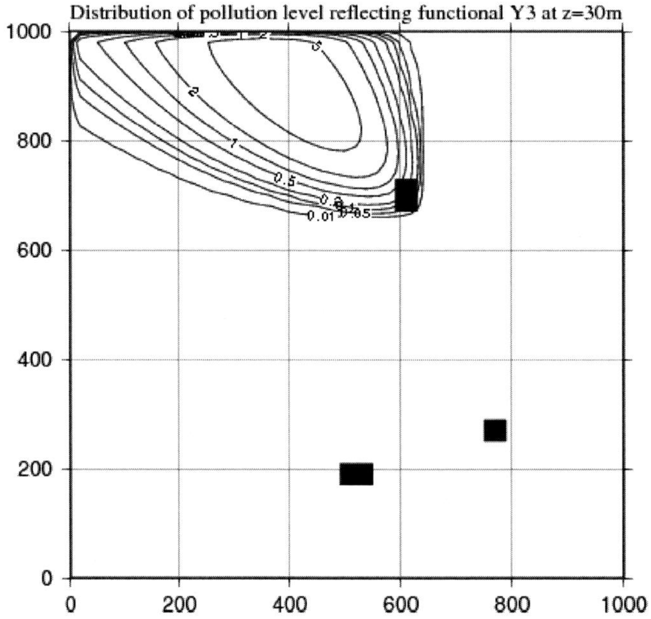

Figure 10: Isogram of pollution level functional J_3 at $z = 30$m.

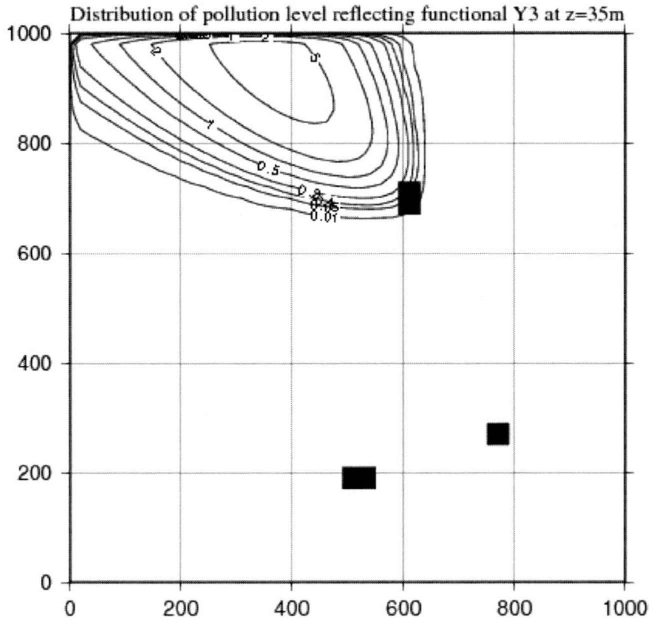

Figure 11: Isogram of pollution level functional J_3 at $z = 35$m.

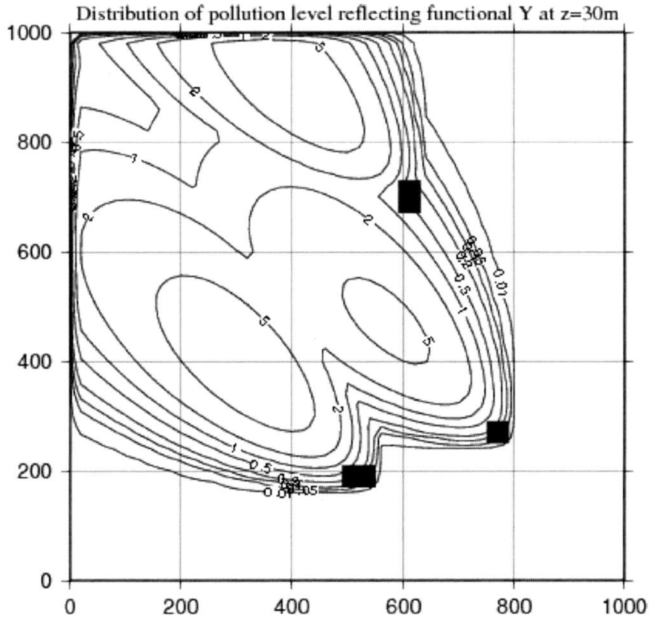

Figure 12: Isogram of pollution level functional J at $z = 30$m.

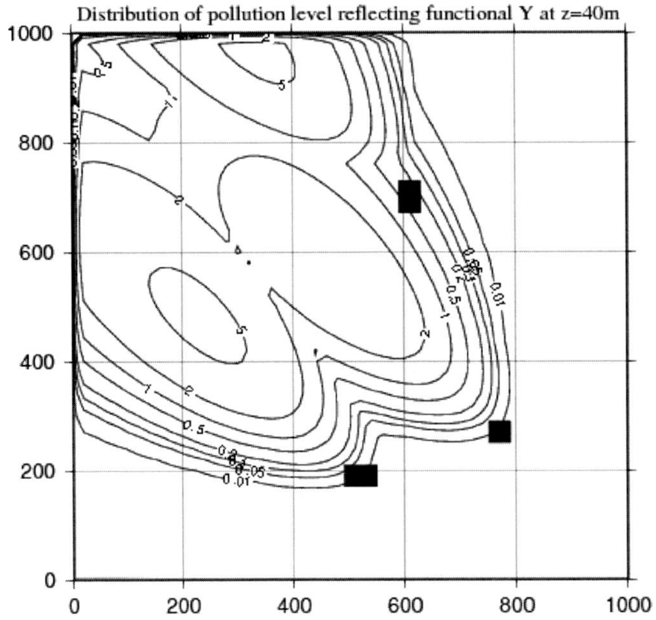

Figure 13: Isogram of pollution level functional J at $z = 40$m.

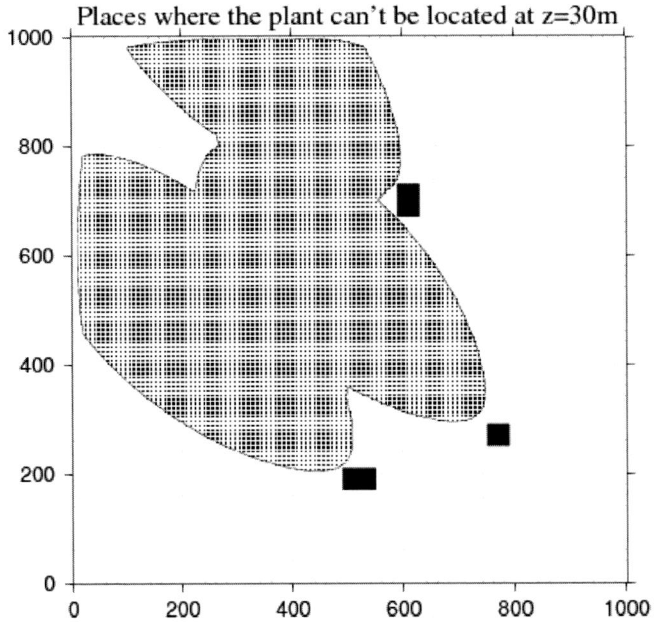

Figure 14: Places where a plant cannot be located at $z = 30$m if $c_1 = c_2 = c_3 = 1$.

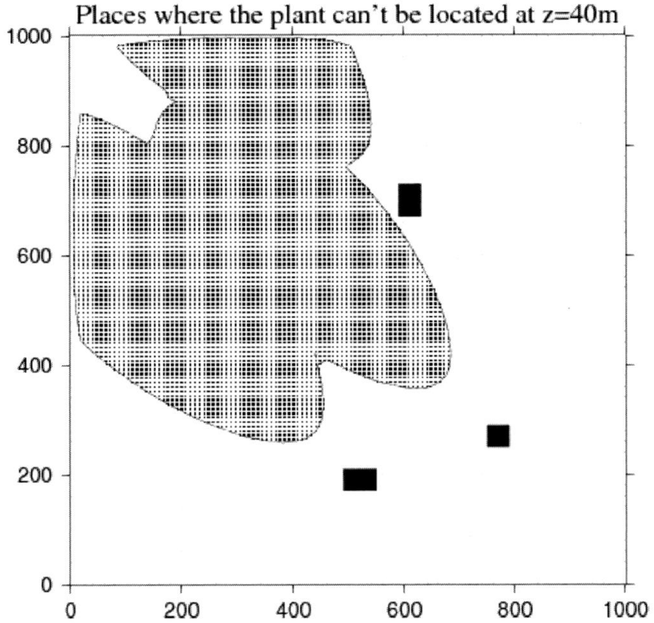

Figure 15: Places where a plant cannot be located at $z = 40$m if $c_1 = c_2 = c_3 = 1$.

A review of 1D problems concerning location of plants and optimization of emissions of operating plants can be found in [8].

5 Optimization of Emissions at Operating Industrial Plants

Assume that n plants A_i are located at points $\mathbf{r_i}$, $i = 1, 2, \ldots, n$ of a assigned area G. The plants emit $\bar{Q}_i$, $i = 1, \ldots, n$ of aerosols per time unit. For simplicity we assume that the emission composition is the same. Furthermore, we suppose that there are m sensitive areas G_k, $k = 1, \ldots, m$ to be protected from the polluted aerosols. The task is to determine affordable planned pollution level emissions rate Q_i for every plant so that the pollution level of the sensitive areas do not exceed the standards c_k under the condition that the total investment into technology to keep up the same production level is minimal.

Now we formulate the above problem as a problem of optimization

$$
\begin{aligned}
I = \sum_{i=1}^{n} \xi_i(\bar{Q}_i - Q_i) &\to \min, \\
J_k \le c_k, \quad k &= 1, 2, \ldots, n,
\end{aligned}
\tag{39}
$$

where ξ_i is the cost for reducing emissions per unit of emission rate. The problem (39) may be reduced to a problem of linear programming. Two different approaches are possible: main and adjoint equations.

i) Main equation approach. The main equation of diffusion and transportation of substances emitted by n plants located at the positions $\mathbf{r_i}$ is

$$\begin{aligned} \frac{\partial\varphi}{\partial t} + L\varphi &= \sum_{i=1}^{n} Q_i\delta(\mathbf{r}-\mathbf{r_i}), \\ \varphi = \varphi_0 \quad &\text{at} \quad t=0, \\ \frac{\partial\varphi}{\partial z} = \alpha\varphi \quad &\text{on} \quad \Sigma_0, \\ \frac{\partial\varphi}{\partial z} = 0 \quad &\text{on} \quad \Sigma_H, \\ \varphi = \varphi_s \quad &\text{on} \quad \Sigma_-, \\ \frac{\partial\varphi}{\partial \mathbf{n}} = 0 \quad &\text{on} \quad \Sigma_+. \end{aligned} \tag{40}$$

We represent the solution of this problem as a superposition of solutions of elementary problems

$$\varphi = \sum_{i=1}^{n} Q_i\varphi_i(\mathbf{r},t) + \varphi_g, \tag{41}$$

where $\varphi_i(\mathbf{r},t)$ is a solution to the problem

$$\begin{aligned} \frac{\partial\varphi_i}{\partial t} + L\varphi_i &= \delta(\mathbf{r}-\mathbf{r_i}), \\ \varphi_i = 0 \quad &\text{at} \quad t=0, \\ \frac{\partial\varphi_i}{\partial z} = \alpha\varphi_i \quad &\text{on} \quad \Sigma_0, \\ \frac{\partial\varphi_i}{\partial z} = 0 \quad &\text{on} \quad \Sigma_H, \\ \varphi_i = 0 \quad &\text{on} \quad \Sigma_-, \\ \frac{\partial\varphi_i}{\partial \mathbf{n}} = 0 \quad &\text{on} \quad \Sigma_+, \end{aligned} \tag{42}$$

and φ_g is a solution to the problem

$$\begin{aligned} \frac{\partial\varphi_g}{\partial t} + L\varphi_g &= 0, \\ \varphi_g = \varphi_0 \quad &\text{at} \quad t=0, \\ \frac{\partial\varphi_g}{\partial z} = \alpha\varphi_g \quad &\text{on} \quad \Sigma_0, \\ \frac{\partial\varphi_g}{\partial z} = 0 \quad &\text{on} \quad \Sigma_H, \\ \varphi_g = \varphi_s \quad &\text{on} \quad \Sigma_-, \\ \frac{\partial\varphi_g}{\partial \mathbf{n}} = 0 \quad &\text{on} \quad \Sigma_+. \end{aligned} \tag{43}$$

If we assume that all the problems (42) and (43) are solved, then we can calculate the pollution level functional for every sensitive area by

$$J_k = \int_0^T dt \int_G p_k \varphi dG = \int_0^T dt \int_G p_k \Big(\sum_{i=1}^n Q_i \varphi_i(\mathbf{r},t) + \varphi_g \Big), \\ = \sum_{i=1}^n a_{ik} Q_i + b_k, \tag{44}$$

where

$$a_{ik} = \int_0^T dt \int_G p_k \varphi_i dG, \tag{45}$$

$$b_k = \int_0^T dt \int_G p_k \varphi_g dG. \tag{46}$$

Now the optimization problem (39) has the form

$$I = \sum_{i=1}^n \xi_i (\bar{Q}_i - Q_i) \to \min, \\ \sum_{i=1}^n a_{ik} Q_i + b_k \le c_k, \quad k = 1,\dots,m. \tag{47}$$

Passing from Q_i to $q_i = \bar{Q}_i - Q_i$ we obtain a linear programming problem

$$\sum_{i=1}^n \xi_i q_i \to \min, \\ \sum_{i=1}^n a_{ik} q_i \ge R_k, \quad k = 1,2,\dots,m, \\ q_i \ge 0, \quad i = 1,2,\dots,n, \tag{48}$$

where

$$R_k = \sum_{i=1}^n a_{ik} \bar{Q}_i + b_k - c_k.$$

Thus, we have to solve in the first approach $n+1$ main problems and one linear programming problem.

ii) Adjoint equation approach. Represent the solution of the main problem (40) as the sum

$$\varphi = \varphi_h + \varphi_g,$$

where φ_g, as above, denotes the solution of the problem (43) and φ_h is the solution of the problem

$$
\begin{aligned}
\frac{\partial \varphi_h}{\partial t} + L\varphi_h &= \sum_{i=1}^{n} \delta(\mathbf{r}-\mathbf{r_i}),\\
\varphi_h = 0 \quad &\text{at} \quad t=0,\\
\frac{\partial \varphi_h}{\partial z} = \alpha\varphi_h \quad &\text{on} \quad \Sigma_0,\\
\frac{\partial \varphi_h}{\partial z} = 0 \quad &\text{on} \quad \Sigma_H,\\
\varphi_h = 0 \quad &\text{on} \quad \Sigma_-,\\
\frac{\partial \varphi_h}{\partial \mathbf{n}} = 0 \quad &\text{on} \quad \Sigma_+.
\end{aligned}
\tag{49}
$$

Let φ^*_{hk} be the solution of the adjoint problem

$$
\begin{aligned}
-\frac{\partial \varphi^*_{hk}}{\partial t} - \operatorname{div}(\mathbf{u}\varphi^*_{hk}) - \mu\Delta\varphi^*_{hk} - \frac{\partial}{\partial z}\left(\nu\frac{\partial \varphi^*_{hk}}{\partial z}\right) + \sigma\varphi^*_{hk} &= p_k,\\
\varphi^*_{hk} = 0 \quad &\text{at} \quad t=T,\\
\frac{\partial \varphi^*_{hk}}{\partial z} = \alpha\varphi^*_{hk} \quad &\text{on} \quad \Sigma_0,\\
\frac{\partial \varphi^*_{hk}}{\partial z} = 0 \quad &\text{on} \quad \Sigma_H,\\
\varphi^*_{hk} = 0 \quad &\text{on} \quad \Sigma_-,\\
\mu\frac{\partial \varphi^*_{hk}}{\partial n} + u_n\varphi^*_{hk} = 0 \quad &\text{on} \quad \Sigma_+.
\end{aligned}
\tag{50}
$$

Then we have the duality relation

$$
\begin{aligned}
\int_0^T dt \int_G p_k\varphi_h dG &= \int_0^T dt \int_G \sum_{i=1}^{n} Q_i\delta(\mathbf{r}-\mathbf{r_i})\varphi^*_{hk} dG\\
&= \sum_{i=1}^{n} Q_i \int_0^T \varphi^*_{hk}(\mathbf{r_i},t)dt = \sum_{i=1}^{n} a^*_{ik}Q_i,
\end{aligned}
\tag{51}
$$

where we abbreviate

$$
a^*_{ik} = \int_0^T \varphi^*_{hk}(\mathbf{r_i},t)dt. \tag{52}
$$

Therefore, the pollution level functional for the sensitive areaΣ_k may be calculated as follows

$$
J_k = \int_0^T dt \int_G p_k\varphi dG = \int_0^T dt \int_G p_k\varphi_h dG + \int_0^T dt \int_G p_k\varphi_g dG = \sum_{i=1}^{n} a^*_{ik}Q_i + b_k,
$$

where b_k is computed by (46).

Hence, the optimization problem (39) becomes

$$
\begin{aligned}
&\sum_{i=1}^{n} \xi_i(\bar{Q}_i - Q_i) \to \min,\\
&\sum_{i=1}^{n} a^*_{ik}Q_i + b^*_k \le c_k, \quad k=1,2,\dots,m,
\end{aligned}
\tag{53}
$$

where a_{ik}^* are given by (52) and $b_k^* = b_k$.

Let us note that

$$a_{ik}^* = \int_0^T \varphi_{hk}^*(\mathbf{r_i},t)dt = \int_0^T dt \int_G \delta(\mathbf{r}-\mathbf{r_i})\varphi_{hk}^* dG$$
$$= \int_0^T dt \int_G p_k \varphi_i dG = a_{ik}.$$

As in the main equation approach, we transform the problem (53) by letting $q_i = \bar{Q}_i - Q_i$ to the form

$$\begin{aligned} &\sum_{i=1}^{n} \xi_i q_i \to \min, \\ &\sum_{i=1}^{n} a_{ik}^* q_i \geq R_k^*, \quad k = 1,2,\ldots,m, \\ &q_i \geq 0, \quad i = 1,2,\ldots,n, \end{aligned} \tag{54}$$

where

$$R_k^* = \sum_{i=1}^{n} a_{ik}^* \bar{Q}_i + b_k^* - c_k.$$

This is again a linear programming problem.

It should be noticed that in the adjoint approach before solving the linear programming problem it is necessary to solve m adjoint problems (50) and one main problem (42) while in the main equation approach we have to solve $n+1$ main problems. Thus, the choice of a method depends on the relation of m and n, namely, if $m > n$ it is better to use the adjoint approach, otherwise to use the main equation approach.

We remark that here the problem of optimization of emissions of operating plants is considered in 3D case. The 2D case was studied by Tran G.L. and Pham T.N. [19], where they carried out many experiments with the use of an up–wind difference scheme for main and adjoint problems in order to verify qualitative properties of the problem. Below we report some numerical results of the problem under consideration for two examples.

Numerical Example 1. Consider the optimization problem of emissions of plants with the following data:

– domain $G = [0,1000] \times [0,1000] \times [0,50]$ (in meters)
– step sizes $dx = 20$m, $dy = 20$m, $dz = 5$m
– time of simulation $T = 4000$ sec, time step $dt = 4$ sec
– velocity vector $\mathbf{u} = (1,0,0)$, falling velocity $w_g = 0.1$
– diffusion coefficients $\mu = 2, \nu = 0.2$
– coefficient of transformation $\sigma = 0.005$
– 3 sensitive areas marked by rectangles:

$\Sigma_1 = [24.5,26.5]dx \times [25.5,26.5]dy$ with $c_1 = 10$,
$\Sigma_2 = [39.5,40.5]dx \times [33.5,35.5]dy$ with $c_2 = 30$,
$\Sigma_3 = [39.5,40.5]dx \times [14.5,16.5]dy$ with $c_3 = 20$

– initial and boundary conditions are homogeneous.

– 4 sources of emissions marked by circles: the first source at point $(200,500,20)$ with rate 100, the second source at point $(300,700,30)$ with rate 70, the third source at point $(300,300,25)$ with rate 50 and the fourth source at point $(600,500,15)$ with rate 90.
– costs for reducing one unit of emission rate of plants are 1.2, 1.4, 1.0 and 1.5, resp.
The result of the numerical computation is given in the Table 5.

Optimal emission rates in Example 1.

Source	Planned emission rate	Cost for reducing emission rates
1	88.326	14.009
2	60.670	13.062
3	36.259	13.741
4	90	0

The pollution level functionals of the sensitive areas before and after reducing emission rates are $(11.322, 34.613, 27.579)$ and $(10, 30, 20)$, respectively.

Next, the following Figures 16 and 17 show the average (in time) concentration distribution of aerosols at $z = 0$m before and after optimizing emission rates of sources, respectively.

Numerical Example 2. The data in this example are the same as in the previous example except for the following changes:
– velocity vector $\mathbf{u} = (1, -0.5, 0)$
– approved pollution standards are $c_1 = 1, c_2 = 2, c_3 = 15$.
The result of calculations is given in the following Table 1.

Table 1: Optimal emission rates in Example 2.

Source	Planned emission rate	Cost for reducing emission rates
1	84.288	18.927
2	63.055	9.723
3	50	0
4	90	0

Here, the pollution level functionals of the sensitive areas before and after reducing emission rates are $(1.120, 2.061, 17.388)$ and $(1, 1.856, 15)$, respectively.

The following Figures 18–19 present the average (in time) concentration distribution of aerosols at $z = 0$m before and after optimizing emission rates of sources, respectively.

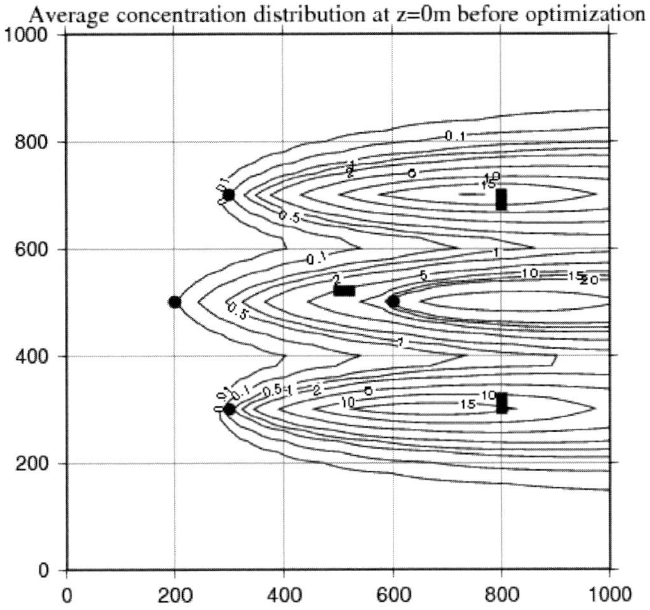

Figure 16: Average concentration distribution at $z = 0$m before optimizing sources in Example 1.

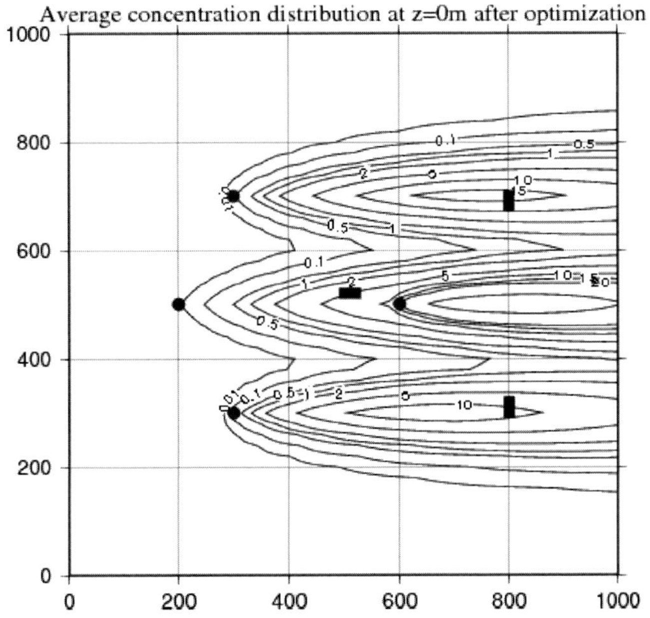

Figure 17: Average concentration distribution at $z = 0$m after optimizing sources in Example 1.

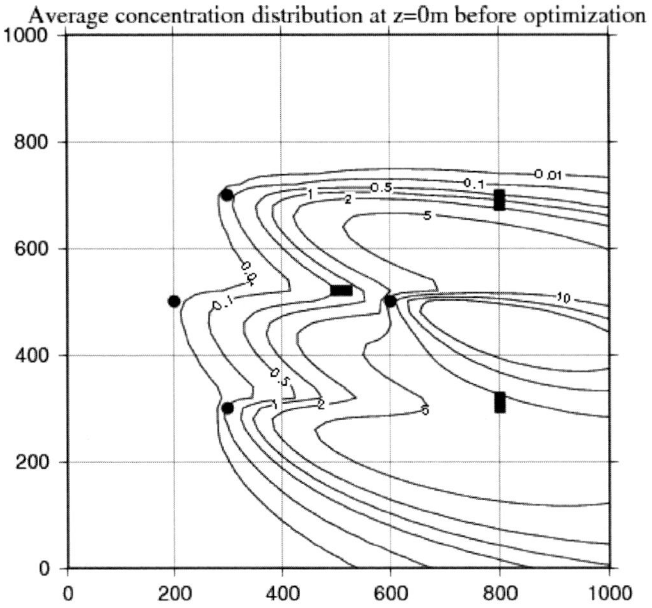

Figure 18: Average concentration distribution at $z = 0$m before optimizing sources in Example 2.

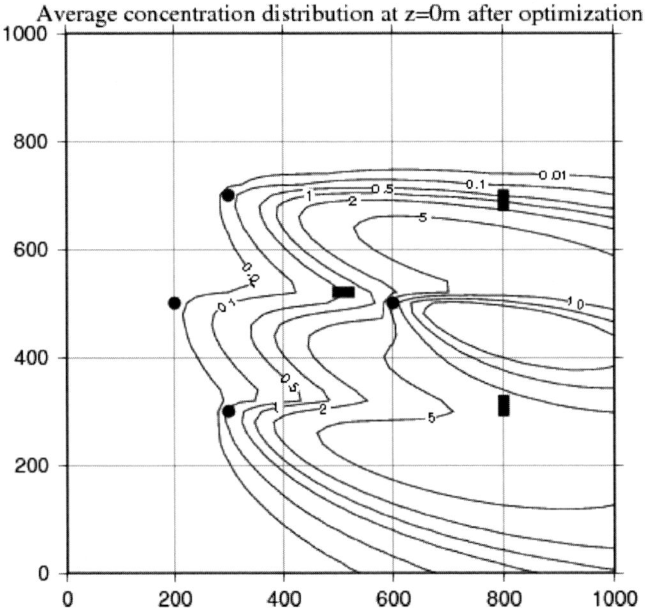

Figure 19: Average concentration distribution at $z = 0$m after optimizing sources in Example 2.

6 Determination of the Coefficients of Diffusion and Transformation of Pollutants

The coefficients of diffusion μ, ν and the coefficient of transformation σ of aerosols are important parameters of mathematical models for the study of air pollution in the atmosphere. These coefficients depend on many factors. Following Kartvelishvili [11], in general, these coefficients may depend on the concentration of aerosols. But as an approximation we suppose that they are constants and shall find them under the assumption that the concentration field of pollutants is known from measurements at grid points and at some times. Discretizing the equation (1) and considering difference equations for those grid points at those time moments where and when measurements are made, we obtain a system of equations of the form

$$\sum_{j=1}^{3} a_{ij}\lambda_j = b_i, \quad i = 1,\ldots,N,$$

where we use the notation $(\lambda_1,\lambda_2,\lambda_3) = (\mu,\nu,\sigma)$. This is a system of many equations with 3 unknowns. For finding them it is possible to use the method of least squares, which gives an approximation of μ, ν, σ with minimal error. This may lead to negative solution. In order to avoid this drawback, it is better to solve the following quadratic programming problem

$$\begin{gathered}\sum_{i=1}^{N}(\sum_{j=1}^{3} a_{ij}\lambda_j - b_i)^2 \to \min, \\ \lambda_i \geq 0, \quad i = 1,2,3.\end{gathered} \tag{55}$$

Numerical Example. For illustrating the ability of the above method we performed a numerical experiment, where the concentration field is obtained from the solution of the problem (1), (2) with the following data:

– domain $G = [0,1000] \times [0,1000] \times [0,50]$ (in meters)
– step sizes $dx = 20$m, $dy = 20$m, $dz = 5$m.
– time step $dt = 4$sec
– velocity vector $\mathbf{u} = (0,0,0)$
– falling velocity $w_g = 0$
– diffusion coefficients $\mu = 2, \nu = 0.2$
– coefficient of transformation $\sigma = 0.005$
– source of emission $Q = 10$ at position $(100,100,25)$
– initial and boundary conditions are homogeneous.

We use the data of the concentration field obtained in the grid points
(x_i,y_j,z_k), $i = istart,\ldots,iend$; $j = jstart,\ldots,jend$; $k = kstart,\ldots,kend$
at times t_m, $m = mstart,\ldots,mend$, which for brevity we write as $(istart : iend, jstart : jend, kstart : kend, mstart : mend)$.

Below we present in Table 2 the results of computing numerically (μ,ν,σ) for various grids.

Table 2: Computed values of (μ, ν, σ) for input data in various grid domains

Grid domain	(μ, ν, σ) found
(0:8, 0:8, 2:6, 99:101)	(1.959, 0.205, 4.643e-4)
(0:8, 0:8, 2:6, 499:501)	(1.957, 0.205, 4.849e-4)
(0:8, 0:8, 2:6, 999:1001)	(1.957, 0.205, 4.858e-4)
(6:14, 6:14, 0:4, 99:101)	(2.021, 0.197, 6.214e-4)
(6:14, 6:14, 0:4, 499:501)	(2.072, 0.206, 5.376e-4)
(1:49, 1:49, 0:4, 499:501)	(2.012, 0.204, 4.976e-4)

Conclusion

In this chapter we have presented a monotone difference scheme for solving the 3D problem of diffusion and transport of aerosols in a bounded domain which ensures high accuracy and especially ensures the positivity of the solution. The realization of the scheme is reduced to solving 1D problems that require economical computational amount. Using the solution of the main problems of propagation of aerosols and their adjoint problems we solve the problem of locating new plant and the problem of optimization of emissions rates of operating industrial plants in order to keep the pollution level in given sensitive areas under a prescribed level. They are very important problems to be solved to achieve sustainable development, especially in developing countries, such as Vietnam, when foreign investment increases from year to year. A simple way for determining several parameters of the air pollution model also is proposed and experimentally studied for its applicability.

Future work will be concerned with problems in unbounded 2D and 3D domains, in domains with complicated geometry, the propagation of active aerosols, problems with complex environmental standards.

Acknowledgements

The first author would like to thank the International Buro of the BMBF (Germany) for a starting grant in 2005 that initiate this German–Vietnamese cooperation. The Vietnamese authors are supported in part by National Basic Research Program in Natural Sciences.

References

[1] V. Abrashin, On a variant of alternating direction method for solving multidimensional problems of mathematical physics. *Differential Equations* 26 (1990), 243–250.

[2] M.E. Berljand, *Moderne Probleme der atmosphärischen Diffusion und der Verschmutzung der Atmosphäre*. Berlin, Akademie-Verlag, 1982.

[3] J.A. Cunge, F.M. Holly and A. Verwey, *Practical Aspects of Computational River Hydraulics*. Pitman Pub. Inc., 1980.

[4] Dang Quang A and Ngo Van Luoc, Exact solution of a stationary problem of air pollution. *Proc. of the NCST of Vietnam* 4 (1992), 39–46.

[5] Dang Quang A and Ngo Van Luoc, Numerical solution of a stationary problem of air pollution. *Proc. of NCST of Vietnam* 6 (1994), 11–23.

[6] Dang Quang A, Monotone difference schemes for solving some problems of air pollution. *Advances in natural Sciences* 4 (2002), 297–307.

[7] Dang Quang A and M. Ehrhardt, Adequate numerical solution of air pollution problems by positive difference schemes on unbounded domains. *Math. Comput. Modelling* 44 (2006) 834–856.

[8] I. Dimov and Z. Zlatev, Optimization Problems in Air Pollution Modeling. In P.M. Pardalos and M.G. Resende (Eds.), *Handbook on Applied Optimization*, Oxford University Press, 2002.

[9] M. Ehrhardt, Discrete transparent boundary conditions for parabolic equations. *Z. Angew. Math. Mech.* 77 (1997), no. 2, 543–544.

[10] M. Ehrhardt, *Discrete Artificial Boundary Conditions.* Ph.D. Thesis, Technische Universität Berlin, 2001.

[11] N.A. Kartvelishvili, *Unsteady open flows.* Hydromet, Leningrad, 1968.

[12] K. Kathirgamanathan, R. McKibbin and R.I. McLachlan, Source release-rate estimation of atmospheric pollution from a nonsteady point source- Part 1: Source at a known location. *Environmental Modeling and Assessment* 9 (2004), 33–42.

[13] G.I. Marchuk, Mathematical modelling in environmental problems. *Studies in mathematics and its applications* 16, North–Holland, 1986.

[14] G.I. Marchuk, *Adjoint Equations and Analysis of Complex Systems.* Kluwer Acad. Publishers, Dordrecht, 1995.

[15] Ngo Van Luoc, Dang Quang A and Nguyen Cong Dieu, Analytic and numerical solution of some problems of air pollution. *SEA Bull. Math. Special Issue* (1993), 105–117.

[16] Nguyen Tat Dac, One-Dimensional practical substance transport problems and their numerical procedures – Some applications. *Proc. of the 4th National conference on Gas and Fluid Mechanics*, Hanoi, 1996, pp. 202–206 (Vietnamese).

[17] A.A. Samarskii, *The Theory of Difference Schemes.* Dekker, New York, 2001.

[18] J.C. Strikwerda, *Finite Difference Schemes and Partial Differential Equations.* Wadsworth & Brooks/Cole, 1989.

[19] Tran Gian Lich and Pham Thanh Nam, Optimization of emissions of aerosols from operating plants for ensuring environmental standards. *Proc. of the 7th National Conference on Mechanics*, Hanoi, 12/2002, 261–268 (Vietnamese).

[20] Tran Gian Lich, Pham Thanh Nam and Pham Ngoc Vinh, Determination of the plant locations for ensuring some environmental criteria. *The 2th Seminar on environmental science and technology issues related to the urban and coastal zones development*, Ha Long, Vietnam, 28-29/09/2004.

[21] H.Q. Wang and M. Lacroix, Optimal weighting in the finite difference solution of the convection–dispersion equation. *Journal of Hydrology* 200 (1997), 228–242.

[22] N.N. Yanenko, *The Method of Fractional Steps.* Springer–Verlag, 1971.

In: Environmental Chemistry Research Progress
Editors: P. Robinson and R. Gallo
ISBN 978-1-60741-055-3

Short Communication

MANGANESE DOPED NANO-TiO_2: SYNTHESIS AND PHOTOCATALYSIS FOR ACID RED 3R DYE

Wang Li-yong, Qin Jiang-lei*, Ding Zhe, Li Lan-fen, Lu Shu-fang
College of Chemistry and Environment Science,
Hebei University, Baoding 071002, China

ABSTRACT

Manganese doped nano-TiO_2 was synthesized with easy and available inorganic salts by hydrothermal methods. TEM results show that the sample was sphere shaped, XRD patterns show that it was anatase structured. Photo-degradation of acid red 3R was performed under high pressure Xe lamp with the photo-catalyst. The effects of pH value, dopant concentration, catalyst's dosage, crystal structure, calcinations on the photo-catalytic reaction were studied. The results indicated that nano-particle doped with 0.5 at. % Mn has higher activity than pure TiO_2, and mineralization percent of organic molecules approaches 100% within 120min.

Keywords: photo-catalysis, manganese doping, nano-TiO_2

1. INTRODUCTION

The principles of photo-catalysis of organic pollutants are well known: by illumination of the catalyst electron-hole pairs are generated. These charges carrier can diffuse to the surface of the catalyst and initiate redox reactions. Since the band gap of the most efficient photo-catalyst, titania, is about 3.2ev, wavelengths below 380nm are necessary for the process [1-4]. But TiO_2 has the shortcomings of low photo-catalytic efficiency and utilizing rate of quantum, this ascribes to the recombination of electrons and holes except the separation of them, and if the recombination is suppressed, the separation process will be prolonged, then the lifespan of photo-generated charges prolonged also, and this will enhance the transportation rate on the interfaces between the semiconductor and electrolyte, so the efficiency of photo-catalysis can be increased [5]. One of the effective methods of increasing photo-catalytic rate is doping nano-TiO_2 with trans-metal ions, and its mechanism is dopants diffuse well-proportioned in the matrix of TiO_2, and the crystal structure will be aberrational to form oxide-bridge structure, it due to the increase of defects and active surface of compound particle, and it will help transfer the photo-generated carrier, then the rate of photo-catalysis increased accordingly [6]. It was discovered by Chio, etc. that nano-TiO_2 doped with Fe (β), Mo (β), V (β) have higher activity of photo-catalysis in the degradation reaction of CCl_4 and $CHCl_3$ than pure TiO_2 [7], Y. Q. Wang, etc. research studies discussed the degradation of Rohdamine B problems with different dopants included Fe(β), Cr(β), Ni(β), Co(β) [5]. But the research of photo-catalytic degradation performed with nano-TiO_2 doped with Manganese is very few, so our experiments is done in this background, in this article we have discussed the factor of pH value, dopants concentration, catalyst's dosage, crystal structure, calcinations and starting concentration of dye solution those influenced the photo-catalytic reaction, and dye of the acid red 3R was simulated organic pollutant sunlight and Xe lamp as lamp-house.

2. EXPERIMENT SECTION

2.1. Synthesis of the Catalyst Powder

$TiCl_4$, $MnSO_4{\cdot}H_2O$ and several inorganic reagents were used as starting material. All the reagents were of analytical grade and used without further purification. In a typical experiment, titanium tetrachloride was dripped into $MnSO_4$ solution (1 $D{\cdot}s^{-1}$) to form a solution that the concentration of Ti^{4+} was 0.5 $mol{\bullet}L^{-1}$, and the ammonia solution (1:1 volume ratio) was dripped into this solution until the pH value of the suspension increased to 9.0, then the mixture was hydrothermal treated in a stainless steel autoclave at 180°C for 3h, the product was recovered by filtration, washed with hot deionized water to remove NH_4^+ and Cl^- and then dried at 90°C in drier or calcined at different temperature.

2.2. Charaterization of the Catalyst

The crystal phase of the sample was determined from the X-ray diffraction (XRD) patters, which were obtained with Y-2000 Automated X-ray diffractometer System, the morphologies and particle size of the sample were obtained by transmission electron microscopy (TEM) performed on a JEM-100SX (Japan), the photo-absorbance experiments were performed on UV-1200 spectrophotometer (China).

2.3. Measurement of the Photo-catalytic Degradation Rate of Acid Red 3R

Acid red 3R(10, 25, 50, 75$mg{\bullet}L^{-1}$) solution was simulating organic pollutants, almost 200mL of this solution was extracted, and a amount of catalyst powder was added to form a suspension which catalyst's concentration were 0.5, 1.0, 1.5, 2.0, $2.5 g \cdot L^{-1}$, then the mixture was supersonic treated for 15 min in a dark place to establish a adsorption equilibrium. The pH value was adjusted by hydrochloride solution (2.0 mol/L) and sodium hydroxide solution ($2.0\ mol \bullet L^{-1}$), and then the photoreaction was performed a glass vessel illuminated by high pressure Xe lamp with magnetically stirred. 7.0ml suspension was extracted every 20 min and separated centrifugally, and absorbance of the clear solution was measured by UV-1200 spectrophotometer. The degradation rate was calculated by following formula according to the standard work curves.

$$D\% = [(A_0 - A)/A_0] \times 100\%$$

3. Result and Discussion

3.1. Morphology and Crystal Structure Analysis

The nano-particle was ellipse shape, and particle length was 30~40nm and radius was about 10~15nm, no distinct agglorgation was found from Figure 1, 2, and the particle morphology did not changed after doping with 0.5 at. % Mn element. And the ellipse shape could ensure plenty of active center and enough special surface areas. XRD patterns showed that the sample was anatase structure though doping with Mn^{2+}, and the diffracting peak sharpened after calcinations treatments at 673K and 873K, it was proved that the faces of crystalline grew to integrity. The rutile structure appeared after calcinations at 1073K, and it was showed that anatase structure could transform to rutile structure with calcinations treatment.

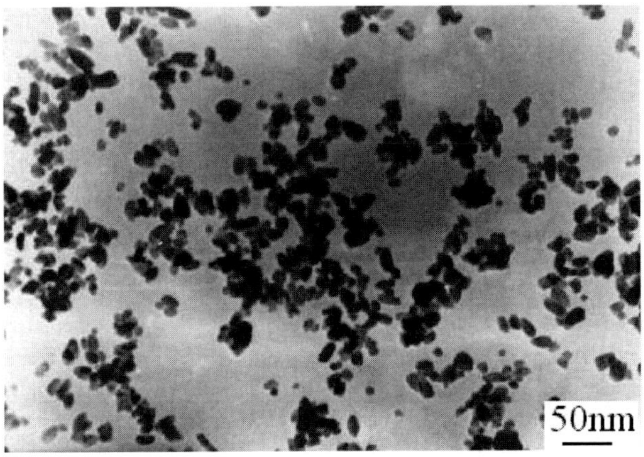

Figure 1. TEM photograph of pure TiO_2.

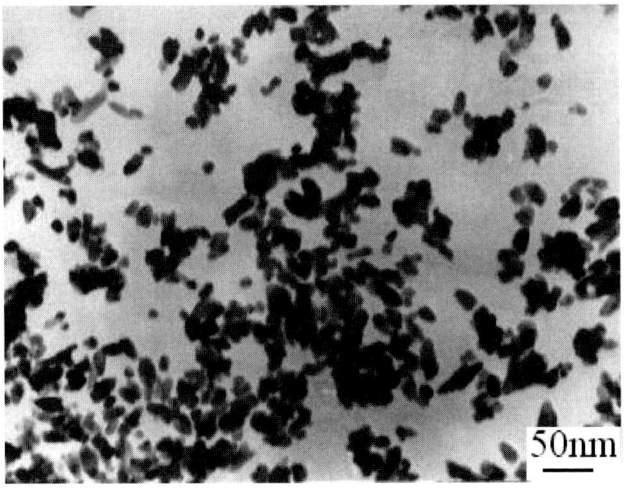

Figure 2. TEM photograph of TiO_2 doped with 0.5% Mn.

Photo-absorbance experiments showed that the photo-absorbance increased on visible and ultraviolet zone and red shift phenomena appeared after doping with Mn^{2+} (Figure 4). The results showed that the spectrum responsive range was broadened, and the electron could be excited by longer wavelength light ascribe to doping energy band [8] and the defects which acts as the active center, so the photo-catalysis efficiency was enhanced [9].

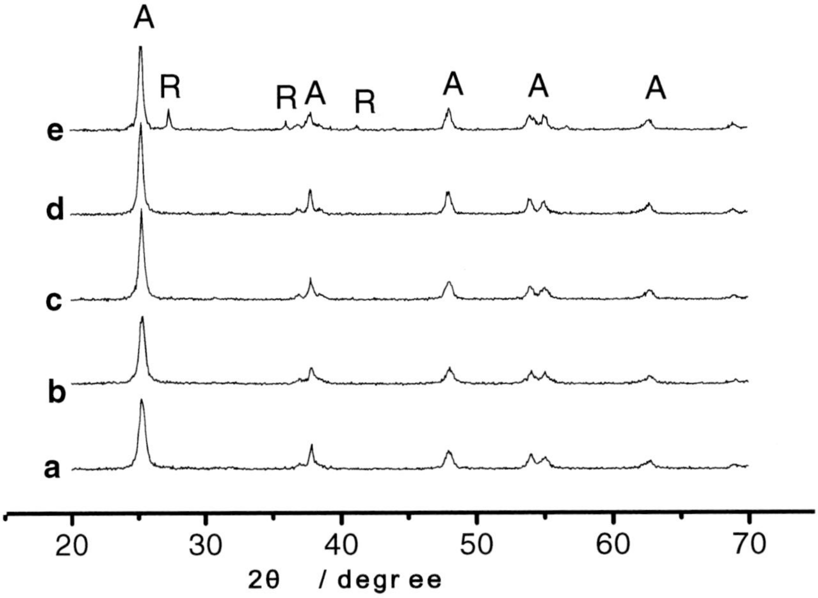

Figure 3. XRD patterns of sample: a: Pure-TiO_2, b: TiO_2 doped with 0.5 at. % Mn, c: TiO_2 doped with 0.5 at. % Mn calcined at 673K, d: TiO_2 doped with 0.5 at. % Mn calcined at 873K, e: TiO_2 doped with 0.5 at. % Mn calcined at 1073K for 1h.

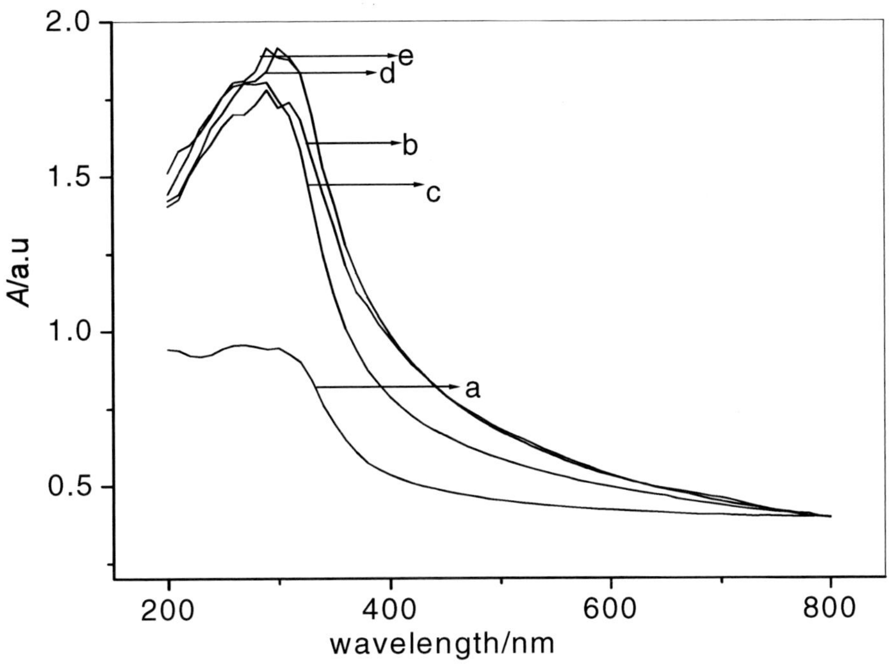

a: commercial bulk TiO_2,b: 0% Mn, c: 0.1 at. % Mn, d: 0.5 at. % Mn e: 0.7 at. % Mn.

Figure 4. Uv-vis spectrum of TiO_2.

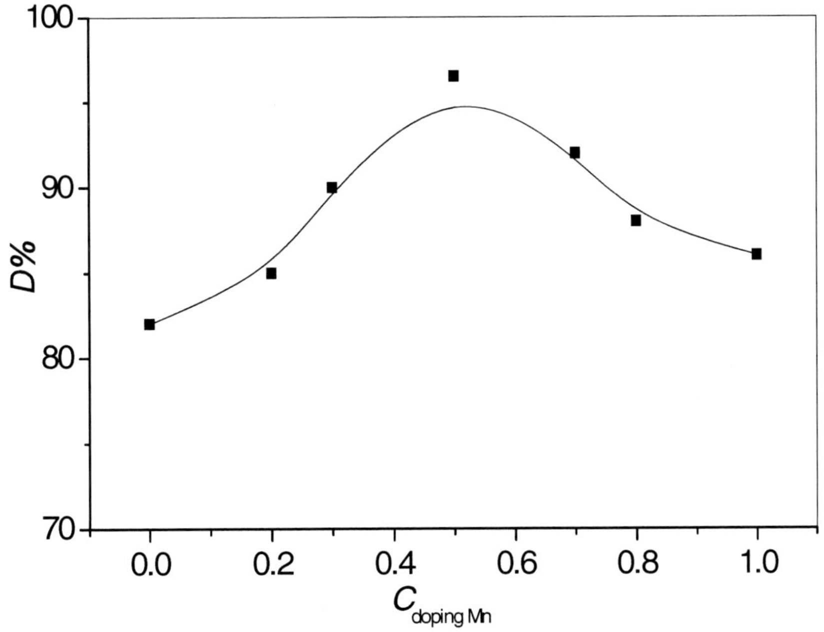

Figure 5. Degradation curve of dye with different doping concentration.

3.2. Effects of Dopant

Higher photo-catalytic efficiency was observed when the catalyst was nano-TiO_2 doped with Mn^{2+} comparing with pure nano-TiO_2, and doping excessive due to decrease of degradation efficiency. The possible reason was that for TiO_2 doped with Mn^{2+}, the presence of Mn^{2+} in the matrix of nano-crystalline TiO_2 increased the photo-reactivity when Mn^{2+} concentration is blow 1.0 at. %. The optimal Mn^{2+} concentration was 0.5 at. % for the photo-degradation. Mn^{2+} doping become detrimental when the dopant concentration is higher than 1.0 at. %. Since the solubility of the manganese in the anatase is below 1.0 at. %, the distribution of the dopant in TiO_2 nano-crystalline is homogeneous when the manganese concentration is below the limit. The photo-catalytic results indicated that a homogeneous distribution of the dopant in nano-crystalline TiO_2 improved the photo-reactivity. When the Mn^{2+} doping is beyond its solubility limit in TiO_2, it is likely to form MnO_2 on the surface of the TiO_2 particles. The photo-chemically inactive MnO_2 will reduce the photo-activity of the nano-crystalline TiO_2.

3.3. Effects of the pH Value

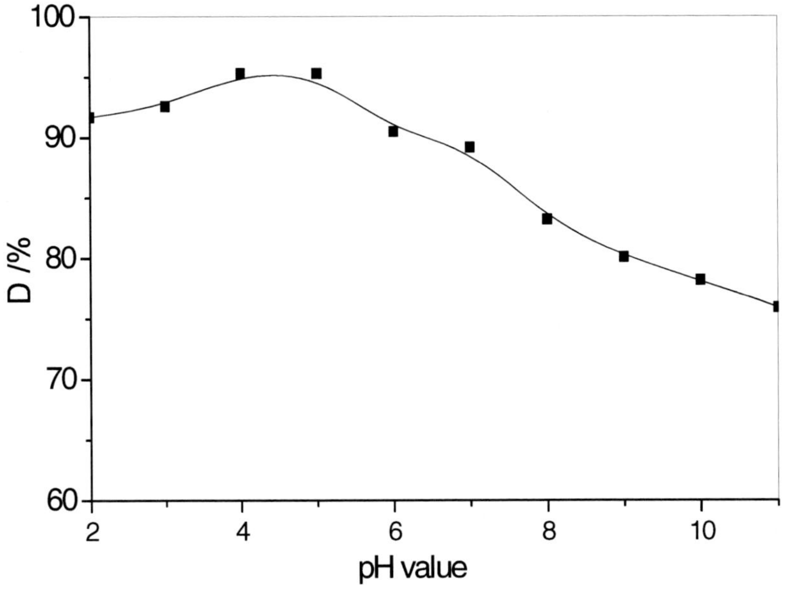

Figure 6. Effects of pH value.

Figure 6 demonstrated that the optimal pH value of the photo-catalytic reaction was 5.0 illuminated by the high pressure Xe lamp, when the starting concentration of dye solution was 25mg•L^{-1} and the dosage of the catalyst was 1.5g•L^{-1}. It could ascribed to the equilibrium of the number surface hydroxyl and the amount of the adsorbed dye molecules on the surface of the TiO_2 nano-particles. At lower pH value the number of the surface hydroxyl groups was higher and the amount of the adsorption of acid red 3R on the surface was more because the dye is negative charged dye but TiO_2 nano-particles are positive charged at low pH (pH<6.6) and attached to each other at low pH, so the reaction parameters were smaller at low pH. At high pH value, they were repulsive to each other and the amount of adsorbed the dye on the TiO_2 nano-particles was little, the number of the surface hydroxyl of the TiO_2 nano-particle was small and the reaction parameters were also small at high pH value. The IR diffuse reflectance spectra of the TiO_2 nano-particles showed that the surface hydroxyl was greater for samples treated at pH 5.0 than treated at pH 10.0. Therefore, the optimal pH value to make the reaction parameters largest was 5.0.

3.4. Effects of the TiO_2 Amount

Photo-catalytic degradation rate enhanced with the increase of the dosage of the catalyst when the concentration of the dye solution was 25mg•L^{-1} and pH value was about 4.0.The results showed that the optimal dosage of the catalyst powder was 1.5 g•L^{-1}, and the degradation efficiency reduced beyond the dosage. The cause was surplus catalyst molecule suspend in the system of reaction would retard illumination, and dye molecules absorbed on the catalyst surface could not contact photo-generated electrons, so the photo-catalytic reaction will be retarded.

Figure 8 demonstrated that different degradation curve of dye at different starting concentration, and degradation rate was near 100% at low concentration, and it reduced with increase of dye solution concentration. This indicated that degradation rate linked with the starting concentration of dye, and higher concentration was due to more absolute amount of degradation.

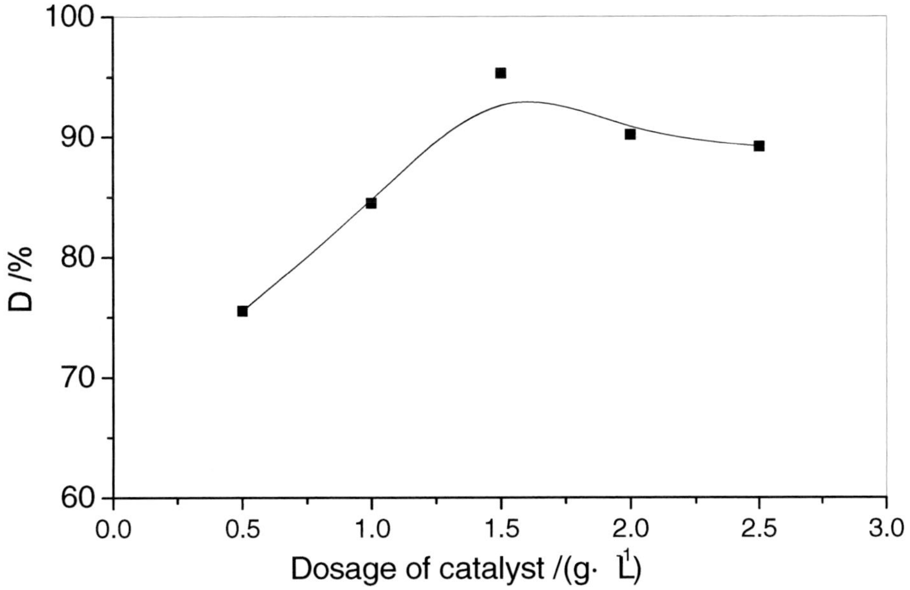

Figure 7. Effects of catalyst dosage.

3.5. Effects of the Starting Concentration of the Dye

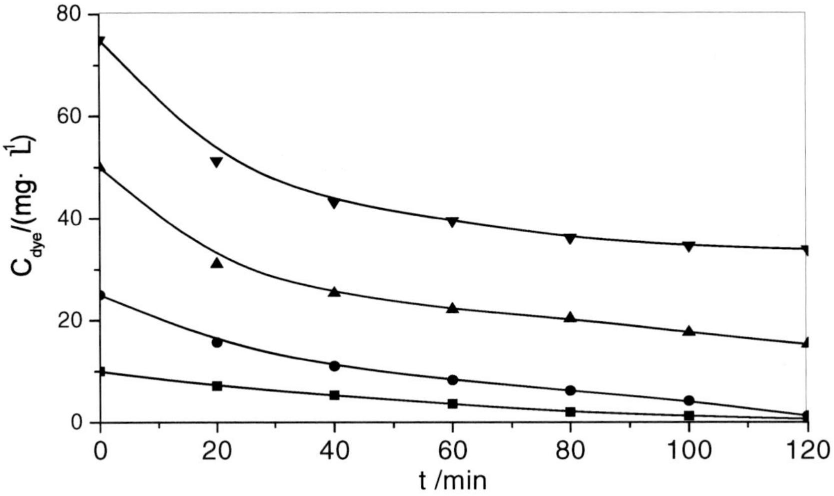

Figure 8. Effect of the starting concentration of dye solution.

3.6. Effects of Calcinations

XRD patterns showed that mainly diffraction peak sharpened as the temperature of heat treatment increased, and crystal lattice constants increased also [10]. Crystal grains size decreased and surface increased, so the efficiency of photo-catalysis was promoted, when the sample was calcinated at 673 K, and nano-crystal particles melted together and grew to form large particles when calcinated at 873K, so the particle's surface decreased, and the photo-catalytic reaction will be retarded [11].

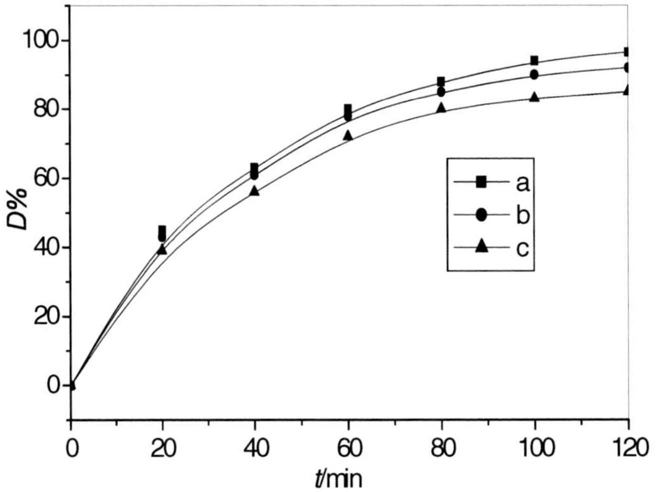

a: Without calcinations, b: Calcinations at 673K, c: Calcinations at 873K.

Figure 9. Effects of calcinations.

4. Conclusion

With nano-TiO_2 doped with Mn^{2+} as the photo-catalyst, high pressure Xe lamp as the illumination resource, and acid red 3R as simulating organic pollutant, the photo-catalytic reaction was studied. Some factors that influenced the degradation reaction included pH value, starting concentration of dye solution, concentration of dopant, dosage of catalyst, etc.

Acknowledgements

We are graceful to youth fund of Hebei University of China by their finance support (No:2006Q11).

References

[1] Ting C. C., Chen S. Y., Liu D. M. *J. Appl. Phys.*, 2000, 88 (8): 4628~4632.
[2] Asahi R., Taga Y., Mannstodt W., Freeman A. J. *Phys. Rev. B.*, 2000, 61 (11): 7459~7465.
[3] Yu J. C., Lin J., Lo D., Lan S. K. Langmuir, 2000, 16 (18): 7304~7308.
[4] Ding Z., Lu G. Q., Greenfield P. F. *J. Phys. Chem.,* 2000, 104 (19): 4815~4820.
[5] Y. Q. Wang, L. Zhang, H. M. Cheng, J. M. Ma, Chem. *J. Chinese Univ.*, 2000, 21 (6) p958-960.
[6] Y.H. Xu, G.B. Gu, X. Q. Cheng, X. J. Li, *J. Sorth China Tech. Sci. Univ.,* 2001, 29 (11) p76-80.
[7] Chio W. Y., Angew, *Chem.,* 1994, 33:1091~1103.
[8] L.D. Zhang, J.M.Mou. *Nanomaterialand Nanostructure.*Beijing: Science Press 2001, 99.
[9] Dana Dvoranováa, Vlasta Brezová, Milan Mazúr, Mounir A. Malati I *Applied Catalysis B: Environmental* 2002, 37, 91~105.
[10] L. Gao, S.ZHeng, Q. H. Zhang Nano-TiO_2 photocatalyst and Application, Beijing, *Chemical technology* Press2002, 121~122.
[11] S. W. Qian, Z. Y. Wang, M. Q. Wang *J.Mater.Sci.& Eng.* [J] .2003, 21(1):48~52.

INDEX

A

B

C

D

E

F

G

H

I

J

K

L

M

N

O

P

Q

R

S

T

U

V

W

X

Y

Z